D1058464

Forest soils: properties and processes

K.A. ARMSON is a professor in the Faculty of Forestry and Landscape Architecture, University of Toronto.

This is a comprehensive study of forest soils for foresters, wildlife and park managers, ecologists, and others interested in forest soils. It provides a valuable text for introductory and more advanced forestry courses.

The first ten chapters deal with basic soil information: texture, structure, and porosity; colour, temperature, and aeration; water; organic content; biological organisms and processes; chemistry; fertility; classification; and surveys. The last six chapters consider the components of the forest soils system as related processes, discussing roots, fire, and water and nutrient cycles as they exist in natural forests and as they are modified by man. Professor Armson examines the processes of forest soil development, and the place of soil as a part of a continuously changing landscape from both the historical and ecological viewpoints. An appendix describes procedures for soil profile description and sampling. Full bibliographical references are supplied.

K.A. ARMSON

Forest Soils: properties and processes

UNIVERSITY OF TORONTO PRESS
Toronto Buffalo London

Toronto Buffalo London
Reprinted 1979
Printed in Canada

Canadian Cataloguing in Publication Data

Armson, K.A., 1927–
Forest soils

Includes index.
ISBN 0-8020-2265-0

1. Forest soils. I. Title.

SD390.A74 631.4 C77-001419-4

Contents

Preface

This book was written for foresters, wildlife and park managers, ecologists, and others with an interest in forest soils. The arrangement of the text should make it suitable both for an undergraduate introductory course in forest soils and for use in a more advanced program. The first ten chapters deal with basic soil information such as physical and chemical properties, soil water, organic matter, soil biology, fertility, soil classification and survey; examples illustrating these properties are drawn chiefly from forest soils. Chapters 11 to 16 consider the components of the forest-soils system primarily in terms of related processes. Tree roots, fire, water, and nutrient cycles as they exist in natural forests are examined along with the implications of their modification by man. The processes of forest soil development are discussed and the place of soil as a part of a continuously changing landscape is treated from both a historical and ecological viewpoint.

While the writing of a text which may be used for both introductory and advanced students presents a hazard to the author, I believe that because of the wide range of interests that may be served it is important to provide the opportunity to pursue a particular topic in some depth and also serve as a reference for both students and those in professional practice. It is my experience that the most useful college texts are those which are referred to long after the particular course of study has been completed.

Each chapter begins with a concise statement, as free as possible from technical jargon, introducing the subject matter. A detailed treatment of the individual topics within the subject area follows. Stress has been placed on the interrelationships existing in forest soils and on the methods of measurement of many soil properties, since a knowledge of the reliability of quantitative information is necessary if conclusions about soil data are to be soundly based.

No attempt has been made to ensure a comprehensive and complete listing of the literature. Rather, those studies are cited which reflect and substantiate the relevant statements in the text and which are, for the most part, more readily available. Inevitably, a bias of the writer in terms of interest and acquaintance with the literature must show through. This book has been written with emphasis primarily on North American forest soils. However, I have not hesitated to include results of studies from elsewhere in the world when it seemed appropriate to do so.

I wish to thank the many students and colleagues who, over the years, have stimulated both my awareness and knowledge of many aspects of forest soils; in particular I am indebted to Dr R.J. Fessenden. I am grateful for support in the publication of this book to: The Faculty of Forestry and Landscape Architecture, University of Toronto, and Cominco Ltd. Dr David C.F. Fayle most capably drafted many of the illustrations in the text.

K.A.A.

Forest soils: properties and processes

Introduction

Man cannot exist without soil. His history is to a very large degree the story of its use and abuse; civilizations that have flourished have always been based on an intensive use of soils for food production. Thus, from historic times knowledge and experience about soils have been related primarily to agriculture. The importance of soils in supporting natural forests has been tacitly understood, but man's use of such forests has been in general an exploitative one. Forests have provided shelter and a source of game, wood for fuel, construction, and industrial purposes. Forests are an integral part of the cycle of shifting cultivation and their role in the water cycle, amelioration of harsh environments, and recreational use has long been recognized.

Systematic observations and study of forest soils occurred first in the nineteenth century, particularly in relation to the forest management practices developing at that time in Europe, especially in Germany. Gustav Heyer's *Lehrbuch der forstlichen Bodenkunde und Klimatologie* (1856) is an early example of such records. The emphasis at this time was largely on the physical aspects of soils. During the latter part of the century, studies by Ebermayer (1876) and others focused attention on the forest litter and the chemical contribution it made to the soil. The impetus for such studies came from a large-scale removal of this litter for bedding of farm animals and the intense use of fallen branch material for fuel - practices that resulted in reduced forest soil fertility. In 1887, *Studien über die natürlichen Humusformen und deren Einwirkung auf Vegetation und Boden* by a Dane, P.E. Müller, was published. This represented a major change in approach to the study of forest soils since it viewed them primarily as zones of biological activity. The terms mull and mor were defined and the use of line drawings of soil profiles, together with a set of colour plates, establishes this work as a classic in the literature of forest soils. The last few decades of the century saw intense development of forestry practices in Europe and in 1893 Ramann's

Forstliche Bodenkunde und Standortslehre brought together information about the physical, chemical, and biological properties of forest soils in a manner that had not been done before and dealt with the application of forest soils' knowledge to certain forest practices. Another general text which appeared was Henry's *Les Sols forestiers* (1908). Henry placed considerable emphasis on the amount and composition of forest litter and dealt with the importance of soil properties on silvicultural practices. His book is notable for its treatment of water in the soil and the role of forests in affecting certain parts of the hydrologic cycle. Henry also presented a synopsis of the forest soils of France together with a tabulation of chemical properties of certain soils.

It might have been expected that such works would have provided the base for an expanding area for scientific study. But such was not to be the case! An agricultural revolution in part spawned by the technological developments of the industrial revolution and by the tremendous expansion of arable land, particularly in North America, together with the ability to transport large quantities of raw materials cheaply, resulted in increasing emphasis on the study of soils for agricultural use. At the same time, the natural forests in both hemispheres were exposed to exploitation for wood to satisfy domestic and industrial demands on an increased scale. In such a climate the study of forest soils, particularly as it related to forest management and silvicultural practices, could only languish. An exception was Bornebusch's (1930) study of the fauna of forest soil. This work was supported financially by the Carlsberg Fund and heralded the development of a new area of soil study - soil biology - some 20 to 30 years later. It is remarkable that Bornebusch dealt with soil organisms not only as numbers and mass within the soil but also in terms of their activity both in an ecological sense and as a metabolic function.

Following World War I (1914-18) and into the depression and drought era of the 1930s a renewed interest in forest soils occurred, largely in relation to developing reforestation practices to provide national sources of wood in times of emergency, as in the United Kingdom, or to rehabilitate eroded lands damaged by improper farming, as in North America. Knowledge and experience were primarily species orientated and related to man-degraded soils - the heathlands and abandoned farmlands which, if they had supported forests decades or centuries before, now had few if any of the attributes of natural forest soils. The study of forest soils proceeded slowly and with a marked emphasis away from natural forests.

Following World War II (1939-45), two American books on forest soils were published, one by Wilde (1946) who was largely instrumental in developing the study of forest soils in the Lake States and relating soils' information to silvicultural practices; the other by Lutz and Chandler (1946) was the first comprehensive attempt in North America to summarize the state of forest soils' knowledge. Between the two world wars, a small group of foresters in Scandinavia had proceeded with forest soils' work; Hesselman, Romell, and Tamm are noteworthy. Tamm's *Den nordsvenska skogsmarken* (1940) was translated into English by M.L. Anderson and appeared as *Northern Coniferous Forest Soils* (1950). In 1958, Wilde's *Forest Soils, Their Properties in Relation to Silviculture* was published (also in a German edition *Forstliche Bodenkunde* (1962)). Aaltonen's *Boden und Wald* (1948), although it dealt primarily with Scandinavian conditions, contained much that was generally relevant but never received the attention it deserved outside of Europe. Duchaufour's *Précis de Pedologie* (1960), although a general soil science text, presented much information on forest soils.

As a result of the dislocations associated with World War II, the global nature of the conflict, and the demands on scientists and professionals from many disciplines to provide new knowledge, our understanding, particularly of subtropical and tropical forests and their soils, was expanded. This extension of knowledge was further developed as new countries began their own development and exploitation of natural resources. Beginning largely in the 1950s and at an accelerating rate through to the 1970s there has been an increasing need for a more comprehensive and deeper understanding of forest soils for several reasons:

1 / In areas of the world where natural forests have been exploited on a wide scale and where demand for raw material is high, there is an increasing intensity of forest management for wood production. To a large degree this management employs conifers and silvicultural practices involving site preparation and subsequent tending treatments. Fire, chemicals, and cultivation are the tools most frequently used in these practices and it is necessary to understand the processes in the forest-soil system, not only in relation to what may be the short-term objectives of management but also in relation to the possible longer-term effects of these practices on the soil.

2 / In many highly developed industrial countries there has developed a use of forests for recreational and aesthetic purposes, often related to natural scenic values and to wildlife populations. Up to the 1950s, with some exceptions, such uses were of low intensity, but as this type of use has increased there has been a need for a greater appreciation of the properties of forest soils in relation to the management of such areas.

3 / Forested watersheds are often of prime importance for the supply of potable water to many cities and towns. Forest hydrology is concerned with the properties and interrelationships between forests and the soils in such watersheds. The management of a watershed to produce water necessitates an understanding of the implications that any management practice may have on the yield, and the chemical and biological attributes of the water.

An increasing demand in many parts of the world for fresh water, together with the problem of organic waste disposal results in a need for greater knowledge of the forest soil properties in relation to such demands.

4 / As the world population grows there are fewer areas of forested soils remaining which have not been subjected to major modification by man. For scientific reasons alone, if for no other, the properties and natural processes associated with these soils should be studied. They can provide a measure or scale against which the effects of man in modifying similar forest soils may be examined.

A study of forest soils relies not only upon soil science as a basis, but also upon knowledge, understanding, and techniques from forestry, agronomy, geology, biology, and other disciplines. Conversely, besides the forester, wildlife or park manager who may require some knowledge of forest soils in their professional life, there are many others such as ecologists, landscape architects, anthropologists, and archaeologists who to a greater or lesser degree will find a knowledge of forest soils rewarding.

1

Forest soil: what it is and how to describe it

Green plants use energy from the sun and by photosynthesis convert carbon dioxide and water into carbohydrates. Soil provides anchorage for the plants and much of the raw materials - water, nutrients, and oxygen - necessary for plant growth. Thus, the scientific definition for soil is 'the unconsolidated mineral material on the surface of the earth that serves as a natural medium for land plants' (Soil Science Society of America, 1973); by extension, forest soils are those which support forest vegetation.

The soil is not just a storehouse of raw materials used by plants; it is a space in which not only the roots of plants can grow but where a host of other organisms, plant and animal, large and small, can live and die. The supplies of the raw materials are not constant but change with the seasons and time and with the growth and development of vegetation supported by the soil. Such changes are also reflected by the populations of soil organisms. Soils therefore are dynamic bodies. Indeed, it is their capacity to change and be changed which has made soils so amenable to agricultural and forestry practices. If the properties and natural processes occurring in a soil are known and understood, then intelligent, non-destructive practices are more likely to be employed than if soil is viewed as a static inanimate part of the earth which can be used or misused at will.

At a time when there is concern about supplies of energy in a world of ever-increasing population, it is appropriate to consider the energy relationships of forest soils. Soil can be viewed as an energy reactor. It supports plants which utilize solar energy and within the soil the transformation of organic matter, whether it be that of the surface accumulation of leaves or dead roots beneath, involves energy changes; the chemical and physical weathering of soil mineral particles also involves such changes. In natural soils, seasonally and over long periods of time the rates and nature of the energy transformation processes will alter. The rate of vegetation litter fall varies as the development of the forest changes. The regimes of soil moisture and temperature differ seasonally throughout the year and also with changes in activity of the vegetation and therefore affect biological activity and physical and chemical weathering processes. If the soils or the vegetation they support are altered or manipulated by man, then many of the soil processes and the pathways of energy transformation change. If the use or alteration is superficial, then the natural soil-vegetation system will usually rebound easily to its original state, but if the exploitation is great or if intensive cultural practices are introduced, then the soil itself will be modified. The culture - silviculture - of forests represents an effort by man to make a more efficient use of a tree's use of solar energy so that it is in the form he desires, as shade, wood for use, or browse for his animals. In so doing, he manipulates the forest but consciously or unconsciously he also changes the soil to a greater or lesser extent.

A look at soil

A soil is a three-dimensional body; it has depth and width and extends over a part of the earth's surface. It is examined by exposing it in vertical cross-section. Usually this is done by digging a soil pit, but often road cuts or excavations made for other purposes may be used. When the exposed cross-section is viewed, one or more bands or layers are usually seen, each distinguishable primarily by differences in colour and form. These layers are often horizontal (paralleling the surface) and are called *horizons.* The vertical sequence of horizons which are observed for any soil is termed the *soil profile* and it is the nature and properties of the profile which form the basis for most classifications of soils. The field description of soils is one of the most critical abilities a student of soils can develop. It demands keen observation and objective judgment. Paradoxically, although the importance of the dynamic nature of soil processes has been emphasized, an examination of the soil profile can record only the tangible results or by-products of these processes. Thus the amount and type of surface organic layer can reflect something of the nature of decomposition processes. From the presence of a zone of accumulation of calcium carbonate at depth in the profile, inferences may be made concerning the weathering and movement of calcium in the soil. The soil, in fact, provides clues to the keen observer; it is then up to him to rationalize from such evidence a story of the processes that have been and are taking place. Laboratory analyses of soils can provide more detailed and quantitative documentation of both properties and processes. It is even possible to measure directly some processes - annual surface organic accumulation, the rate of water movement through a profile, or the movement of chemical compounds using radioactive tracers. These techniques are usually both involved and expen-

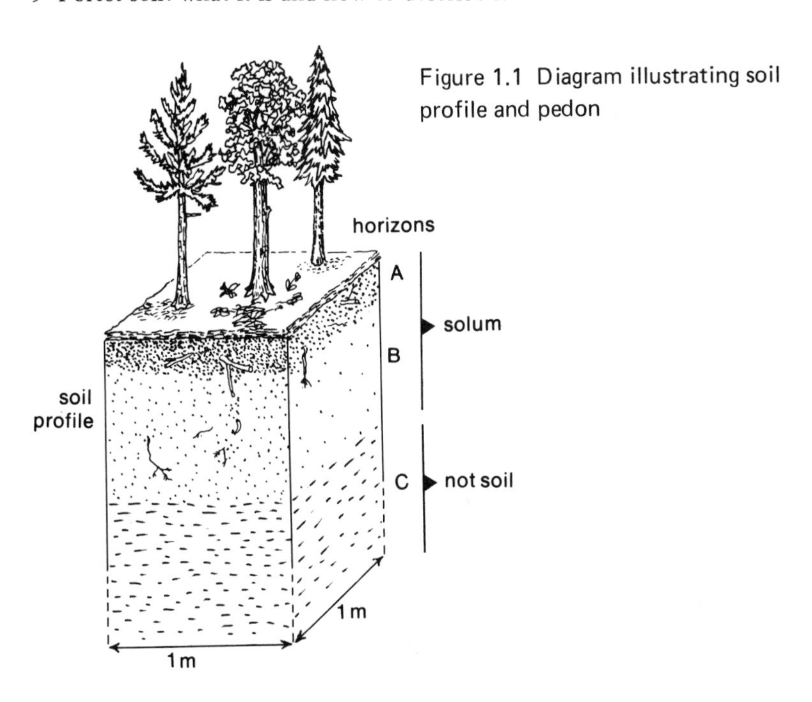

Figure 1.1 Diagram illustrating soil profile and pedon

sive and in the field one must rely primarily on a macroscopic examination of the soil for evidence of its characteristics.

The description of a soil

When a soil profile (Figure 1.1) is examined, it is common for the horizons to show variation in thickness, colour or other qualities. Also, the differences between horizons may show variation. Some horizons may be so indistinct in terms of their boundaries that they merge or blend from one to the other. Examples of certain forest soil profiles are shown in Figure 1.2. Thus, in setting out to describe a profile, questions arise. What is the proper width of soil profile to be described? How deep is the soil to be examined? The answers to these questions can only be somewhat arbitrary, because different soils will vary to different degrees, but soil scientists generally agree that the unit of soil for which a soil profile description is usually made is termed a *pedon*. It has been defined as follows (Soil Survey Staff, 1975):

A pedon has the smallest area for which we should describe and sample the soil to represent the nature and arrangement of its horizons and variability in the other properties that are preserved in samples. ... It has three dimensions. Its lower limit is the somewhat vague limit between the soil and the "not-soil" below. Its lateral dimensions are large enough to represent the nature of any horizons and variability that may be present. A horizon may vary in thickness or in composition, or it may be discontinuous. The area of a pedon ranges from 1 to 10 m^2, depending on the variability in the soil.

In practical terms, the minimum width of the soil profile face of a pedon is one metre and may be as great as three and a half metres. The vertical extent is often less easily decided upon. The influence of soil weathering processes associated with organic matter and living organisms becomes less with depth, but usually there is a zone above which the materials can be termed soil and below which is essentially geological material. This is best termed *not soil*. Previously, this geological material was referred to as parent material.*

The horizons located above the geological material (not soil) make up the *solum*. In many forest soils, roots may descend vertically several metres below the general zone of the solum. Although the vertical extent of the soil profile is still defined by the overall solum depth, efforts should be made to describe and quantify, as far as possible, the extent of such deep roots and the nature of the materials in which they occur.

When a profile is being described, it is customary to determine first the general sequence and number of horizons including the nature of the not soil and to identify them in terms of a standard system of notation (Table 1.1). In mineral soils there are four master horizons:

1 / Surface organic layers. In forest soils these comprise the surface accumulations, mainly of foliage, twigs, and other plant debris. Usually they can be separated into zones, vertically, according to degree of decomposition, with the least decomposed on the top and the most decomposed on the bottom. In some soils, little accumulation may occur because of rapid decomposition. The layers are designated:

* Organic matter is always a component of soil and because it determines or modifies so many properties it could logically be termed parent material. Organic soils such as peats do not have lower geological materials which could be referred to as parent materials. Further, in many soils there is increasing evidence that the upper surface layer horizons have often been developed in surficial geological materials which, while similar in many properties to the underlying geological materials, do differ from them. Therefore the term *not soil*, although somewhat awkward, is certainly more valid than the term *parent material*.

Figure 1.2 Illustrations of various forest soil profiles (n.b. divisions of black and white rod are in decimetres)

(a) profile in calcareous clay loam till under deciduous forest, Southern Ontario

(b) deeply weathered profile in loam under conifers, North Carolina piedmont

(c) profile in shallow till over bedrock (at 55 cm) under mixed forest, central Ontario

(d) profile in acidic outwash sands under boreal forest, northern Quebec

(e) bisequum profile in calcareous sandy loam under mixed forest, north Alberta

PHOTOGRAPH: (a) courtesy R.J. Fessenden; (e) courtesy Canada Department of Agriculture; (b), (c), and (d) by author

L - slightly decomposed material, readily identifiable as to origin.
F - partially decomposed, in which the origin is somewhat difficult to discern.
H - usually amorphous, in which the origin is unidentifiable and the lower portion is often mixed in with the mineral soil. In the USA these layers are designated as O.

2 / Mineral horizons at or near the surface. These are characterized by some degree of organic matter accumulation, usually in amorphous form (humus) from the surface organic layers, and also by the loss of clay, iron, and aluminium. The degree to which accumulations of organic matter or losses of other constituents occurs will vary considerably. These horizons are designated A. Typically in forest soils it is these horizons L, F, H (O) and A which reflect or are modified to the greatest degree by disturbances such as fire, windthrow, logging, flooding, cultivation, and grazing.

3 / Mineral horizons beneath those of 1 and/or 2 which are characterized by accumulation of one or more of silicate clay, iron, aluminium, or humus. They may also be typified by residual materials formed by hydrolysis, reduction, or oxidation which serve to distinguish it from horizons above or below. These mineral horizons are designated B.

4 / Mineral horizons (excluding bedrock) which may be like the horizons above (the solum). Although they lack the characteristics of the solum they may have an accumulation of carbonates and soluble salts; exhibit cementation or induration; show evidence of either permanent or temporary saturation. These are identified as C horizons. Underlying consolidated bedrock is designated by R.

Horizons in which the dominant processes of soil weathering result in a net removal of materials in suspension or solution are said to be *eluvial.* Horizons identified as A horizons are an example. B horizons commonly are zones of accumulation of products precipitated from solution or deposited from suspension, and are referred to as *illuvial* horizons.

In soils which are saturated with water for part or all of the time, these conditions will cause the reduction of certain minerals, especially those that contain iron, so that the soil takes on a blue-gray colour or is mottled with gray-yellow-orange-red colours. Such conditions are termed *gleys.*

Usually the master horizons (A,B,C) are accompanied by suffix letters which serve to indicate the particular manner in which the horizon has been modified. For example, Ah indicates a horizon characterized by humus accumulation; Bt designates one of clay accumulation. Lithologic discontinuities in the profile are designated by prefixing the horizon notation with a roman numeral. Thus if the

TABLE 1.1
A selection of soil notations for the World,[1] United States,[2] and Canada[3]

World	USA	Canada	Remarks
O1 Of Oh	Oi Oe Oa	Of Om Oh	Refer to organic layers developed under poorly drained or saturated conditions; they may be present in soils which have been drained
O1f O1h Ofh	O1 O O2	L-F L-H F-H	Organic layers developed under dominantly aerobic conditions
A	A	A	
Ah	A1	Ah	Humus accumulation
Ah-E	A1	Ahe	Humus accumulation and eluviation
Ap	Ap	Ap	Plow layer, usually 15-18 cm thick
E	A2	Ae	Eluviation dominant
AB or EB	A3	AB	Gradual transition horizon
BE or BA	B1	BA	Gradual transition horizon
A/B	A & B	A & B	Interfingered horizons
B	B	B	
Bt	B2t	Bt	Horizon of clay accumulation
Bfe	Bir	Bf	Horizon of iron enrichment; criteria of form and origin of iron vary with the system
Bfeh	B2hir	Bfh	Horizon of iron and humus enrichment
Bhfe	B2hir	Bhf	Horizon of humus and iron enrichment
Bgfe	B2gir	Bgf	Horizon in which part or all of the iron accumulation may be derived from ferrous iron; associated with gleying as evidenced by mottles
Bh	Bh or B2h	Bh	Horizon of humus accumulation
Bna	B2(natric)	Bn	Horizon with prismatic structure and significant level of exchangeable Na
Bs(cambic)	B2(cambic)	Bm	Horizon slightly altered by hydrolysis, oxidation, or solution, or all three to give a change in colour and/or structure
Bg	Bg or B2g	Bg	Horizon exhibiting strong reducing conditions; mottling is prevalent
C	C	C	Relatively unaltered unconsolidated geological materials
R	R	R	Consolidated bedrock

TABLE 1.1 continued

World	USA	Canada	Remarks
Commonly used suffixes			
b	b	b	Buried horizon
	p	p	Horizon altered by plowing or other forms of cultivation
m	m	c	Horizon irreversibly cemented or indurated
cn	cn	cc	Cemented concretions
g	g	g	Horizon with characteristics of gleying
ca	ca	ca	Horizon with carbonate accumulation
sa	sa	sa	Accumulation of salts excluding gypsum (note that Canada includes gypsum)
x	x	x	Fragipan
–	f	z	Permanently frozen horizon

SOURCES:

1 International Society of Soil Science, Commission 5, Working Group. ISSS Bull. No. 31 (1967).

2 Soil Classification: A Comprehensive System, Seventh Approximation. Soil Survey Staff, Soil Conservation Service. U.S. Department of Agriculture, 1960.

3 The System of Soil Classification for Canada. Canada Department of Agriculture, 1974.

A and B horizons were developed in a windblown sand, but the C horizon was a waterlaid sand, then the C horizon would be designated as II C. The numeral I is understood for the uppermost material and is not written.

Frequently it is desirable to subdivide horizons. This is indicated by using arabic numbers. In a system of notation where letters are used as suffixes to the master horizon, the arabic numbers follow the latter suffix. If a Bt could be subdivided into two horizons, they would be designated from the uppermost down as Bt1, Bt2. In a notation where a master horizon has arabic numbers as suffixes primarily to designate the character of the horizon, as in the United States, any subdivision employs a second arabic number. If an A2 can be subdivided into two horizons, they would be noted as A21 and A22. In the United States, letter suffixes follow the arabic numbers, as in B21ir, B22ir. Table 1.1 presents the horizon designations suggested for world use by the International Soil Science Society, together with those used in the United States and Canada. In any soil profile a set or sequence of an eluvial (A) and illuvial (B) horizon is termed a *sequum.* Sometimes in one soil profile two sequences may be described and the soil is then termed a *bisequum.*

Further advice and information on the preparation of a soil pit, the description of a profile, the tools required and suggestions on sampling soils are given in Appendix 1.

2
The architecture of soil: texture, structure, and porosity

When a soil profile is viewed in the field the most obvious features are those associated with the solid particles, but on closer examination it will be seen that the living components such as roots occur mainly in the spaces between the solid particles. It is also in these spaces that the water and air of the soil are present. Describing a soil's texture and structure is very much like describing a building by observing the kinds of construction materials - wood, bricks, stone, and how they have been assembled, how large the building is, and the number of stories. In other words texture and structure describe the building materials or matrix of the soil. But, as in buildings, although the construction materials are important, the arrangement of space inside the building is more critical. So it is with soils. From the standpoint of plant growth, soil organisms, and the vast majority of soil processes, the holes or spaces in the soil are of utmost importance. Since the characterization and measurement of the soil's pores is an involved procedure which cannot be done completely in the field, we have to rely largely upon data about the solid phase of the soil and infer from this what the space patterns may be.

The geological origin of mineral particles is of interest because many of the physical and chemical properties of the soil may be directly related not only to the mineralogical nature of the particles but also to the type of geological weathering and/or transportation. For example, in areas which have been glaciated as in northern North America, Europe, and Asia, much of the unconsolidated, surficial deposits consists of fragmented rocks and minerals laid down by the advancing or retreating ice sheet as moraines or tills. These materials show little or no evidence of sorting and the particles have sharp to subrounded edges. Although distinctions can be made between materials deposited as the ice sheet advanced (they are more compacted) and those laid down as it retreated (they are looser, more rounded, and may show local pockets of sorting by sizes), they contrast markedly with fluvial and lacustrine materials deposited in water. These latter deposits will usually show striking patterns or bandings of different-sized particles. The size of the particles is related to the rate of movement of the water in which they were deposited and the densities of the rocks and minerals. Particles of larger size and greater density are associated with greater rates of water movement than are the smaller, less dense particles. The same physical principle which results in differential sorting in nature is utilized in the quantitative determination of particle sizes in the laboratory.

In the field, therefore, one of the most useful steps that can be taken before any soil descriptions are begun is to obtain information about the geology of the area and, in particular, the landforms and nature of the surficial deposits in which the soil profiles are located. With this as a background a much more effective study and understanding of the physical nature of the soil materials is possible.

Soil texture

Soil texture refers to the relative proportions (by weight) of different-sized particles. Arbitrarily, particles larger than 2 mm in diameter are not included. This does not mean that particles larger than this have no importance in the soil but rather that the particles less than 2 mm determine other soil properties, such as water retention and movement, aeration and certain aspects of fertility, to a much greater degree.

Two commonly used classifications distinguish between particle sizes. The first is the International System and the second is that of the United States Department of Agriculture. The categories for each are shown in Table 2.1. Each category is termed a *soil separate*. It will be seen that the International System has fewer separates, particularly in the sand sizes, than the American system. This is a disadvantage since it places together soils with a very wide range of particles and serves to emphasize the rather arbitrary nature of such divisions. Texture is one of the more stable properties of a soil, since it is little, if at all, modified by cultivation or other processes over relatively long periods of time.

DETERMINATION OF SOIL TEXTURE

In the field it is possible to determine the texture class of soils with reasonable accuracy using 'rule of thumb' guides. These guidelines may vary from region to region; Table 2.2 lists the main texture classes with field recognition features found useful in temperate North America.

For any detailed study, however, it is necessary to determine texture by particle-size analysis of a soil sample in the laboratory. The methods used depend largely on the size of separates. Sieving removes particles larger than 2 mm and can be used to fractionate size classes to a minimum of 0.05 mm. Sedimentation

TABLE 2.1
Particle size-classes – USDA and International Systems (size limits in mm)

USDA	2–1	1–0.5	0.5–0.25	0.25–0.1	0.1–0.05	0.05–0.002	<0.002
	Very coarse sand	Coarse sand	Medium sand	Fine sand	Very fine sand	Silt	Clay
International		2–0.2		0.2–0.02		0.02–0.002	<0.002
		Coarse sand		Fine sand		Silt	Clay

TABLE 2.2
Distinguishing features useful in field determination of soil texture

Texture class	Features
Sand	Loose and single-grained. Individual grains can be seen and felt. Squeezed in the hand when dry, it will fall apart when pressure is released. Squeezed when moist it will form a cast which will crumble when touched.
Sandy loam	Contains enough silt and clay to make it somewhat coherent. Sand grains are readily seen and felt. Squeezed when dry the cast will readily fall apart. Squeezed when moist the cast will bear careful handling without breaking.
Loam	Mellow, with a somewhat gritty feel, yet fairly smooth and plastic when moist. Squeezed when dry, the cast will not break if handled carefully. When moist a cast can be handled freely without breaking.
Silt loam	When dry it may appear cloddy but the lumps are easily broken. When pulverized it feels soft and floury. When wet the soil puddles. Casts formed of either dry or moist soil can be readily handled without breaking. When moistened soil is squeezed between thumb and finger it will not 'ribbon,' but will form flat 'pastry flakes.'
Clay loam	When dry it forms hard lumps or clods. When moist it can be squeezed to form a thin ribbon, which will break readily, barely sustaining its own weight. When moist the soil is plastic and will form a cast that will take much handling.
Clay	Forms very hard aggregates when dry. When wet it is plastic and sticky. Moist clay can be pinched out between thumb and finger to form a long flexible ribbon. Note that some clays are friable and lack plasticity in all moisture conditions.

techniques serve to determine particles down to clay size (<0.002 mm). Within the clay separate, centrifugation methods are used. Whatever the procedure, the soil samples are prepared usually by drying, and then the soil aggregates are reduced to primary particles. This involves treatments to remove organic matter, binding agents, and finally the dispersion of flocculants.

Sieving is done with dry samples, but the sedimentation technique involves determination of settling velocities in a liquid – usually an aqueous solution. In this procedure Stokes's Law is utilized to determine the velocities: $v = g(P_s - P_l)d^2/18n$, where v is the particle setting velocity (cm/sec), g is the acceleration due to gravity (980 cm/sec^2), P_s is the particle density (frequently assumed as 2.65 g/cm^3), P_l is the liquid density (if water, it is often assumed to be 1 g/cm^3), n is the viscosity of liquid (if water, it is 0.01005 poises at 20°c), and d is the equivalent diameter of soil particles, assuming each is spherical (cm). This formula to be valid makes several assumptions. One is that the particles are large in comparison to the liquid molecules so that Brownian movement will not affect the rate of fall. This assumption does not hold for the finer colloidal-size particles of the soil. Another assumption is that the particles are spherical, smooth, and of uniform density – hardly valid for most soil particles. It is more appropriate, therefore, to use the term *equivalent diameter* or to present results in terms of particle settling velocities when using this formula to determine particle sizes.

The *Bouyoucos hydrometer* and *pipette* are instruments commonly used for determining particle sizes by sedimentation. The results of particle size analysis may be presented in a number of ways. Most commonly the proportions of the sand, silt, and clay fractions are expressed as a per cent of soil dry weight. Often a textural triangle (Figure 2.1) which relates the amounts of these three fractions to specific descriptive names is used. In different parts of the world textural triangles with quite different subdivisions may be used, even though the textural names may be the same; thus the same words used to describe soil texture would refer, quantitatively, to different combinations of particle sizes. Further it should be noted that the terms sand, silt, and clay have two meanings. One refers to the diameter limits of the soils as separates; for example, silt (International) is particles whose diameters are in the range 0.02 to 0.002 mm. The second refers to silt as textural class; that is, it contains more than 80 per cent silt-size particles and less than 12 per cent clay particles.

In detailed studies of soils, particularly where the geological origin of the soil materials or the weathering processes are of interest, a graphical method of presenting results is often used. The amounts of different-sized particles are plotted cumulatively against their respective equivalent diameters or settling velocities. Examples are shown in Figure 2.2. The textural curves in Figure 2.2a are similar in form for all horizons, indicating that they are of the same geological origin.

Figure 2.1 Textural triangle used in North America

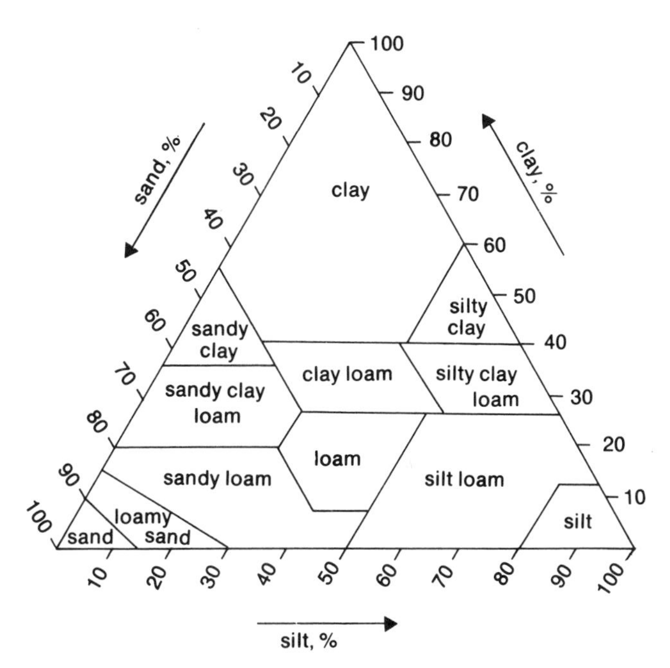

By far the largest proportion of particles is within the range of 0.1 to 0.2 mm (fine sand), a fraction that is readily windblown; in fact, this soil profile was situated in an area of dunes. In contrast, Figure 2.2(b) shows the particle-size distribution for horizons developed in a glacial till in which there is a general distribution of all size-classes. The marked accumulation of clay in the B2 horizon is apparent and it will be seen that there is an absence of particles in the A1 and A2 horizons in the size range 0.04 to 0.22 mm. A visual inspection of the minerals in this size range in the C horizon showed most of the particles to be of shale. It may be deduced, therefore, that much of the clay accumulation in the B2 horizon is derived from the weathering of these shale particles.

Another manner in which particle size data may be used is illustrated by Arnold and Cline (1961). They plotted the frequency distributions for particle sizes in the range of 3.9 to 2000 μ for horizons in two profiles and showed clearly the presence of windblown deposits over an underlying till (Figure 2.3). Quantitative measurement of soil texture can therefore be used to characterize the soil matrix, and also may form a set of evidence which will identify possible geological origins and sometimes indicate the site and nature of certain soil weathering processes.

Figure 2.2 Cumulative per cent texture curves:

(a) soil profile in windblown sand

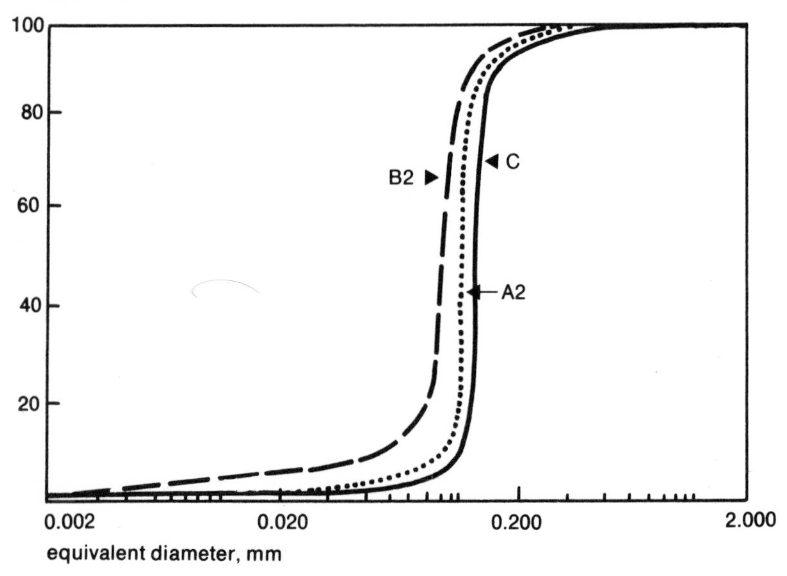

(b) soil profile in clay loam till

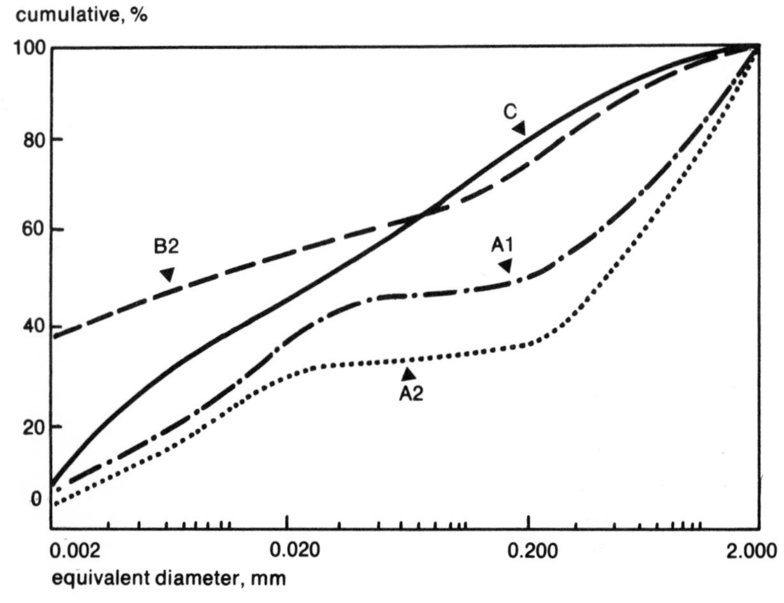

Figure 2.3 Frequency distributions for particle sizes in the range 3.9 to 2000 μ for a soil with wind deposits over till (Potsdam loam) and a uniform till (Gloucester stony fine sandy loam). From R.W. Arnold and M.G. Cline in *Soil Science Proceedings of America* (1961), by permission Soil Science Society of America

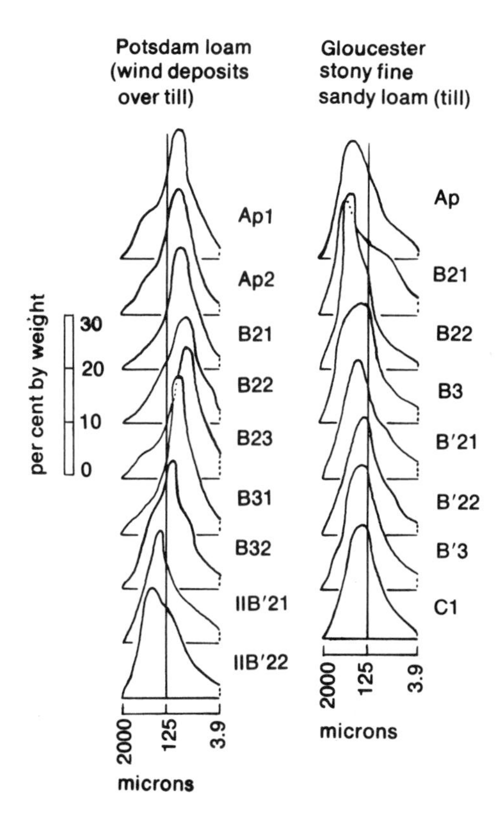

STONES

In many forest soils, stones occupy a considerable portion of the soil volume and thus reduce the volume that can be occupied by particles of 2 mm or less (fine earth) and pore space. The volume and distribution of stones in forest soils have not been adequately documented. Lyford (1964) using various sampling techniques found that for a forested Gloucester sandy loam, coarse fragments (i.e. >2 mm diameter) averaged 60 per cent by weight or 50 per cent by volume. The distribution of coarse fragments by size classes in one of these soils is given in Figure 2.4.

Figure 2.4 Distribution of coarse fragments by size classes (per cent by weight) in a Gloucester sandy loam. Size class width is proportional to size class diameter range. From W.H. Lyford (1964), by permission

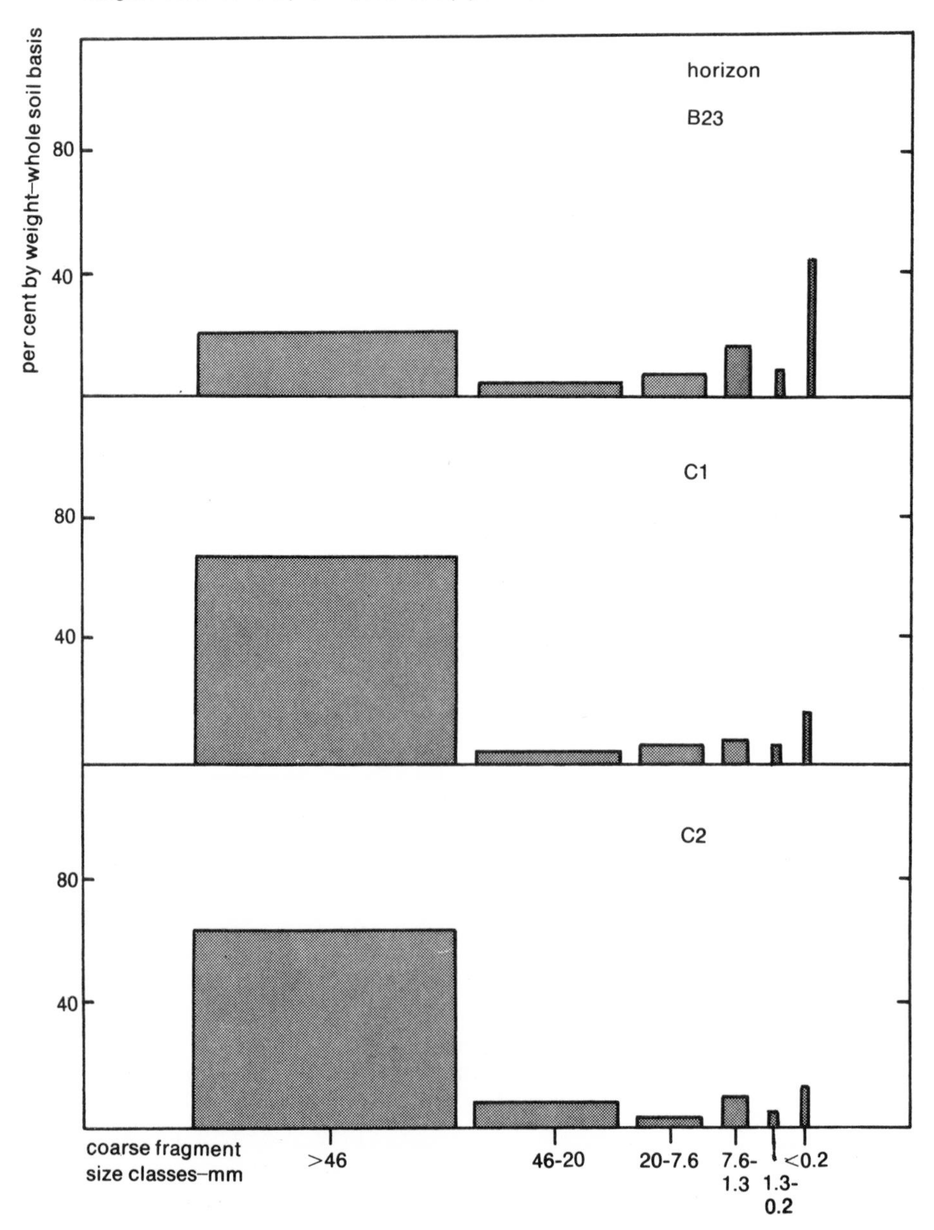

Soil structure

In many soils the primary particles which are described in soil texture do not occur singly but are often aggregated or bound together. This is particularly evident in soils which contain appreciable amounts of clay and/or organic matter. Other chemical compounds may also act as binding agents in the soil. Each aggregate is called a *ped* - a unit of soil structure formed by natural processes. The combination and arrangement of primary particles into peds constitutes the *soil structure.* In the assessment of soil structure it is the size, shape, arrangement, and distinctness of the peds which are noted particularly.

If the primary particles are not aggregated, they are described as *structureless* and may be single-grained as in a loose sand or massive as often occurs in deposits of unweathered materials, predominantly clay. Essentially the terms used to describe the type of structure are based on the geometry of the aggregate:

Block-like. These are peds in which the particles are arranged about a central point such that the ped is isometric. The faces or sides of the ped may be flat or curved.
*Spheroidal.** The aggregates are bounded by rounded or irregular surfaces, but are essentially isometric. Those which are porous are described as having a *crumb* structure, whereas those which are not porous are termed *granular*.
Prism-like. This describes peds which are vertically orientated in the soil, reflecting the arrangement of the particles about a vertical axis. The sides are typically flat. If the tops of the peds have rounded caps, they are said to be *columnar*, but if not rounded they are described as *prismatic.*
Plate-like. These are peds which are horizontally orientated with relatively flat horizontal surfaces. If the peds are very thin they are often described as *laminar*.

Figure 2.5 illustrates the main types of structure and size and shape of soil peds.

In classifying the structure of soils, it has been recognized that some peds when removed from the soil retain their individual distinctness, whereas in other soils the peds are weakly formed. The term *grade* is used to describe the degree of distinctness of aggregation:

* The terminology for describing structure is not uniform, even in North America. In the United States, the four main types of structure are Block-like, Spheroidal, Prism-like, and Plate-like, but in Canada the Spheroidal type is grouped with Block-like, since both describe isometric aggregates and the structureless category is listed as a type of structure, i.e., the Canadian system recognizes Structureless, Block-like, Prism-like, and Plate-like types of structure.

Figure 2.5 Types, kinds, and classes of soil structure. From *The System of Soil Classification for Canada* (1974), reproduced by permission of Information Canada

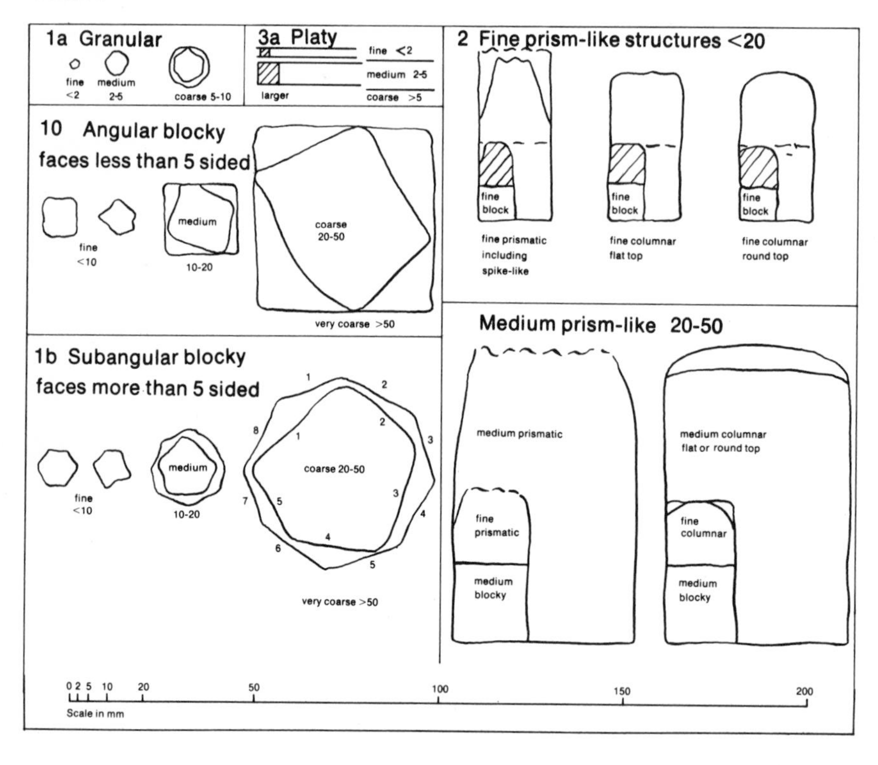

strong – the peds are distinct and retain their form when removed from the soil.
moderate – the peds are distinct, but when removed from the soil a small proportion break into fragments.
weak – the peds are only weakly developed in the soil itself.

The grade of structure is usually very dependent on the moisture condition of the soil at the time of examination. It is, therefore, not a stable property of the soil and will vary considerably with the time of year that the soil is viewed. The stability of the peds in a soil can be important. Aggregates especially at the surface, which resist the force of rain drops or disruptive actions of cultivation, ensure a more stable system of pores for the movement of water and air in the soil. This is usually beneficial for plant growth and tends to minimize surface erosion.

ORIGIN OF SOIL STRUCTURE

The development of structure in soils is usually the result of physical and chemical changes in the soil, often in association with biological factors. The major causes of structure are:

1 / *Chemical reactions.* In many soils the nature of adsorbed cations on the clay surfaces affects both the degree of aggregation and stability. In moist clays where monovalent Na^+ ions are prevalent, the clay particles are dispersed, whereas if divalent cations such as Ca^{2+} dominate, aggregates are usually present and relatively stable. Hydrous ions and aluminium oxides, perhaps sometimes combined with some form of organic matter, also can cement particles together. Many of the block-like and prism-like peds found in soils with relatively high clay contents, will have clay coatings or skins over the ped surface. These clay skins have resulted from a movement of clay in suspension followed by its deposition. The skins will usually give the ped a smooth, shiny surface.

2 / *Organic matter and soil organisms.* Many organic compounds are capable of aiding in the formation of aggregates. Humus materials commonly have high exchange capacities and, as a result, a high ability to bond ions at exchange sites. In many soils the addition of organic material is not only as litter to the surface but also as additions and secretions by roots in the soil itself. It is frequently observed that there is usually some aggregation of soil particles about root surfaces although whether the aggregation is related to the root itself or more likely to the activity of the root surface microorganisms is not usually clear. Many of the larger soil animals such as earthworms, millipedes, and arthropods comminute organic material and the peds are in effect their faecal remains. These may in turn be colonized by various microorganisms. Bacteria, actinomycetes, and fungi can cause aggregation in the soil, either by the production of gums and resins which may bind primary soil particles or by a mechanical 'net' effect such as is produced by fungal hyphae. The overall effect of organic matter, both living and dead, in creating structure within the mineral soil, especially in forest and range soils is evident when a profile is examined.

3 / Wetting and drying cycles are particularly effective in soils which contain clay minerals that will change volumetrically with changes in moisture content.

4 / Freezing and thawing cycles result primarily in the breakdown of larger aggregates or in the initial development of aggregates in amorphous massive structureless materials which become exposed to weathering.

Soil consistence

Soil consistence refers to the ability of a soil to resist deformation or stress and also to the degree of adhesion and cohesion of the soil mass. These properties vary considerably, not only with the soil, but also with the moisture status of the soil. The terms used to describe consistence vary depending on moisture: dry - loose, soft, hard; moist - loose, friable, firm; wet - non-sticky, sticky.

The consistence of a soil is commonly associated with the textural class. For example, when dry, sands may be described as having a loose consistence, but many clays are hard. If water is added, the sand remains loose, but the clay will normally be firm, until, as more water is added it becomes sticky and plastic. Wet sands are non-sticky and not plastic.

Soil porosity

The porosity of a soil is that volume of the soil not occupied by solids. The net effect of soil weathering or formative processes in bedrock or unconsolidated geological materials is an increase in porosity. The importance of the pore spaces in terms of soil organisms, water and air movement has already been mentioned. It is so important that one way of describing soil might be to term it an 'assemblage of holes.' The sizes of the pores will vary from the smallest, a few microns in diameter, in and between clay particles to large spaces several centimetres in width which may be formed by large tree roots or are the burrows of soil animals. Any factor which will affect soil structure will alter the size and arrangement of pores, if not total porosity in a soil. A study of pore spaces can only be made in undisturbed samples or in soils in the field, since pore spaces will be changed drastically when a soil sample is broken or crushed.

MEASUREMENT OF SOIL POROSITY

The measurement of soil porosity is concerned with one or more of the following features:

1 / *Total porosity.* The total pore space may vary not only with depth in the soil, but also with time, as a result of changes in soil structure.

2 / *Pore space distribution.* Two soils may have the same total pore space, yet the distribution of pore sizes may be quite different. One soil may have a preponderance of small-size pores, whereas another may have a few large pores.

3 / *Relative amounts of air and water in pores.* For most soils the amounts of air and water will vary considerably, even over short periods of time. For plant growth the total porosity may be less critical than the proportions of air and water.

One common method of measuring the total pore space of a soil is to push a sample cylinder of known volume into a soil so that the sample completely fills the cylinder, without being compressed or otherwise distorted. The contents of the cylinder are then placed in an oven and dried to a constant weight at 105°c.

The oven dry weight of the soil is determined and expressed on a volume basis. This is the *apparent* or *bulk density* of the sample.

If the density of the solids in the soil sample is also determined by means of a pycnometer, then the total pore volume of the soil sample can be calculated. For example, the oven dry weight of a soil was determined as 125 g in a sampling cylinder of 100 cm^3 volume; the bulk density = 125/100 = 1.25 g/cm^3. The density of the soil materials was determined as 2.65 g/cm^3, thus the proportion of the sample cylinder occupied by solids was 1.25/2.65; therefore (2.65 - 1.25)/2.65 represents the proportion of the sample volume occupied by space. This calculation may be made using the general formula

$$\text{Pore space (per cent)} = \left(1 - \frac{\text{bulk density}}{\text{particle density}}\right) \times 100$$

and for the example given above

$$\text{Pore space} = \left(1 - \frac{1.25}{2.65}\right) \times 100 = 52.8\%.$$

If the proportion of air and water in the pore space is to be determined, the procedure is similar except that, when the soil is sampled in the field with the cylinder, care is taken to ensure that no moisture escapes from the sample, and the fresh weight of the soil is determined before the sample is oven-dried. The difference between the fresh and dry weights of the sample represents the amount of water it contained and it is assumed that one gram of water occupies one cubic centimetre of volume. In the above sample, if the fresh weight of the soil was 142 g, then the amount of water was 142 - 125 = 17 g or 17 cm^3. Within the total pore volume, 17 cm^3 was occupied by water and the amount occupied by air was 52.8 - 17.0 = 35.8%. The volume distribution of the three main soil components was: solids 47.2%, water 17.0%, and air 35.8%.

The determination of bulk density and porosity in soils that are stony or contain woody roots is very difficult, if not impossible, with a metal sampling cylinder. A method sometimes used in such soils is to remove an irregular quantity of soil, whose fresh and dry weights can be determined, and then to refill the hole from which the soil came with a material of known volume such as a uniform dry sand. A variation of this procedure is to line the hole with thin, plastic sheeting and then fill it with a measured volume of water.

Another method (Rennie, 1957) involves taking a natural structural aggregate from the soil and by means of a series of weighings in air and in kerosene to calculate the bulk density, real density, porosity air and water volumes of the aggregate.

A method using a gamma-ray attenuation has been used for measuring both water content and bulk density in soil (Gurr, 1962) but the equipment is relatively elaborate and expensive.

TABLE 2.3
Bulk densities for three representative forest soils

Loamy Typic haplorthod[1] (podzol)		Clay loam Typic eutroboralf (gray brown luvisol)		Histosol[2] (peat)	
Horizon	Bulk density (g/cm^3)	Horizon	Bulk density (g/cm^3)	Layer	Bulk density (g/cm^3)
0	0.27	A1	0.84	Moss peat	
A2	1.20	A2	1.31	Live undecomposed	0.020
B21h	0.97	B21	1.23	Undecomposed	0.056
B22ir	1.11	B22t	1.38	Partially decomposed (woody inclusions)	0.153
B23,B24	1.37	C	1.61	Decomposed peat	0.237
B3	1.34			Herbaceous peat	0.125 – 0.156
C1	1.66				

1 Data from Hoyle (1973).
2 Data from Boelter (1964).

For a certain type of soil where bulk density may be difficult to determine, it may be possible to develop specific relationships with another property more readily measured. Curtis and Post (1964) for stony, sandy loam soils in the north-eastern United States developed a relationship between bulk density and loss-on-ignition. It was a curvilinear relationship

$$Y = 2.09963 - 0.00064X_1 - 0.22302X_2,$$

with

$$R = 0.96, \qquad \text{Standard Error } (Sy12) = 0.054,$$

where

$$Y = \log(\text{bulk density} \times 100), \quad X_1 = \log(\%\ \text{loss-on-ignition}), \quad X_2 = (X_1{}^2).$$

Reigner and Phillips (1964) on coastal plain soils found bulk density related to gravel content.

Another and more direct procedure is the use of an air pycnometer (Wooldridge, 1968) which involves subjecting a sample core of air-dry soil to air at constant pressure. The sample core can also be used in conjunction with a moisture pressure membrane to determine pore size distribution.

For forest soils generally, there is a progressive increase in bulk density from the soil surface with increase in depth. Total porosity then shows a converse relationship being greatest in the surface horizon and becoming progressively less with depth. This is illustrated by data given in Table 2.3. These changes in bulk density reflect the fact that amounts of organic matter in a soil are usually greatest in the upper soil layers and the bulk density of organic matter is much less than that for mineral particles. The soil weathering processes, together with the prevalence of soil organisms and roots in the upper part of the soil, tend to bring about greater development of soil aggregates and increased porosity.

When bulk densities of organic soils, especially peats, are determined, Boelter and Blake (1964) have shown that the bulk densities vary considerably, depending on the moisture content and it is preferable to use a bulk density based on 'wet volume.'

The use of the cylinder sampling technique to determine bulk density and total porosity is a convenient and simple method to determine the possible compaction effects of animals, humans, or machinery on soils.

3
Colour, temperature, and aeration

Colour is the most obvious property of any soil and historically the names that have been used to describe many soils do so by colour. For example, red and yellow podzolic, chestnut, terra rossa, black cotton, sols brun acide, and braunerde are names which have been used by soil scientists to describe different types of soils. To a large degree, colour of a soil reflects organic matter content and mineral composition. The amount and form of these materials are in turn affected by such properties as soil moisture, temperature and aeration; these properties control the rate and development of plant growth, the activity of organisms and form of compounds as well as the rate of chemical reactions in the soil. The measurement or description of colour has presented many problems. It was not until the 1940s that there was general use of the Munsell* colour notation system and even with colour charts, there are differences in colour perception among individuals, especially men. Nevertheless, the colour of soil is always noted in any field examination of a profile.

In contrast, soil temperature and aeration are two properties which are very seldom assessed in routine soil descriptions. If the physical properties such as texture, structure, and colour are considered as relatively stable, then those of temperature and aeration are among the most changeable properties of a soil, changing not only seasonally but daily and even from hour to hour in many situations. The temperature of a portion of the soil at any time reflects the balance in terms of heat energy transfers: aeration reflects the state of gaseous exchange, particularly for carbon dioxide and oxygen. The characterization of a soil in terms of its temperature or aeration involves a description of a regime of temperatures or conditions of aeration throughout a relatively long period of time, at least a year. Whereas many soil properties are only indirectly affected by fac-

* Munsell Soil Color Charts, Munsell Color Company Inc., Baltimore, Maryland, USA

tors at the soil surface or above it, soil temperature and aeration are directly affected by changes in heat energy received at the soil surface and the state of the atmosphere above the soil.

Although precise, quantitative documentation of soil temperature and aeration regimes is not possible during a field inspection, there are, nevertheless, objective assessments of texture, structure, bulk density, colour, and the general nature of the profile from which inferences can be drawn concerning both temperature and aeration regimes.

Soil colour

The colour of a soil is usually one of the first properties to be noted in a field description and, as has already been mentioned, soils have been categorized in the past by colour words. In temperate regions of the world, particularly, the darkness or lightness in colour of a soil has been used as an index of level of organic content and, by extension, of fertility. As with many generalizations, there are exceptions to this relationship and a healthy scepticism should be maintained towards such conventional wisdom.

To a varying degree a soil will reflect the colour of the geological or other material present in it. As the weathering processes associated with soil profile development proceed, the greater will be the modification of these intrinsic colours. Organic material tends to impart brown to black coloration to a soil and the amount and degree of decomposition will determine the intensity of the colour. If the organic matter is only partially decomposed it will give more of a brownish colour, whereas if it is well decomposed and is present as a fine humus, it will usually give the soil a blackish coloration.

Soil minerals containing iron are a major source of colour, as weathering takes place. Iron oxide (Fe_2O_3) is red in colour and, since it is one of the main products of weathering, its presence as a coating around other soil particles will give the soil a reddish hue. If the iron is subject to reducing conditions, as for example when aeration is poor, then ferrous forms which are grayish blue will occur. Limonite ($2Fe_2O_3 \cdot 3H_2O$), a hydrated oxide of iron, is yellowish in colour. Manganese compounds on weathering give a dark coloration in well-aerated soils because of the presence of MnO_2.

THE MEASUREMENT OF SOIL COLOUR

Colour descriptions of soils are made using Munsell colour notations. Essentially the colour is described in terms of *hue, value,* and *chroma.* Hue refers to the major colour (only red and yellow are commonly used in describing soils, although green, blue, and purple are other hue colours). The hue is given a desig-

Figure 3.1 Determining soil colour in the field using a Munsell Soil Color Chart (photo by K.A. Armson)

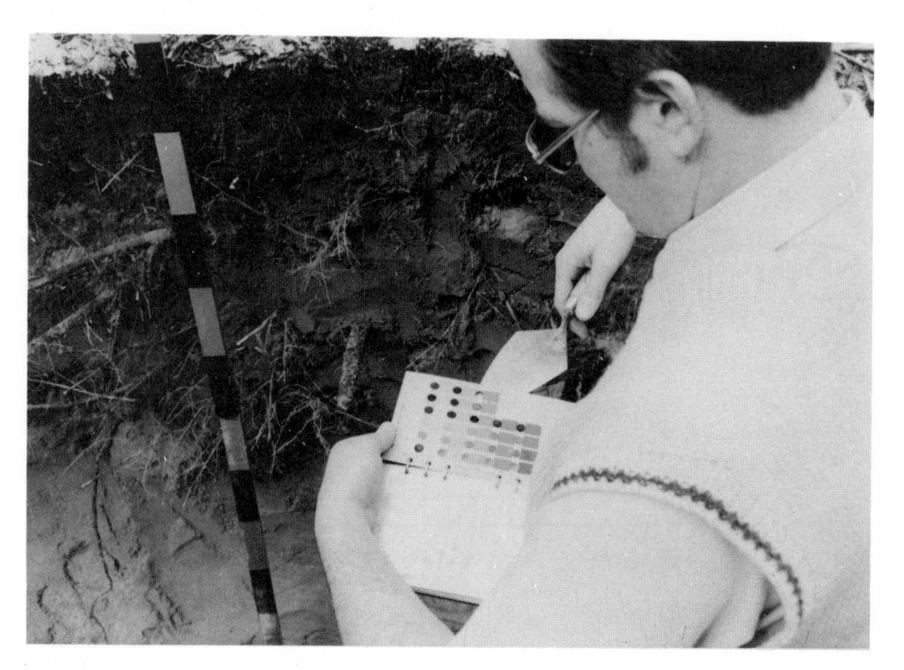

nation such as 7.5R or 10R; the number indicates the intensity of the particular colour, in this instance, red (R). Hue designations may combine colours as in 7.5YR (yellow and red). Value refers to the degree to which a colour is black or white; it is rated on the basis of numbers from 0 representing black to 10 for white. For most soils, values lie between 2 and 8. Chroma refers to the strength or departure of the colour from neutral; it is expressed numerically by numbers from 0 to 20 (0 to 8 for soils), where 0 represents neutral gray – the higher the chroma number, the brighter the colour.

In order to use a colour chart (Figure 3.1), a small sample is held in the openings beneath the colour chips and moved about until the closest matching chip is located. The colour is noted as hue, value/chroma, for example 10YR5/6. The quality of illumination is important in assessing colour. In the field, colour should not be determined in bright sunlight but rather under a condition of diffuse daylight. This will ensure greater consistency in describing soil colours and also minimize the drying out of the soil as it is being moved between colour chips.

Figure 3.2 Relationship between Munsell Soil Color Values and associated soil organic matter contents for reforested dune sand soils in the Netherlands. From J. Schelling, Stuifzandgronden. Inland-dune sand soils (1955), by permission

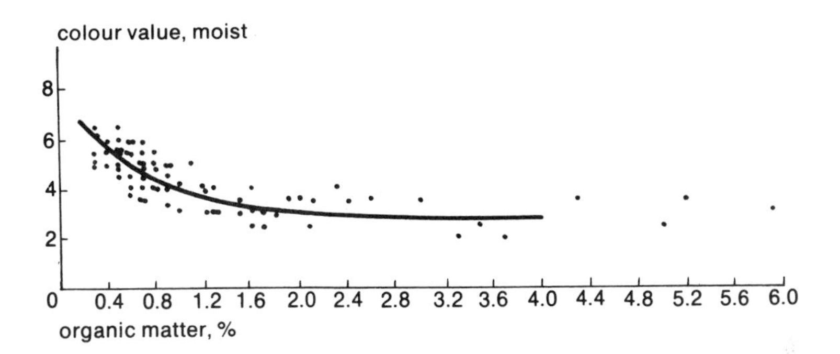

The texture of a soil determines the intensity of colour in relation to a given amount of colouring material. A certain amount of humus or iron oxide coating in a medium sand imparts a greater intensity of colour than the same amount of colouring matter in a silt or clay soil. The reason for this is that per unit volume of soil the total particle surface area increases as the particle size decreases. A soil containing particles with a diameter of 0.5 mm has less than 1/250th the surface area of a soil with particles of 0.002 mm diameter (the upper limit of clay size particles).

As the moisture content of a soil changes, so will soil colour. The drier a soil, the lighter its colour and the more moist it becomes, the darker it will appear. In the Munsell colour system, therefore, a soil in the moist state will have lower number values than when it is dry. This is also true for soils of the same texture which differ only in organic matter content – the greater the organic content the lower the value number. An example of this relationship is illustrated in Figure 3.2 for reforested soils in a limited area in the Netherlands. A close relationship between organic matter content and soil colour value exists over the range 0 to 1.5 per cent organic matter.

The degree to which the colours of red, yellow, and blue-gray, largely reflecting the state of iron compounds, are used to assess the degree of aeration in a soil profile may vary. Where blue-gray colours predominate, it can be inferred that reducing conditions, commonly associated with poor aeration, occur during a greater part of the year. If such blue colours or mottles do not occur it is usually assumed that well-drained, aerated conditions exist. The presence and degree of development of mottles is normally taken as evidence of partial or seasonal conditions of poor aeration.

Soil temperature

Soil temperature reflects the balance between heat energy gains and losses. Incoming energy is received primarily as solar radiation during the day and at the soil surface some part is back radiated, part is used in evaporation, and part moves into the soil along an energy gradient. At night when no solar radiation is received, the direction of the heat energy gradient in the soil reverses as there are net radiation losses from the soil surface, losses due to evaporation of water and also a heating of the soil itself. Cochran (1969) has outlined a number of soil thermal properties which to a large degree control soil temperature.

Thermal conductivity
This is a measure of the rate of heat transfer and is measured as the amount of heat flowing through a 1 cm^3 of a substance in 1 sec when one face of the cube is 1°c cooler than the face opposite to it. The units are cal/cm/sec/°c. Low values indicate slow rates of heat transfer at a constant temperature gradient and high values indicate rapid rates of heat transfer.

Volumetric heat capacity
This is the quantity of heat which must be supplied to a 1 cm^3 volume to raise its temperature 1°c. Its units are cal/cm^3/°c. The smaller the heat capacity the greater the temperature change with unit additions or gains in heat energy.

Thermal diffusivity
This is the ratio of thermal conductivity to volumetric heat capacity and thus is an index of the change in temperature with time and depth in the soil. The thermal conductivity, volumetric heat capacity, and consequently the thermal diffusivity of a soil will vary considerably with differences in other soil properties such as:

Organic content. The thermal conductivity and volumetric heat capacities for most organic materials are lower than for inorganic materials. For most mineral soils the variation in content of intermixed organic material probably has little or no effect since under field conditions other properties such as soil moisture content will dominate. Surface soil organic layers, common in many forest soils will result in less heat transfer to or from a soil depth and because of the lower heat capacity there will be a greater temperature change with heat gains or losses. On a 24-hour basis a litter mulch will result in greater soil surface temperature variation but less variation within the soil body proper (Figure 3.3).

Texture and porosity. Generally, thermal conductivities increase with increasing particle size. Sands have higher conductivities than clays, and denser minerals

Figure 3.3 Daily temperature variations with depth for an unmulched and mulched clay soil where the litter mulch has a much lower thermal conductivity, volumetric heat capacity, and thermal diffusivity than the soil. From P.H. Cochran (1969), by permission United States Department of Agriculture, Forest Service

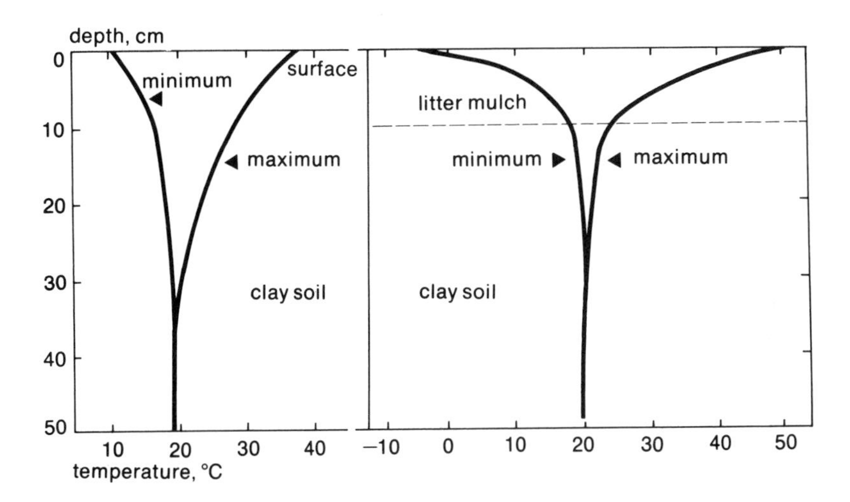

TABLE 3.1
Thermal properties of sand, peat, clay soils, and Lapine A1 and AC horizons (pumice) for a moisture content of 40 per cent by volume

Material	Thermal conductivity (cal/sec/cm/°C)	Heat capacity (cal/cm^3/°C)	Thermal diffusivity (cm^2/sec)
Pumice A1	0.00116	0.55	0.0021
AC	0.00125	0.55	0.0023
Peat*	0.00070	0.52	0.0014
Clay*	0.0038	0.70	0.0054
Sand*	0.0054	0.70	0.0077

SOURCE: Cochran, 1969
* Values taken from van Wijk, 1965.

have higher conductivities than those that are less dense. The thermal heat capacities also are generally less with decrease in particle size, but the degree of difference is considerably less than for conductivities (Table 3.1) Stoniness can affect soil temperatures. Troedsson (1956) found for forested soils in Sweden that summer temperatures were 1-3°C higher in most stony soils than in soils poorer in stones. Bouyoucos (1916), noted that with rapid changes in air temperature, the temperatures of the coarser-textured soils would alter more quickly than those of finer-textured soils or peat. The greater the porosity of a soil, the lower the thermal conductivity. Compaction of soils, especially the surface layers results in a greater increase in soil temperature during net heat gain than in a similar uncompacted soil.

Soil moisture. In most soils, the moisture content of the soil exerts a great influence on the soil temperature regime. The volumetric heat capacity of water is high (1 cal/cm^3°c) compared to that for most soil materials. The heat capacity of a soil increases therefore with increase in moisture content. Thermal conductivity for a soil also increases curvilinearly with increase in moisture content; the increase is greatest at the lower moisture contents. Thus diffusivity of a soil shows a general increase with moisture content to a maximum and then a decline. The moisture content at which the maximum occurs will vary from soil to soil. If the surface of the soil is wet, a large proportion of heat energy may be used to evaporate water.

Colour. The colour of a soil is important in terms of the heat budget only at the soil surface. The lighter the colour of the surface particles, the greater the proportion of incoming heat energy that is reflected, whereas the darker the surface particles, the greater the amount of energy which will be absorbed. The darker surface soils consequently will be warmer than the lighter coloured ones.

The external factors which affect the soil temperature regime most markedly are those which modify the amount of incoming solar radiation. Some are primarily factors of position, time of year, and topography. As latitude increases, the amount of solar radiation per unit area of the earth's surface decreases. This is modified by the seasonal changes in the earth's position with respect to the sun. At any particular latitude the slope of the surface and aspect will enhance or decrease the amount of solar radiation. Thus in the northern hemisphere a south-facing slope will receive more incoming radiation than level ground and a north-facing slope will receive less. The converse is true in the southern hemisphere.

The other main modifying factor of soil temperature is the vegetation which the soil supports. Vegetation intercepts some part of the incoming radiation. The amount of interception will vary with the degree of vegetative cover, the kind of vegetation, and the time of year or stage of development of the vegetation. Evergreen species will exert a more consistent modifying effect than deciduous spe-

Figure 3.4 Annual cycle of mean daily soil temperature for 11 years of data at Hubbard Brook. Forest cover - northern hardwood; soil a coarse loamy Typic haplorthod. From C.A. Federer (1972), by permission United States Department of Agriculture, Forest Service

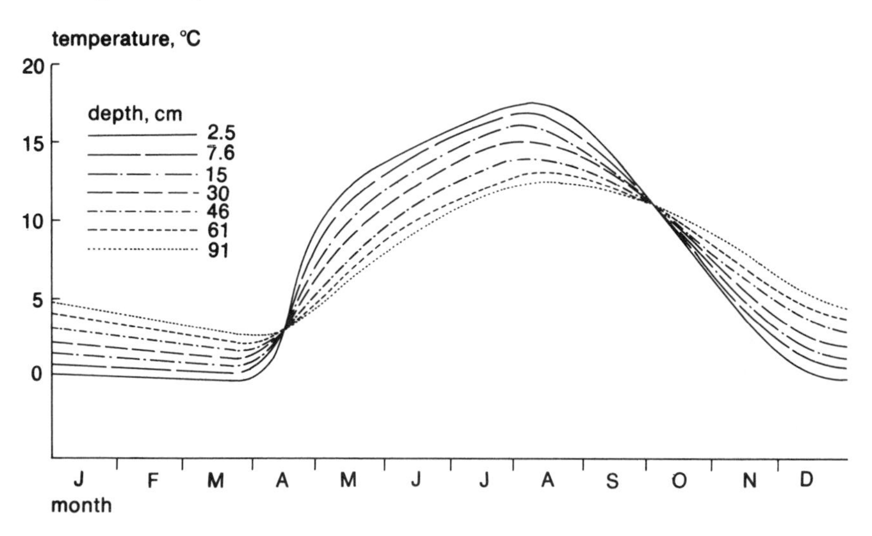

cies. During times when there is a net heat loss from the soil surface (as at night or seasonally), the vegetation reduces the rate of such loss. The effect of vegetation is to reduce both the total amount of radiation received by the soil and the rate at which it may be lost from the soil surface to the atmosphere. Both these actions minimize the extremes of soil temperature that occur. Li (1926) concluded that the effect of the forest cover in raising minimum surface temperatures was primarily due to the litter layer at the soil surface, whereas the effect of lowering the maximum soil temperature was due to the forest cover itself.

SOIL TEMPERATURE REGIMES

There are many techniques for measuring soil temperature. One of the most simple and convenient ways is to use thermistors. These are usually inserted in vertical stacks into the soil at various depths. The annual cycle of the mean daily soil temperature under a northern hardwood forest is shown in Figure 3.4. The soil temperature increases rapidly after snow-melt and reaches a maximum in early August. Isothermal conditions exist twice during the year, once in April during warming and once during early October as cooling takes place. This pattern of soil temperature regime will be modified, depending upon differences in forest

Figure 3.5 Mean daily air temperatures and weekly soil temperatures under Monterey pine in New Zealand. From G.M. Will in New Zealand Journal of Agricultural Research (1962), by permission

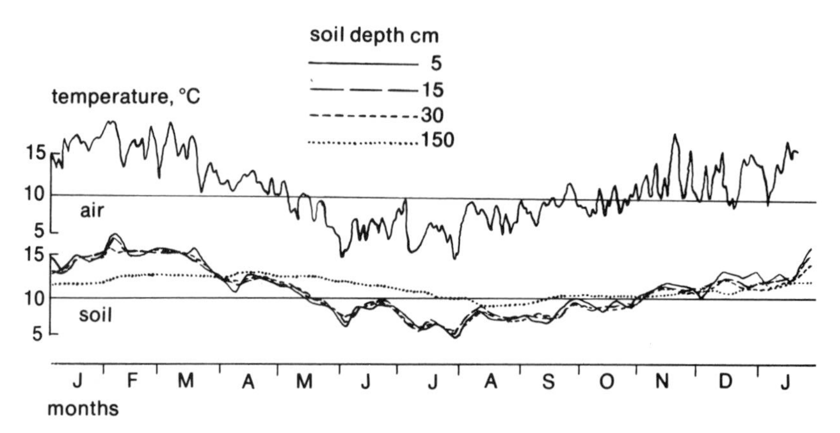

cover and latitude. Figure 3.5 shows the soil temperature regime under Monterey pine growing in New Zealand.

The significance of the level of soil temperature and the regime or seasonal fluctuation are considerable. The activity of roots of higher plants is to a large extent temperature controlled, increasing with increase in temperature to some maximum. In many forest soils the stratification of temperature with soil depth during a major portion of the year may result in parallel sets of rates of both chemical and biological activities. As shown by Figure 3.4, the soil temperatures in June and July at 2.5 cm are at least 5°c greater than temperatures at 91 cm depth. The root systems of many trees are so distributed in soil that, while roots near the surface may be at a temperature which permits rapid growth, roots lower in the soil may be at soil temperatures which limit or even prevent root growth entirely.

It has already been noted that freeze-thaw cycles can affect soil structure. The frequency and depth in the soil to which these occur can alter soil physical conditions to a considerable degree and by doing so the porosity and opportunity for change in biological and other soil weathering processes.

SNOW AND FROST

In areas where below-freezing air temperatures persist for extensive periods of time and where precipitation in the form of snow is common, the freezing and thawing process, depth of frost penetration, and amount of snow all have hydrologic importance. Soil temperatures are affected by snow cover in particular.

TABLE 3.2
Soil temperatures and their variation for snow covered and bare plots – New Hampshire (data of Hart and Lull, 1963)

Depth below soil surface (cm)	Mean temperature (°C)		Coefficient of variation (%)	
	Snow-covered	Bare	Snow-covered	Bare
0.6	0	-3.2	2	40
7.5	0	-3.6	1	22
15.0	0.9	-2.8	1	17
30.0	1.7	-1.2	2	11
61	2.7	0.4	3	7

Snow moderates soil temperatures, so that they are less under snow cover than when the soil is bare (Table 3.2). The effect is shown first in the upper soil layers; Hart and Lull (1963) noted that soil temperatures at depths of 0.6 and 7.5 cm were affected more quickly than at depths of 30 and 60 cm where there was a three week lag. The type of forest cover will influence the amount of snow at the soil surface; generally the snow cover is greatest under deciduous forest and least under evergreens. For example, Striffler (1959) noted that the depth of frost penetration was greatest in soil under a closed red pine canopy and least under northern hardwoods. Although he related this to the greater snow depth under the hardwoods, he also considered that the nature of the soil horizons exerted an influence. The hardwood soil was a mull with a well-developed A1 (Ah) and was associated with granular or honeycomb frost, whereas under the pine, surface O(L-H) layers were not mixed with the mineral soil and were associated with concrete frost development (Figure 3.6). Pierce et al. (1958) noted in the northeastern United States that frost was observed twice as frequently in open land as in forested soil. They also found that frost depths under hardwood forests were only half those under conifer stands.

Soil aeration

Soil aeration refers to the process whereby air in the soil is exchanged with air from the atmosphere. There are three gases which constitute the greatest proportion of both the atmosphere and soil air. Nitrogen occupies some 79 per cent by volume of each, oxygen 20–21 per cent, and carbon dioxide 0.02–1 per cent. Both in the atmosphere and in the soil there can be wide variations from place to place, particularly in oxygen and carbon dioxide levels. Generally, the levels of oxygen are lower in the soil air than the atmosphere above and the levels of

Figure 3.6 Frequency of frost occurrence under different forest cover in northern Lower Michigan. From W.D. Striffler (1959), by permission United States Department of Agriculture, Forest Service

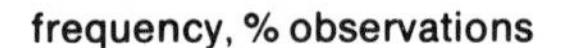
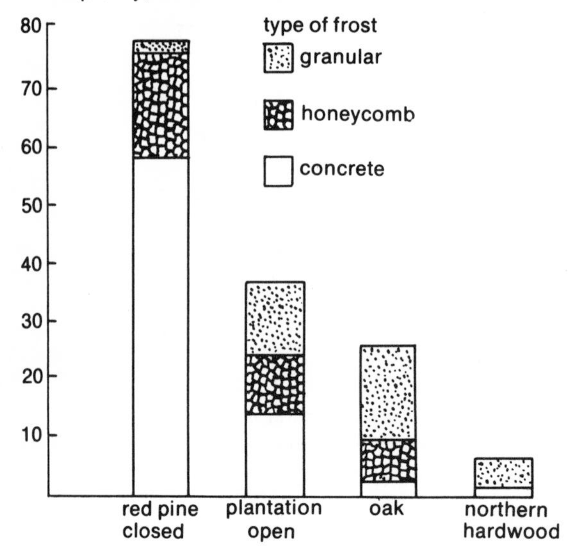

carbon dioxide are higher. This reflects the interrelations between biological activity and these gases; oxygen is required to ensure metabolic functions of aerobic organisms and carbon dioxide is produced as a result of respiration. In soils the concentrations of oxygen and carbon dioxide are inversely related; as one increases, the other decreases.

The total amount of air in the soil is related to the soil porosity and to the moisture content of the soil at any time. As the volume of pore space occupied by water increases, the amount filled with air must decrease, and consequently soils with high moisture contents may be poorly aerated.

The soil water will contain dissolved gases and, depending on the temperature and rate of movement of the water, the amount of dissolved oxygen may be significant for many organisms. The total air space in the soil may be misleading as an index of aeration if the pores at the soil surface are closed, thus preventing gaseous exchange with the atmosphere.

It is generally considered that gas exchange is affected by two mechanisms. *Mass flow* represents a response by gases to gradients in total pressure between the soil and the atmosphere. Temperature will affect the specific volume of a gas and also bring about convection interchange, warm air rising and being replaced by cooler denser air. This may occur in soils with marked diurnal changes in soil temperature, especially near the surface. Changes in barometric pressure will also result in mass flow exchange between the soil air and the atmosphere. Mass flow is of less importance in bringing about gas exchange than a second mechanism, *diffusion.* By diffusion, exchange of a gas takes place as a consequence of partial pressure differences between the gas in the soil and the same gas in the atmosphere. Thus in a soil, particularly during periods of biological activity, there is primarily a diffusion of oxygen into the soil from the atmosphere and carbon dioxide out from the soil. Diffusion of a gas in the soil is related to the following factors:

1 / *Concentration gradient.* The volume of a gas diffused is dependent on the concentration gradient; the steeper the gradient, the greater the volume which will diffuse.

2 / *Density of gas.* The rate of diffusion is inversely proportional to the density of the gas (Graham's Law of Diffusion). Since the density of a gas is directly proportional to its weight, this means that, for a given pressure difference, oxygen will diffuse more rapidly than carbon dioxide and hydrogen will diffuse at four times the rate of oxygen.

3 / *Temperature.* The rate of diffusion varies directly with absolute temperature, increasing with an increase in temperature.

4 / *Free pore space.* There is a relationship between the rate of diffusion and free pore space, diffusion increasing with increase in pore space over a wide range of values. There have been many detailed studies of this relationship since Buckingham's classic work of 1904, but since the 1950s greater emphasis has been placed on *in situ* measurements of diffusion rates, particularly of oxygen.

MEASUREMENT OF SOIL AERATION

Measurement of soil aeration in early studies consisted of extracting samples of gas from the soil and analysing for oxygen and carbon dioxide concentrations. Samples were extracted by various means, syringe, pumping out, or by burying and then extracting sample containers. Lemon and Erickson (1952, 1955) described a technique using a platinum microelectrode as a method for characterizing soil aeration. The advantages of the technique were twofold. First, it enabled oxygen diffusion rates to be measured directly. Secondly, the measurement takes place in a pore but as a result of oxygen diffusing across a water film to an oxygen-reducing surface similar to that of a plant root. The principle employed is that, if a certain electrical potential is applied between a platinum electrode in the soil and a reference electrode, oxygen is reduced at the platinum surface. The electric current flowing between the two electrodes is then proportional to the rate of oxygen reduction, and hence it is possible to determine the oxygen diffusion rate (O_2 μg cm^{-2} min^{-1}). Letey and Stolzy (1964) have reviewed the theory and present details of the construction of equipment. The factors that influence the measurement of oxygen diffusion rates have been reviewed by Birkle et al. (1964).

SOIL AERATION REGIMES

Within a soil the general pattern is for the oxygen concentration to decrease with increasing depth in the soil, and the carbon dioxide concentrations to increase. During a season, the levels of moisture will greatly affect the gas concentrations. Thus oxygen concentrations will usually be considerably less as soil moisture levels increase, whereas carbon dioxide levels will increase. In well-drained coarse-textured soils, diffusion of gases is most rapid, but with increasing fineness of texture there is a tendency for diffusion rates to decrease, largely in relation to an increasing amount of water films through which the gases must pass.

For forest soils most information on aeration relates to drainage practices. For example, Lees (1972) measured oxygen diffusion rates in an afforested peatland in which drainage was affected by plowing. Oxygen diffusion rates were significantly greater in the upper part of the turves and decreased markedly with depth.

The root systems of higher plants and microorganisms in the soil are subjected to a constantly changing regime of aeration, and one of the pronounced difficulties in studying aeration is the dynamic nature of the supply. Most studies of aeration have involved either the experimental maintenance of oxygen supply at a constant rate or have used compacted soils of high bulk density to indirectly assess the possible effects of reduced aeration. This latter procedure, however, does not distinguish between effects due to aeration, moisture, and mechanical impedance. Leyton and Rousseau (1958) studied the root growth of seedlings of a number of conifer species and found that, when oxygen levels were reduced from 20 per cent to 10 per cent, there was a reduction in the rate of root elongation for all species, but that it was least for Scots pine and greatest for Norway spruce; black spruce, jack pine, and Sitka spruce were intermediate. At concentrations of 5 per cent, root growth for all species was very small with little difference between species. Armson and Struik (1969) found for 2-year-old black spruce grown at three oxygen concentrations under simulated outplanting that, although height extension was not reduced at the 10 per cent oxygen level, there was a decrease in root development. Both height and root development were markedly reduced at 4.5 per cent oxygen concentration (Table 3.3).

TABLE 3.3
Black spruce seedlings (2 years old); mean height increment (cm) and seedling root surface area increment when grown in oxygen concentrations of 19, 10, and 4.5 per cent

Oxygen concentration (%)	19.0 ±1.0	10.0 ±2.0	4.5 ±0.5
Height (cm)	8.5	8.5	3.6
Root area (cm^2)	23	14	1.9

4
Soil water: the lifeblood of soil

Water occurs in the pore spaces of soil in one or more of three phases, as a liquid, solid, or gas. Most studies of the movement and retention of water in soils have been concerned with it as a liquid, but water vapour and ice are forms of water which may be significant for shorter or longer periods in certain situations.

The water content of a soil is continually changing during a large part, if not all of the year and these changes in content result from additions and losses to the soil system. Within the soil pores which are not filled with water as a liquid or solid, the soil air is usually nearly saturated with water vapour. Major movement of water occurs in the liquid form.

Concepts of the mechanisms by which water is retained and moves in the soil have relied mainly on the consideration of soil as a physical system – a relatively stable matrix in which water moves in responses to differences in energy state. We explain the movement of water through a pipe from a higher elevation to a lower elevation as due to the higher energy state of the water above, equivalent quantitatively to its vertical distance above the lower position. So we refer to the movement of water in soil as being a result of differences in energy state; water flows from a position of higher to one of lower energy. The origin of this energy is of major concern then in understanding the behaviour of water in soils. In many soils the pore system is not fully filled with liquid water and the extension and movement is more involved for such an unsaturated state than if fully saturated conditions prevail. The soil resembles a complicated plumbing system involving a wide range of pipe sizes in which there are many air blocks. The movement of water vapour is dependent on a contiguous system of open pores and takes place largely in response to partial pressure differences resulting from temperature changes in the soil. Thus water vapour will move from warmer to cooler soil positions; the amount of water moving under these conditions is negligible compared with what will normally move in the liquid phase.

The occurrence and development of different pore systems within a soil, and the heterogeneous nature of the soil along with the presence of organisms and roots, serve to complicate and modify soil moisture movement and retention. To assess or interpret the moisture dynamics of a soil, it is necessary to have a clear grasp of the basic principles that determine the energy state of water at any position and time in the system.

Soil water is important to man not only because of the amounts which he may utilize directly for his own needs or those of the plants upon which he depends, but also indirectly because of the many other biological and chemical processes by which soil sustains life, all of which require the presence of water. Water is the 'lifeblood' of soil.

Energy relationships of water

The energy - termed the *total potential* - that water has in the soil has several origins. These are:

1 / *Gravitational potential.* This is the energy of position or elevation within the soil. Water at one position has greater or lesser gravitational potential depending on its vertical distance above or below the second position.

2 / *Osmotic potential.* One of the attributes of soil water is that to varying degrees it contains solutes. Such solutes lower the vapour pressure of water and thus may affect the diffusion of water vapour in the soil. This is not usually of consequence in terms of the overall movement of soil water. Osmotic potential is of main significance in the absorption of water by living organisms since the cell membranes act as semi-permeable barriers.

3 / *Matric potential.* This is the energy that soil water has, and results from the forces of capillarity and adsorption associated with the soil matrix.

Water molecules are asymmetric in terms of the arrangement of their hydrogen atoms and exhibit polarity, an excess negative charge on the oxygen side and an excess positive charge on the hydrogen side. As a result, water molecules may be held by adhesive forces to the surfaces of soil particles. In coarse-textured soils where the pores are relatively large, the amount of water retained by such adsorptive forces is negligible, whereas in finer-textured soils it may be sizeable. Many of the pore spaces in soils are of a size that allows capillary forces to prevail. The contact of the water with one solid surface is illustrated in Figure 4.1(a). A water molecule at B is subject to its weight acting vertically (w) and forces of cohesion with surrounding water molecules and adhesive forces acting normally to the surface of the solid. At C, no adhesive forces affect the water molecules, but as the solid surface is approached the increasing effect of adhesive forces results in a movement of the liquid surface to A. In capillary-size pores (Figure

Figure 4.1 (a) Diagram of curvature of water in contact with a solid plane surface, *A*, *B*, and *C* are three positions of water molecules and *a* is adhesive force; *w*, force due to weight of molecule, and *c*, resultant of cohesive forces. (b) Diagram illustrating rise of water in capillary tube: *r*, radius of curvature; α, angle of contact; *C′* and *D* represent positions of water molecules separated by vertical distance *h*.

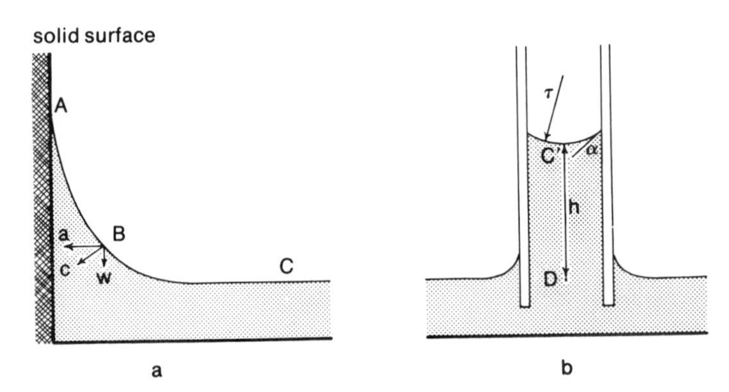

4.1(b)), the rise of water in the tube is represented and the pressure of a molecule at C' is less than atmospheric by an amount represented by $(2T \cos \alpha)/r$ where T is the surface tension of the liquid, α is the angle of contact, and r is the radius of the tube.

The pressure of a molecule at D, on the same level of the surface of the liquid, is atmospheric and the difference in its pressure and that at C' can be expressed by pgh, where p is the density of the liquid, g is the acceleration due to gravity, and h is the vertical distance between C' and D. Thus

$$\text{Pressure difference} = (2T \cos \alpha)/r = pgh.$$

As the pressure difference is inversely proportional to the radius of the pore, the smaller the pore, the greater the pressure difference or capillary force. If the energy level of the free water surface outside the capillary is used as a datum and assumed to be zero, then the energy level of the water beneath the miniscus is less and is a negative value. By analogy, in soils, the water held within capillary size pores is at a lower energy level than that at a free water surface at the same soil position and the term *matric suction* is frequently used rather than matric potential; this use implies that the force of pressure is negative.

The total potential of the soil water at any position is the sum total of the individual potentials, of which three of the major ones have been described. For most forest soils the osmotic potential, in relation to water movement and retention in the soil can be disregarded and the matric potential and gravitational potential are thus the main considerations. The nature of the matric potential is determined by the capillary and adsorptive forces which are a function of the physical surfaces and pore sizes of the soil; the gravitational potential is dependent on the degree of vertical displacement or elevation within the soil. It is therefore possible, from a knowledge of these physical properties, to postulate the probable characteristics of water retention and movement in a soil.

Movement of water in soil

Movement of water takes place in both liquid and vapour states, but because of the quantities involved only liquid movement is discussed.

Initially it is convenient to consider a simple situation of saturated flow. The quantity of flow, q, may be expressed in an equation known as *Darcy's Law*:

$$q = K\Delta H/L,$$

where $\Delta H/L$ is the *hydraulic gradient*, i.e. the difference in hydraulic head per unit of column length. K is the proportionality factor and is termed the *hydraulic conductivity*.

This equation, for saturated flow, has also served as a model in attempts to quantify unsaturated flow, which is the more usual condition in soils. The moving force instead of a hydraulic gradient is the water potential gradient and is largely made up of the matric potential or suction component. The proportionality factor K is modified and represents a conductivity, associated with differences in water potential. Conductivity in unsaturated, as compared with saturated, soil is subject to considerable modification.

1 / In a saturated soil, pores of all sizes are filled with water, but in an unsaturated soil the size of pores in which water moves varies. Water held at high matric suctions moves through pores of smaller radius than water held at lower suctions. According to Poiseuille's Law, rate of flow of water will vary with the fourth power of the radius of the tube or pipe; hence a reduction in radius by one-half results in a sixteenfold reduction in flow-rate.

2 / As water moves from a position of lower matric suction (high energy) to one of greater suction (low energy) the water will move progressively from the larger pores first. As these empty, the pore fills with air, providing a barrier to water moving through smaller pores which lead into the larger air-filled pore. The route of water movement becomes more tortuous and less direct, so that the conductivity is reduced.

Conductivity in unsaturated soils is very much affected by the size and continuity as well as the relative abundance of different-sized pores. If a coarse-textured soil, such as a sand is close to saturation, the conductivity of water in it will be greater than the conductivity for a finer-textured soil holding water at the same low matric suction. As the water moves in relation to increase in matric

Figure 4.2 Conductivities for silty clay loam and a sandy clay loam plotted against soil matric suction. Adapted from D.R. Nielsen, D. Kirkham, and E.R. Perrier in *Soil Science Society of America Proceedings* (1960), by permission Soil Science Society of America

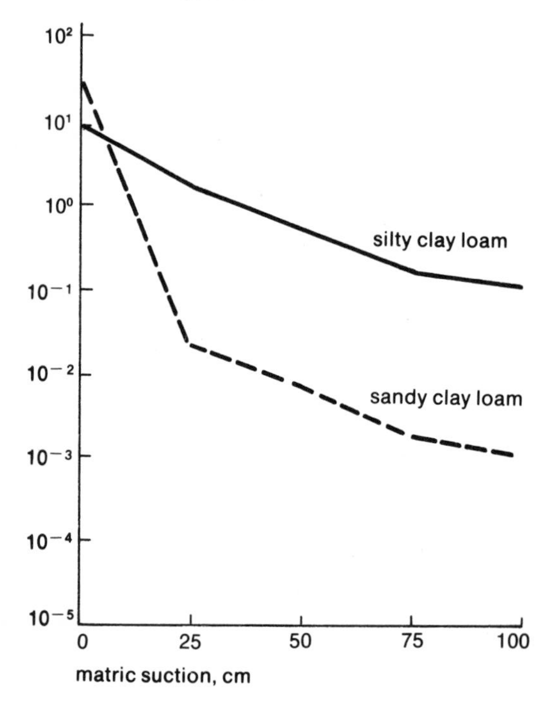

suction, i.e. in response to difference in energy state, the conductivity of water in a coarser-textured soil is reduced to a greater extent than the conductivity in a finer-textured soil. Figure 4.2 illustrates changes in conductivity for two soil materials, a fine-textured-silty clay loam of loessial origin and a coarser-textured sandy clay loam of till origin.

As water moves into a dry soil it does so almost as a curtain falls across a stage; the boundary between the wet and dry soil is not only a moving one as long as water is being applied, but it is usually a distinct one. This boundary is termed the *wetting front.* Commonly the wetting front is observed moving downward, but since the movement is primarily a function of differences in soil water matric potential (suction), the wetting front initially moves in an essentially equidistant manner from the point of water addition. At the wetting front the moist-

Figure 4.3 (a) Increase in soil water content with depth during a 24-hour infiltration period. Soil was a moist clay loam. Broken line indicates water content necessary to saturate profile. (b) Soil water tension versus soil depth during 24-hour infiltration period. From D.R. Nielsen, J.M. Davidson, J.W. Biggar, and R.J. Miller in *Hilgardia* (1964), by permission University of California

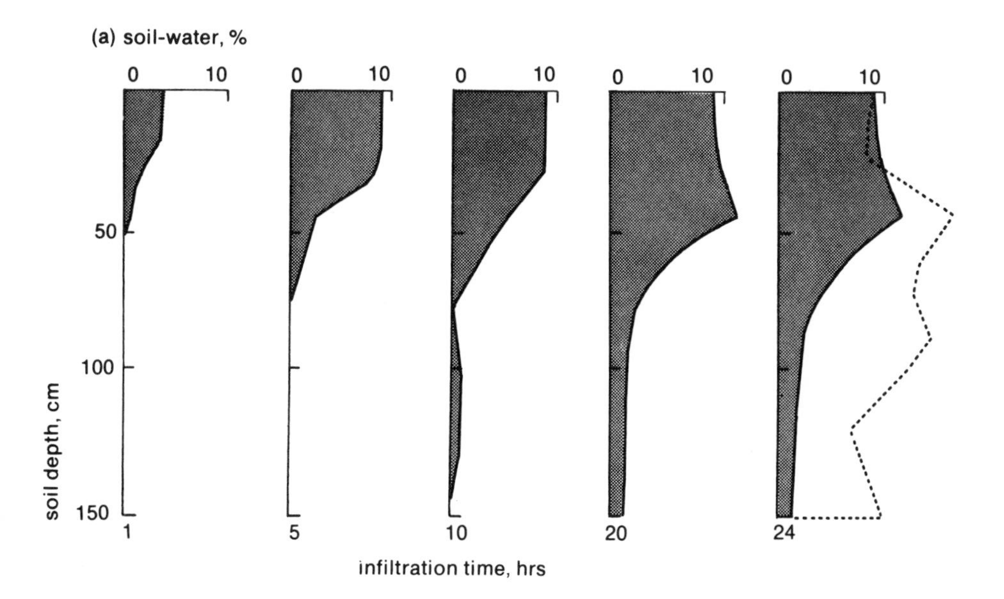

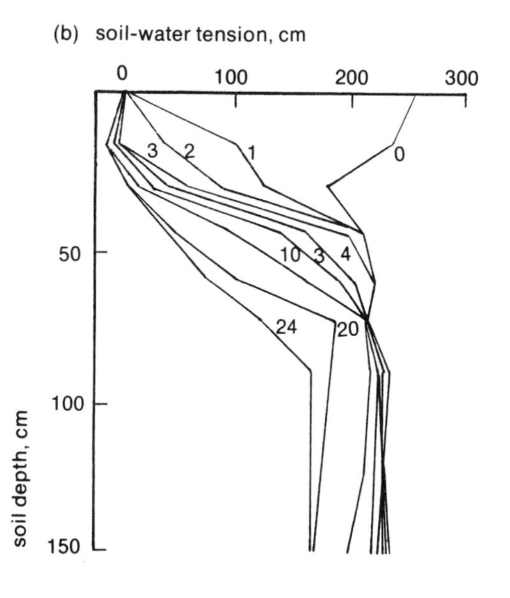

ure gradient is very steep (the reason for the distinct boundary between the dry and wet soil). Immediately behind the wetting front is a band termed the *wetting zone*; in it the gradient of moisture increase is much less steep than at the wetting front. Above this zone is the *transmission zone* in which the moisture gradient is slight and matric potential is highest, except for the point of entry of free water into the soil. Consideration so far has been only of the wetting of a dry soil. Under natural conditions, water is usually moving into soil that is already moist and the wetting front is then more diffuse in relation to a less steep moisture gradient. A soil moisture profile during infiltration is shown in Figure 4.3 (a,b). As long as water is continually being applied to the soil surface, movement of the wetting front will continue. When addition of water stops, the subsequent movement of water in the soil is dependent largely on the texture and structure of the soil. In coarser-textured soils or finer-textured soils with considerable structure, the rate of movement of the wetting front slows until, after one or two days, it becomes imperceptible. In some finer-textured soils the movement of the wetting front becomes less with time but may continue for months. The moisture condition in the soil transmission zone behind the wetting front, after the front has stabilized, is often referred to as the *field capacity* moisture status of the soil. In the past, field capacity has been defined as the moisture status of a soil two to three days after it has been saturated and free drainage has virtually ceased. Such an occurrence is usually infrequent in well-drained soils. For this reason and because field capacity also describes the moisture condition which stabilizes in a soil a few days after it has been wetted by rain or irrigation, the moisture content of a soil described as being at field capacity is not an exact value. Empirically, for many soils it is considered to be the moisture content at matric suctions equivalent to 0.1 to 0.33 atmosphere tension. Obviously, finer-textured soils in which the wetting front continues to move do not have a field capacity.

The moisture status of a soil is normally in a constant state of change. As water is added, either at the surface in the form of precipitation or from upward movement from a water-table, it is redistributed within the soil. During a large part, if not all of the year, water is being lost from the surface by evaporation and from the soil due to absorption by roots. Frequently within the same soil, one part may be on a wetting cycle while another portion is on a drying cycle. A diagram of the relationships between moisture content and associated matric suction is given in Figure 4.4. The decrease in water content with increase in suction is represented by a continuous drying curve. This would occur as water evaporates from the surface or as roots absorb water from a soil. A continuous curve also represents the wetting cycle, but it is different from the drying curve in that over a major portion of the range of suction the soil of the same suction

Figure 4.4 Diagram of relationships between water content and associated soil-water suctions for a soil on drying and wetting cycles.

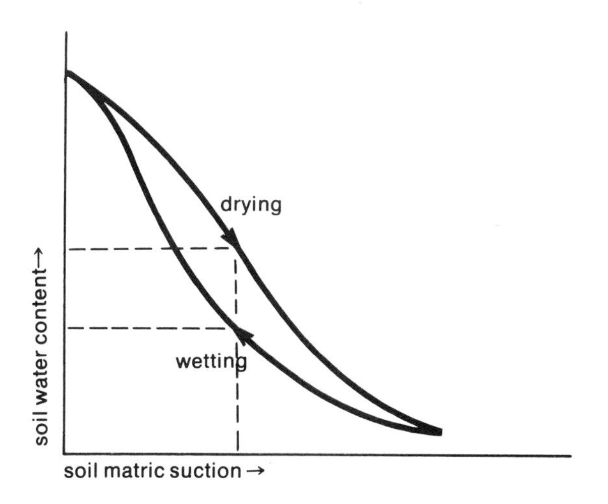

will contain more water on a drying cycle than on the wetting cycle. This is an example of *hysteresis* - the moisture status being contingent on the direction of the energy gradient (soil moisture suction). The continuous curves (Figure 4.4) are termed *soil moisture characteristic curves*, but obviously they are not necessarily unique for a soil; hysteresis prevents that. The main causes for variation are:

1 / As the moisture content of a soil changes, shrinking or swelling may take place. This is particularly evident in many soils with appreciable quantities of clays of certain types and/or organic matter. The colloidal nature of these components allows for volumetric changes with different moisture contents and this can result in altered pore sizes and pore arrangements. Such changes would not account for hysteresis but rather a random difference in the soil water versus suction curve. Coarse-textured soils of low organic content will show such changes only to a slight degree.

2 / Air may be entrapped in the soil as the wetting process proceeds and this results in the wetting curve water contents being lower at a given suction than those in the drying curve.

3 / A phenomenon which Hillel (1971) termed the 'ink bottle' effect. This refers to the fact that at an interface, movement of water will take place when the suction at the interface exceeds that of water in the bounding pore with the

Figure 4.5 Illustration of 'ink bottle' effect: (a) Removal of water from pore R is dependent on matric suction at small pore r; (b) Movement of water into empty pore R will proceed when matric suction in R exceeds that at r.

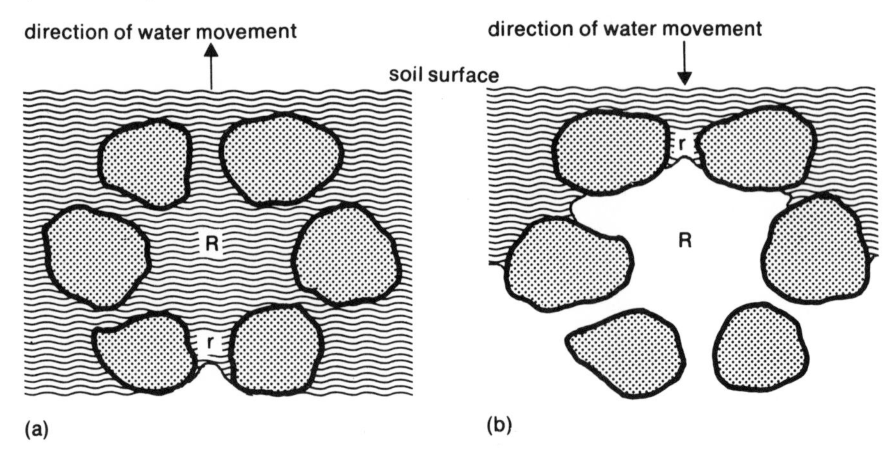

smallest neck or radius. Conversely if the bounding pore is empty, refilling of the pore system from the interface takes place in relation to the suction of the maximum radius of the pore (Figure 4.5).

There are certain general differences in moisture curves for soils. Coarse-textured soils have most of their water held at low suctions (high matric potentials) and finer-textured soils tend to a somewhat more uniform distribution in water content across the range of suctions and a greater total water capacity.

MOVEMENT IN LAYERED SOILS

So far, the discussion of water movement has considered soils as uniform or homogeneous in their pore distribution. A majority of forest soils are heterogeneous in their structure, their pore space, and pore distribution, as a result of soil horizon differentiation. In many soils, textural discontinuities are often present as a result of geologic differences in both materials and mode of deposition. Miller (1969) reviewed the general principles involved and used both models and field profiles to illustrate the effect of underlying coarse-textured layers on water movement. Such underlying coarse layers result in a finer-textured soil above retaining considerably more water than if it were uniform and of a similar depth. Miller and Gardner (1962) measured rates of infiltrating water into stratified soil (fine over coarse) and found that the rate depended on the nature of transmitting pores, their water content, and the water potential gradients existing and developed as flow takes place. The effects of layering on infiltration rates

Figure 4.6 Effect of subsurface layers of sand on infiltration rates into standard soil as a function of time for downward wetting. Layers were 0.5 cm thick at a depth of 8 cm. Diameter size classes of sand in mm. From D.E. Miller and W.H. Gardner in *Soil Science Society of America Proceedings* (1962), by permission Soil Science Society of America

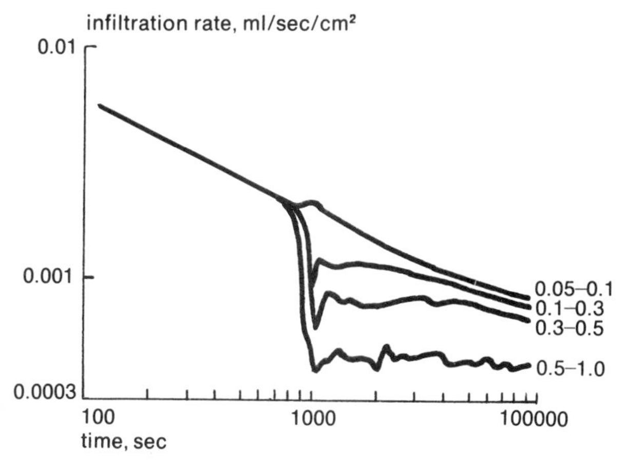

are shown in Figure 4.6. The rate decreases with increases in particle size and hence in pore size. The amount of water held by a layered soil increases as the underlying particles become coarser; it is greatest in the upper layer adjacent to the soil-layer interface. Water movement (conductivity) from larger to smaller pores is practically unrestricted at the contact zone if the pore volume of both is about the same and the pore size difference is not extremely great (Eagleman and Jamison, 1962). Such would be the case for a fine sandy loam over a silt loam. If the pore volumes and sizes are considerably different as for a sand over a silt loam, Eagleman and Jamison (1962) found that, following saturation, the conductivity at the place of contact remained fairly constant but, after the large pores in the sand were emptied, the conductivity was virtually zero since there were only a small number of small pores in the sand in contact with a large number of small pores in the silt loam.

Soil water measurement

The measurement of soil water usually involves estimates of two parameters. One is the quantity of the water in the soil and the other is its energy status. These measurements may be obtained in a number of ways, some of which involve direct, and others, indirect measurement. Wilson (1970) has presented a general review of the advantages and disadvantages of the techniques used to measure soil water.

Gravimetric technique

A sample of soil is weighed moist (W_w); it is then dried in an oven at 105°c until it is at constant weight (W_D). The moisture content (per cent) is then calculated as:

$$\text{Moisture content (per cent) weight basis} = \frac{W_w - W_D}{W_D} \times 100$$

If the bulk density of the soil is also determined, and this is usual if core samplers are used, then the moisture content is best expressed on a volume basis:

$$\text{Moisture content volume basis} = \frac{\text{MC\% (weight basis)}}{100} \times \text{bulk density.}$$

If one gram of water is assumed to occupy 1 cc, then the moisture content on a volume basis is equivalent to surface depth of water per unit soil depth. For example, the moisture content (weight basis) was determined as 16.5% and the bulk density as 1.20 g/cm^3. Then moisture content (volume basis) = (16.5/100) $\times$ 1.20 g/cm^3 = 0.198 cm^3/cm^3; because the units cancel out, this value has been termed the *water ratio* (Richards and Richards, 1957). This amount of water is equal to 0.198 $\times$ 1 cm = 0.198 cm per centimetre of soil depth. If the soil horizon for which these data (moisture content and bulk density values) apply was 8.5 cm thick, then the amount of water in that horizon would be 0.198 $\times$ 8.5 cm = 1.68 cm. This type of conversion is a very useful one since it enables the quantity of water in the soil to be expressed in the same terms used to measure precipitation. The water ratio also has the advantage that it is independent of units of soil thickness measurement. For example, if the soil horizon depth was 8.5 inches, the amount of water in the horizon would be 0.198 $\times$ 8.5 inches = 1.68 inches.

The gravimetric technique is most useful in homogeneous, inorganic materials of relatively coarse texture. It entails disturbance of soil and if repeated samples are to be taken, this can pose serious difficulties. Drying at 105°c does not usually remove all adsorbed water and in some soils, higher in certain types of clay, the amount of adsorbed water may be significant.

Generally, bulk densities should be determined for the soil at the same time as moisture content so that the amount of water may be expressed on a volume basis. This is of utmost importance in soils with any organic matter and/or content of stones.

Figure 4.7 Tensiometers used to measure soil moisture. Two tensiometers, carrying case (far left) and soil probe (second from right) used with portable tensiometer (second from left). Photo courtesy Ontario Ministry of Natural Resources

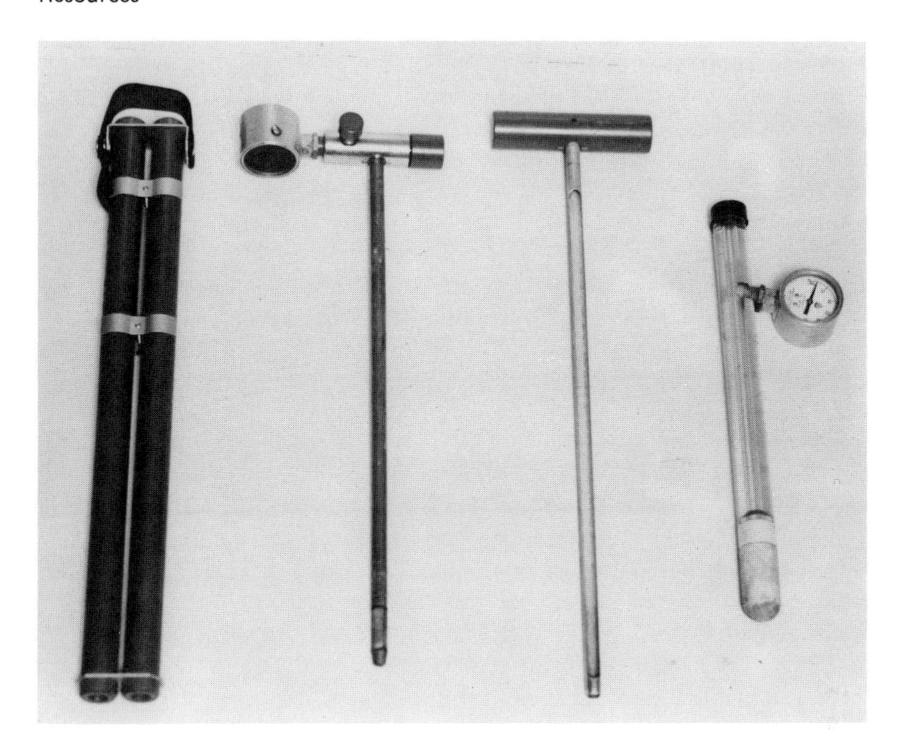

Tensiometers

The matric potential or suction of soil water is most readily measured using a tensiometer (Figure 4.7). It consists of a ceramic container filled with water and connected to a manometer or vacuum gauge. The tensiometer is installed so that the ceramic portion is in complete contact with the soil. If the soil pores are saturated with water, there will be no net flow of water from the tensiometer through the ceramic cup to the soil. If the soil is unsaturated and this means that the soil water held by it is at a lower energy state (higher suction), then water will flow from the tensiometer to the soil and the resultant suction force will be transmitted to and registered on the manometer or vacuum gauge. The range of matric suctions over which a tensiometer can normally be used is 0 to 0.8 bar. Although this may appear to be a small range compared with that over which plant roots are considered to be capable of removing soil water (0 to 15 bars), it includes the major portion of their soil water content for coarser-textured soils. At suctions beyond 0.8 bar the ceramic cup 'leaks' air and this, rather than the fact that the highest reading theoretically possible is 1 bar, is the limiting factor.

Tensiometers are convenient in coarser-textured soils where irrigation may be practiced. They are relatively cheap, sturdy, and easy to instal at different soil depths. Since tensiometers do not measure osmotic potential, their readings are unaffected by changes in concentrations of soil solutes associated with fertilizer additions. A low cost recording-type tensiometer has been constructed for remote field conditions (Walkotten, 1972).

Pressure plate apparatus

The relationships between the matric potential and the amount of water in the soil are determined by use of the pressure plate apparatus. There are various forms of this apparatus but they consist basically of a pressure chamber in which a sample of soil may be placed on top of a ceramic plate or cellulose membrane. These allow water but not air to pass through them. Air or nitrogen is applied to the chamber at some predetermined pressure and any water held at suctions less than the applied pressure will pass through the bottom plate or membrane until an equilibrium is reached when no more water will pass. The water retained by the soil is then held by matric suction forces equal to or greater than the applied pressure. At equilibrium, either the amount of water which has passed through the membrane is measured, or, more usually, the soil sample is weighed and the dry weight of the soil sample determined at the end of a series of equilibrium weighings. The moisture contents at each pressure equilibrium can then be calculated. In routine determinations only the drying sequence values are determined. The continuous curve of soil moisture content against matric suction is the soil moisture characteristic curve. The range of matric suctions over which these curves are commonly determined is 0 to 15 bars and marked differences may be noted for soils of different textures (Figure 4.8).

Electrical resistance units

The principle on which these are based is that the resistance to an electrical current in the soil increases with the decrease in soil water content. Two electrodes encased in gypsum (Bouyoucos and Mick, 1940) or in fibreglass (Colman and Hendrix, 1949) are placed in the soil and an electric current is passed between them. The electrical resistance is measured. The electrical resistance values must then be related to soil water content and/or matric suction readings for a particular soil. The units of gypsum, although somewhat bulkier than the fibreglass units, buffer the electrodes against salts in the soil, whereas the fibreglass units are very sensitive to electrolytes in the soil solution. The gypsum units tend to

Figure 4.8 General relationships between soil moisture content and soil matric suction for soils of two contrasting textures - sand and clay.

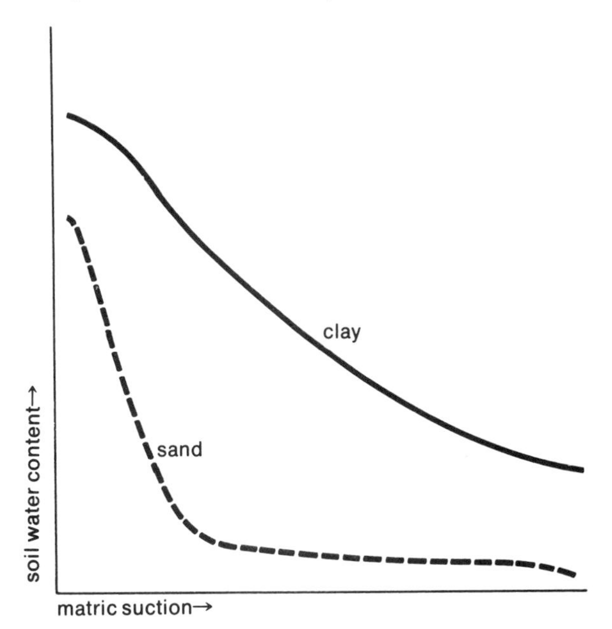

disintegrate during the course of a season and the readings may become erratic if the breakdown of the gypsum block is great enough. The fibreglass units have a long life and can be used over periods of up to several years.

An example of the type of relationships which can occur between electrical resistance, matric suction, and soil water content is shown in Figure 4.9. For this particular soil, the sensitivity of the meter to moisture changes is not very great when the soil is wet (low matric suctions of less than 0.33 bar).

Neutron scattering
When a source of fast moving neutrons is placed in the soil, the speed of the neutrons is reduced when they collide with the other nuclei, especially the hydrogen nuclei of water. The probe of the neutron scattering device is lowered into the soil through access tubes, usually made of aluminium. In the probe there is a detector of slow neutrons as well as the emitter of fast-moving neutrons. The source of fast-moving neutrons is usually radium-beryllium or amercium-beryllium. The flux of slow-moving neutrons is measured by a rate meter or scaler. An

Figure 4.9 Soil moisture characteristic curve and related electrical resistance readings for a sandy loam. From K.M. McClain, (1973)

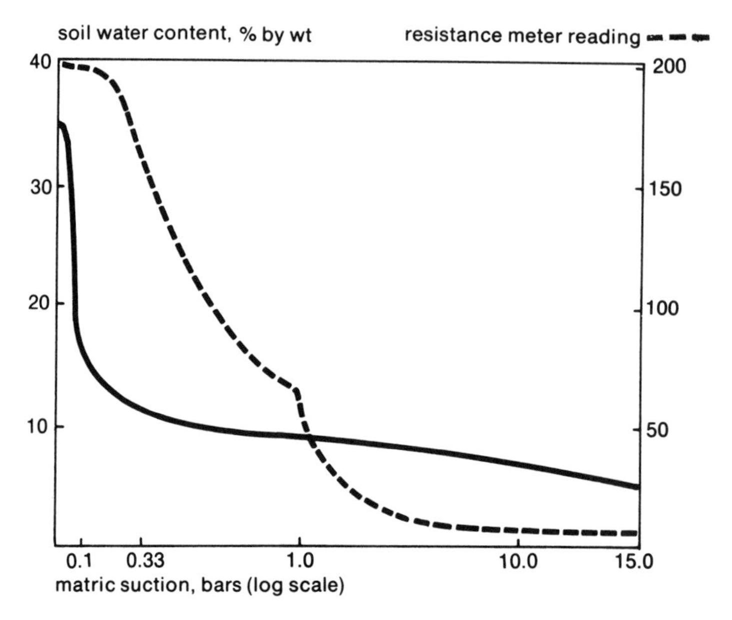

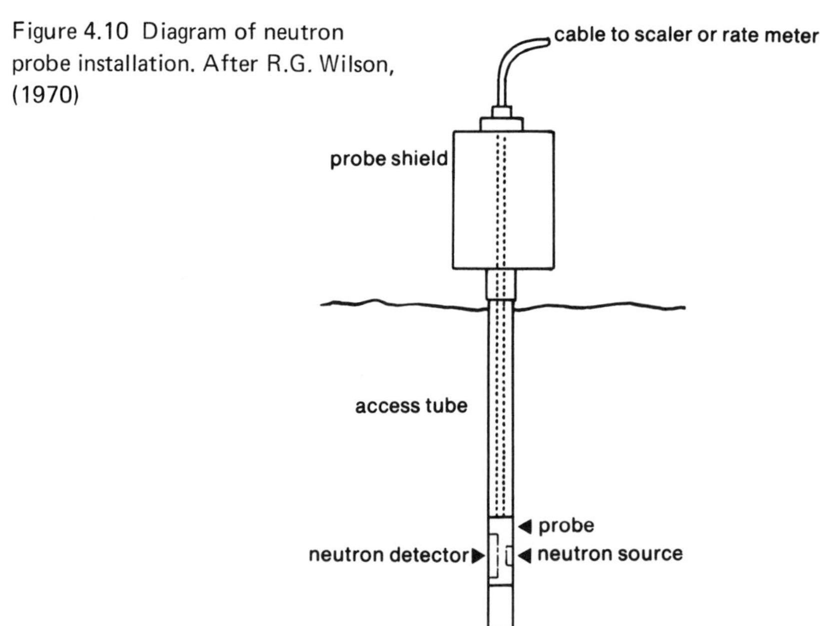

Figure 4.10 Diagram of neutron probe installation. After R.G. Wilson, (1970)

Figure 4.11 Comparison of soil water contents (% by volume) estimated gravimetrically and using a nuclear probe. From R.S. Sartz and W.R. Curtis (1961), by permission United States Department of Agriculture, Forest Service

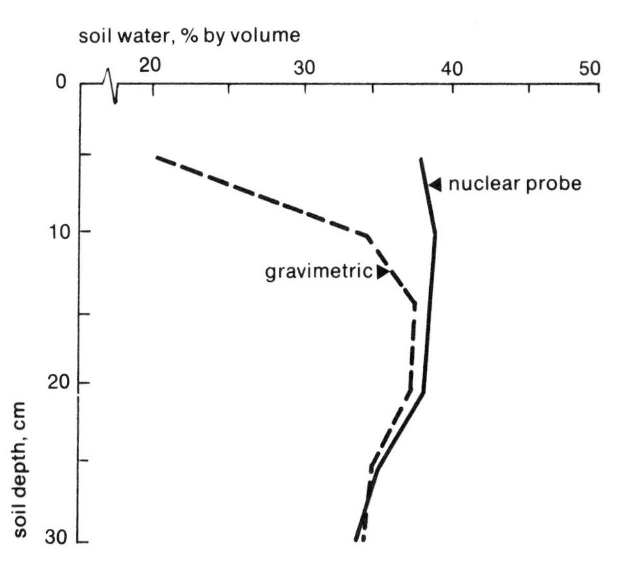

increase in flux is related to a decrease in the presence of hydrogen nuclei and hence of water. A diagram of a neutron probe installation is shown in Figure 4.10. The sphere of neutron activity varies with change in soil moisture content and in dry coarse-textured soils may be up to a radius of 50 cm. Calibration of the instrument is important because of the presence of hydrogen ions in forms other than water, as in organic matter and clay minerals. Near the surface the neutron probe is not satisfactory (Figure 4.11) because of loss of neutrons from the surface. Pierpoint (1966) suggested that the use of a polyethylene shield at the soil surface would prevent neutron escape. Visvalingam and Tandy (1972) give a comprehensive review of the neutron method.

As with most field techniques for soil water measurement, calibration for the soils to be studied and a concern for sources of error and variation are matters of importance.

Gamma-ray attenuation

In this radioactive technique, cesium 137 is used as a source of X-rays and the scattering or adsorption of the rays is dependent on the density of the matter in their path. The source and the detector of the rays are placed separately (Reginato and van Bavel, 1964). Gurr (1962) discusses the general theory of the method. It is not a procedure that is in general use.

Ultrasonic energy

Mack and Brach (1966) described a technique in which low energy ultrasonic waves were propagated at specific megacycle frequencies through soil samples of various water contents. They found that changes in texture and organic matter content had no significant effect on propagation at higher frequencies. This method has not been adapted to field use.

Soil moisture regimes

The changes in soil moisture regimes that occur in forest soils will depend on many factors – precipitation, texture, vegetation, drainage, and others. A general pattern is illustrated in Figure 4.12 for a deeply weathered ultisol supporting a 15-year-old pine forest in South Carolina. Precipitation during March results in an increase (recharge) of soil water at the deepest level (137–168 cm). During the remainder of the time from April to October there is a continual withdrawal by evapotranspiration from the soil. In the upper zone (0–38 cm) and to a lesser extent in the 38–76 cm zone, additions of water from precipitation offset these losses to a degree. During the April to June period there is a removal of water from the two lowermost depths (107–137 and 137–168 cm) which can be attributed to absorption by tree roots, since the authors (Metz and Douglass, 1959) found little to no loss in soil at this depth when the surface was either barren or supported broomsedge. At a given time, May 1 for example, the upper part of the soil is being wet while the lower portion is undergoing removal of water.

Much of the consideration of soil water has been in relation to plant growth and it is customary to express the amount of water in soil not as total water, but as '*available water*.' Usually, the water held in a soil between an upper matric suction of 15 bars and a lower one of 0.33 or 0.1 bar is the quantity measured as 'available water.' The 15-bar value is considered to approximate the *permanent wilting percentage* (Hendrickson and Veihmeyer, 1945) and 0.33 or 0.1 bar, field capacity. This concept of water views it as a more or less static store which is drawn upon by plants. In reality, however, the demand by the vegetation on a soil for water is a changing one and stress develops only when the rate of water supply does not equal the demand rate. As Hillel (1971) has pointed out, it is the dynamics of the movement of water or *flux* – the amount of water flowing per cross-sectional area per unit of time which is the critical factor. The assessment of a soil's capacity to supply water to vegetation under a given set of climatic factors should, therefore, be based on the following main observations:

Figure 4.12 Total soil water (cm) by soil depth under young pine forest during period March to October - South Carolina. From L.J. Metz and J.E. Douglass (1959), by permission United States Department of Agriculture, Forest Service.

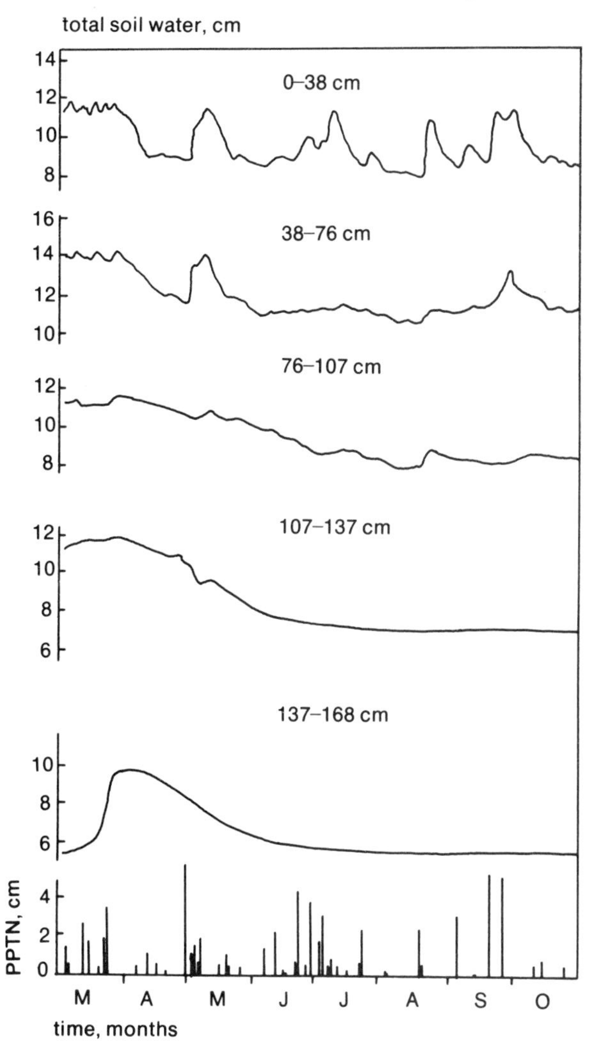

1 / The texture and structure of the soil, which determines the pore pattern and therefore the general nature of the soil moisture characteristic curve.

2 / The presence of textural boundaries and the degree of difference in texture and the depth or position of these interfaces in relation to the solum and

root distribution should be noted. When plants grown in containers of soil or with balled root systems are planted out in natural soils, a textural boundary between the container soil and the surrounding material is created. This can create a marked effect not only on soil water movement but also on the ability of roots to penetrate outside the container soil (Grover et al., 1964).

3 / The degree to which water moves laterally within the soil or upwards from a water table may be important in providing water to the vegetation's roots.

4 / The stage of development of the vegetation and in particular the development, density, and distribution of the roots are highly significant factors if the flux of water per unit area of absorbing root surface is considered. In the juvenile stage of vegetative development, root systems while small are usually still expanding and therefore capable of exploiting new zones of soil water. Under these conditions, reductions in conductivity of water in the soil will have a minimal effect in reducing the total flux. The effect of root density is essentially one of degree of exploitation - the greater the density the greater the ability of the plant to absorb water from a unit volume of soil in a given time. Finally the root distribution, especially the vertical extent, enhances the ability of a species or vegetation to utilize water at depth in a soil.

5
Soil organic matter

One of the most important characteristics of soils, and, in fact, a diagnostic feature, is the presence of organic matter and organisms. Certain soils such as peats are almost entirely organic material, whereas in soils of arid areas the organic content may be small.

The green plants which a soil supports convert solar energy to organic matter by means of photosynthesis. The overall efficiency of the conversion is low; for example, Odum (1971) has estimated that the efficiency of conversion to net primary production is 0.1 per cent. This value is an average for the biosphere; similar estimates for crops grown under intensive cultivation indicate efficiencies of the order of 3 to 5 per cent. The organic matter produced by such conversion not only modifies the surface environment of the soil but also affects its physical, chemical, and biological properties by providing the basic food source or substrate on which many soil organisms depend.

The amount of organic matter which may be measured in or on the soil at any given time is the conventional way of characterizing the soil's organic status. Yet this is analogous to characterizing the financial status of a business by using the amount of money in its bank account on a particular day as a measure of the firm's activity. Quite obviously it is the rate and nature of the business transactions that serve as the important attributes. Similarly with organic matter, it is the rate of addition, form, and transformations which exert the greatest influence on soil. As might be expected, the gains in organic matter are directly related to the nature and abundance of the vegetation and factors which affect plant development will inevitably influence the amount of organic material which may be added to the soil. The type of vegetation also affects the location of organic addition. In most forests the bulk of the organic addition is added to the soil surface, yet the annual or periodic addition by dead roots can be important because of their location in the soil. Bray and Gorham (1964) stated that the net production of below ground parts is about two thirds of the weight of leaves produced in temperate forests but only about one third for tropical forests. Losses of soil organic matter are primarily those related to its decomposition. In the long run, gains from vegetation and losses from decomposition, although variable, will tend towards a set of levels in any particular situation. Typically with forests, however, there are the infrequent catastrophic gains resulting from natural causes such as windstorms, and major losses occurring from wildfire or erosion. These must be considered in the dynamics of soil organic matter.

Chemical constituents of organic matter are important in the maintenance of soil fertility and this is particularly true for nitrogen, since the majority of green plants must rely on nitrogen in a form derived from complex organic compounds by microbial processes. Although organic matter has long been known to influence soil properties, it is increasingly apparent that certain chemical constituents, especially of fresh origin, can exert a profound influence on processes of soil weathering and on soil profile development.

Rather than viewing organic matter only as a component modifying soil properties, we can consider it also as a material which may determine the nature and intensity of soil processes.

The type and amount of organic matter

The surface additions of freshly fallen leaves, twigs, stems, flowers, fruits, and bark are referred to as *litter.* In forest soils litter is the predominant form of annual organic addition and is most readily measured. Bray and Gorham (1964), in a comprehensive review, considered that the contributions of various litter components on a dry weight basis were of the following order: foliage 60–76%; non-leaf litter 27–31% and understory litter variable, averaging 9% but reaching as much as 28%.

Methods of measurement of litter range from those in which a container of known area is used to a procedure in which the litter is removed in a soil block and then physically separated from the underlying materials. This latter method is less satisfactory than the use of containers unless the objective is to measure the amount of the *forest floor* – that is, all dead organic matter including litter and unincorporated humus of the mineral soil surface. The type and size of container used to collect litter varies considerably; a plastic box-type with low sides and a mesh bottom to allow for drainage is satisfactory, particularly for periodic measurements to determine seasonal variation. Trap sizes range widely, but are in the order of 0.05–0.5 m^2. Except on very level uniform soil surfaces, the main source of variation not related to sample size or sampling intensity is the influence of microtopography. This is illustrated in Table 5.1. The increased accumu-

TABLE 5.1
Mean annual leaf fall accumulation (kg/ha) in relation to surface topography (data of Hart, Leonard, and Pierce 1962)

	Mean accumulation - kg/ha (over-dry wt)		
	1959	1960	1961
Mounds	2,076	1,056	1,839
Slopes	2,876	2,979	2,827
Depressions	4,693	7,333	6,518

Figure 5.1 Annual production of litter in relation to latitude. Means for climate zones – +. After J.R. Bray and E. Gorham in *Advances in Ecological Research* (1964)

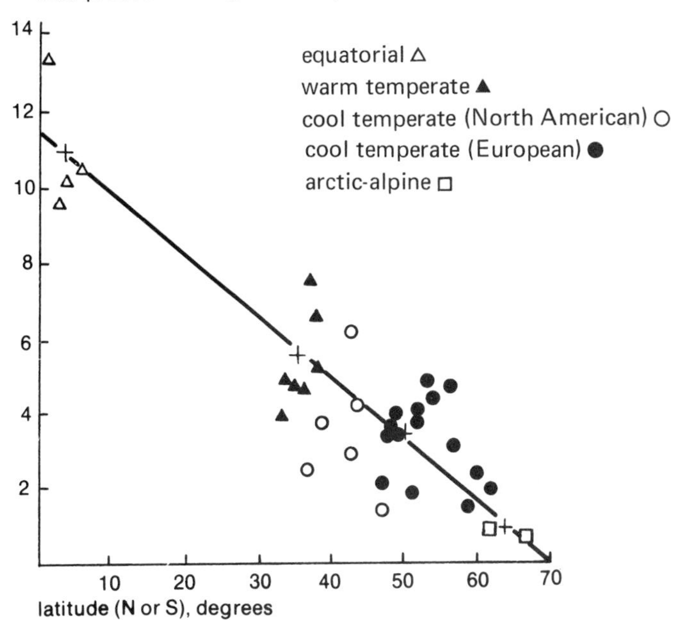

TABLE 5.2
Annual addition of forest tree litter ($\times 10^3$ kg/ha)

Species	Age (years)	Litter ($\times 10^3$ kg/ha)	Reference
Ponderosa pine	100	0.80	Tarrant et al., 1951
Ponderosa pine	350	0.61	Tarrant et al., 1951
Western white pine	–	1.35	Tarrant et al., 1951
Douglas-fir	100	0.92	Tarrant et al., 1951
Douglas-fir	350	2.15	Tarrant et al., 1951
Western hemlock	–	1.05	Tarrant et al., 1951
Norway spruce	60	1.90	Mork, 1942
Norway spruce	60-90	2.89	Ebermayer, 1876
Scots pine	50-75	3.03	Ebermayer, 1876
European beech	60-90	3.40	Ebermayer, 1876
White pine type (eastern USA)	–	1.79	Scott, 1955
Hardwood type (eastern USA)	–	2.23	Scott, 1955
Sugar maple, basswood	60	2.20	Alway et al., 1933
Jarrah (E. marginata)	36	3.10	Hatch, 1955

lation in the depression occurs not only because of its effect as a catch basin but also because the leaves do not dry out in the depression and blow as dry litter frequently does.

For many kinds of forests the annual litter addition ranges from 1,000 to 6,000 kg/ha; the amount will vary with the type of forest and in particular the climatic zone in which it occurs. Bray and Gorham (1964) related values of annual litter production to latitude (Figure 5.1) and, although at any latitude considerable variation exists, there is a general trend towards increase in litter quantities with decrease in latitude. The paucity of data from the lower latitudes is evident. Scott (1955) concluded from a review of factors affecting litter production that the amount is closely correlated with the productivity of the site. The variation in annual litter fall between species within a particular locality appears to be small; for example, the annual litter varied only between 0.90 and 1.03 $\times 10^3$ kg/ha for various stands of white pine, red pine, and jack pine in Minnesota (Alway and Zon, 1930). In a comparison of the weight of the litter beneath 16-year-old stands of white, loblolly, virginia, and shortleaf pines in Virginia, Metz et al. (1970) found that only the weight of litter under loblolly was significantly greater than that under the other species. Annual litter weights for selected forest species are given in Table 5.2.

In forests with full canopies the amount of litter production does not show an inherent tendency to decrease or increase with age (Bray and Gorham, 1964). It could be expected in the early stages of stand development that, as crowns increase in size, litter fall would show a corresponding increase. Zavitkovski and Newton (1971) found that for stands of red alder in Oregon the amount of annual litter increased with stand age up to 10 years, but became fairly constant at a rather high level of 7.4 to 7.8 $\times$ 10^3 kg/ha/yr. Conversely as a forest stand grows older and crowns become more open, the amount of forest tree litter may decline, although if lesser vegetation increases with opening of the overstory, this may reduce the decrease in litter fall. The change in litter production with forest succession is not well documented. Hurd (1971) presents data for four stands in Alaska. Stand 1 (*Alnus-Salix*) occupied an area deglaciated about 30 years prior to the study. Stand 2 (*Populus-Picea*) occupied an area deglaciated for 40–45 years. Stand 3 (*Picea*) stood on an area which had not been glaciated within the past 200 years and Stand 4 (*Tsuga-Picea*) was 132–163 years of age and just outside the Mendenhall Glacier valley. Over a three-year period litter production for the four stands averaged 2,850 kg/ha (oven dry weight basis) and, although variations occurred from year to year and among stands, no real differences were apparent. Within closed canopy forests, litter production appears little affected by differences in stand density. Heyward and Barnette (1936) found no direct influence of stand density on litter production for longleaf pine. Changes in density brought about by thinning are, however, likely to result in differences in litter production. Reukema (1964) found for Douglas-fir over a 13-year period following thinning that litter fall decreased generally with reduction in stand density but year-to-year fluctuations were often considerable.

Probably the most significant factors affecting year to year variations are those growing conditions which determine foliage production. As an example of the magnitude of such variations Ebermayer (1876) found that over a seven year period the leaf fall of Norway spruce ranged from 2.17 to 6.34 $\times$ 10^3 kg/ha/yr (air-dry weight basis). Weather conditions, resulting in drought, excessive winds, and abnormal occurrence of wet snow may influence both the amount of litter and the time of fall.

Studies of the periodicity of litter fall show marked differences between species. Deciduous species in the northern cool temperate zone drop the bulk of their foliage in the autumn; however, Scott (1955), quoting data for New Hampshire, indicated that for gray birch, over 75 per cent of the foliage was shed in July-August. Mork (1942) found for Norway spruce that there were two periods of major leaf fall, one in early summer (May-June) and another in the autumn (September-October). Occurrence of nutrient stress or deficiency may affect litter fall pattern. Potassium deficiency symptoms in red pine are often strongest in early summer when height growth takes place and there is usually an associated removal of potassium from older-age class needles to the newly developing foliage. These older needles die and are cast in early summer rather than in autumn.

While there is an extensive literature concerned with tree litter production, far fewer studies deal with the contribution to total litter by the lesser or subordinate vegetation. Bray and Gorham (1964) review data which indicate that amounts of 0.1 to 2.0×10^3 kg/ha/yr or 3 to 28 per cent of the total forest litter production may be attributed to the understory vegetation. Scott (1955) found for white pine and mixed hardwood stands in the northeastern United States that the understory contributed 15-16 per cent of the annual litterfall.

THE FORM AND COMPOSITION OF LITTER

Form

The form of the litter component is determined by the species from which it originates. The long, stiff foliage of certain pines forms a criss-cross pattern creating an open porous litter. If the rate of decomposition is slow, the litter layer may have a pronounced physical effect by allowing rapid entry of water into the soil as well as facilitating gas exchange. It can also withstand a greater degree of traffic, especially by humans, than can litter made up of short, flexible foliage. Coniferous foliage, especially the larger, longer needles does not alter appreciably with changes in moisture content. The maintenance of a consistent physical structure ensures a beneficial mulching effect and minimizes surface water movement and associated erosion. Leaf fall from deciduous species is variable in form and, in locations where the initial decomposition will proceed rapidly, the physical form of the litter is probably of little consequence. In cool temperate climates, where litter accumulation in the autumn is followed by low temperatures and considerable snowfall, the physical nature of the litter can be of significance. For example, where the litter is predominantly large leaves of species such as sugar maple, the leaves form a flat overlapping litter which becomes more flattened and compacted by the winter snow. In the early spring this litter mat may act, somewhat like shingles on a roof, to facilitate the surface flow of water from snowmelt. Deciduous litter made up of smaller leaves (ca. 2-8 cm in length) will not overlap or mat together in the same manner as larger leaves and will curl up and change dimensions in response to moisture changes to a much greater degree.

Chemical composition

The composition of plant material which comprises the bulk of litter additions is generally considered to be made up of seven constituent groups; six of these are organic and one mineral: (a) cellulose - varies from 15 to 60 per cent of the dry weight: (b) hemicelluloses - 10 to 30 per cent of the dry weight; (c) lignin - 5 to 30 per cent; (d) water-soluble fraction - simple sugars, amino acids, and other compounds - 5 to 30 per cent of the dry weight; (e) ether and alcohol-soluble compounds such as fats, oils, waxes, resins, and pigments; (f) proteins; (g) mineral constituents - 1 to 13 per cent of the dry weight.

For forest litter, the first three components - cellulose, hemicelluloses, and lignin constitute the bulk of the litter weight. Although it has been customary to consider and determine the composition of plant material only after it has become litter, it is apparent that the composition and associated biochemical processes of the component while still on the plant may be critical in terms of what happens to it when it becomes litter (Handley, 1954). The chemical composition of plant material destined to become litter is of interest, primarily for the following reasons:

1 / Certain compounds, especially those that are water-soluble, may be removed from the foliage or other organs by washing. Simple nitrogen compounds and elements may be absorbed by lower vegetation (Tamm, 1953). Other materials may undergo conversion by chemical and biological processes to form compounds which may have various effects in the soil. For example, the non-toxic hydroxyjuglone which occurs in the leaves and fruit of black walnut, when washed to the soil, may be oxidized to juglone (5-hydroxy-1,4-naphthaquinone), and in this form can inhibit the growth of various herbaceous and woody plants. Another example is the effect of certain polyphenolic compounds in fresh litter which have the ability to facilitate the weathering and movement of certain metals, such as iron, in the soil (Coulson et al., 1960a,b).

2 / The major components of litter are carbohydrates and lignins which provide energy for the growth and development of many soil organisms as well as carbon for incorporation in their bodies. The water-soluble compounds such as simple sugars are utilized rapidly by the soil population, but the decomposition or breakdown of the long-chain molecules of other carbohydrates proceeds more slowly in relation to their size (Figure 5.2). The long-chain molecules of cellulose and the hemicelluloses are broken up by enzymic hydrolysis to simple compounds (sugars) which may be absorbed by microorganisms. Initially, hemicelluloses are considered to undergo more rapid hydrolysis in the soil; Alexander (1961) concludes that more microbial species are active in destroying hemicelluloses than cellulose. Lignin is more resistant to decomposition and chemically more variable than the other carbohydrates. Its degree of resistance to decomposition seems to be greater, the larger the number of methoxyl ($-OCH_3$) groups. Fungi are prominent in the breakdown of the large lignin molecules.

Since the initial breakdown of these large carbohydrate molecules takes place outside of the microorganisms and in the soil, many soil properties will frequently

Figure 5.2 Diagrams of (a) unit of cellulose molecule: (b) polyglucuronic unit; and (c) xylan units of hemicellulose molecules

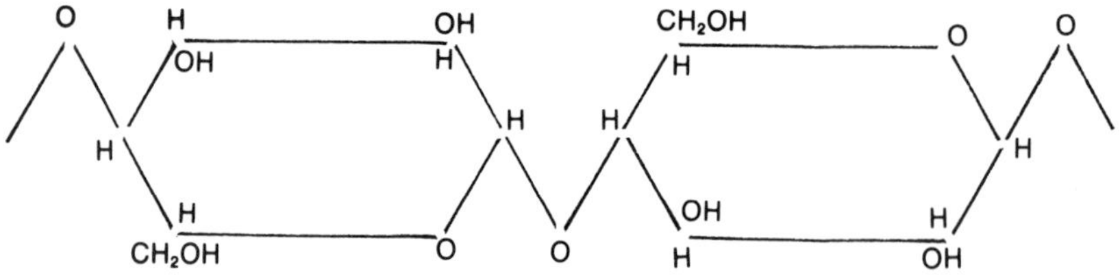

(a) unit in cellulose molecule

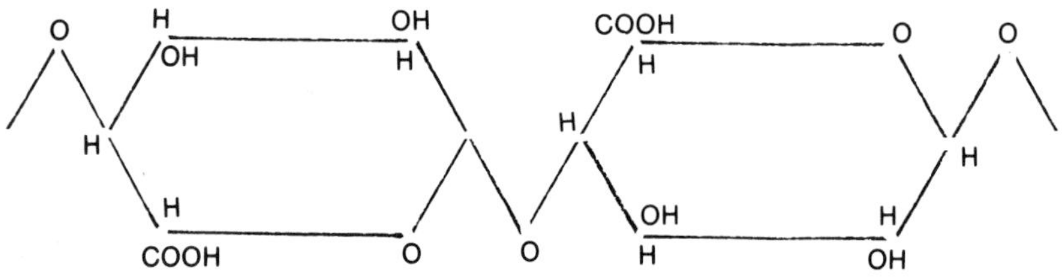

(b) polyglucuronic unit

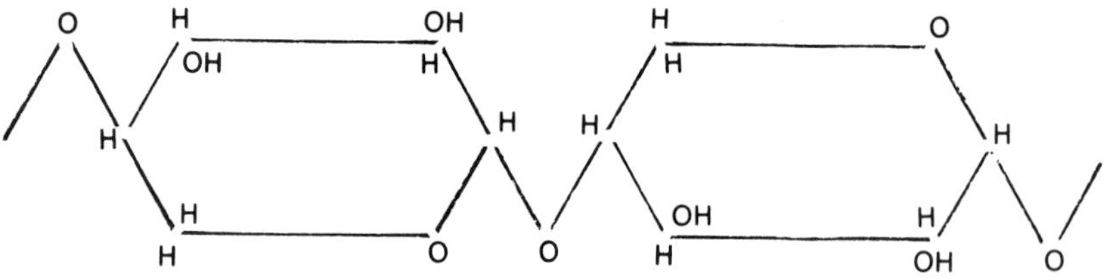

(c) xylan unit

Figure 5.3 Diagram of units in a protein molecule. R′, R″, R‴ represent organic radicals

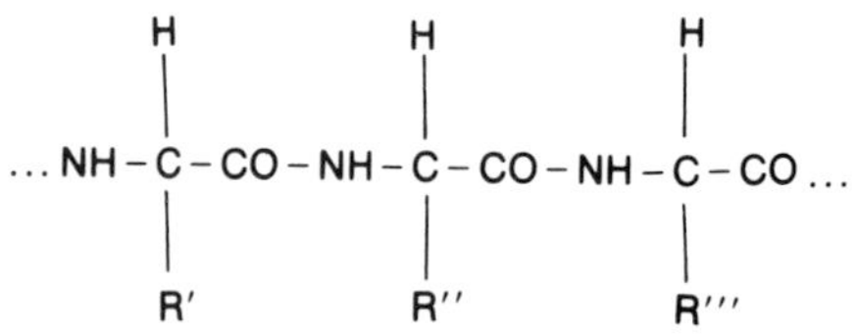

$$\cdots NH-\underset{\underset{R'}{|}}{\overset{\overset{H}{|}}{C}}-CO-NH-\underset{\underset{R''}{|}}{\overset{\overset{H}{|}}{C}}-CO-NH-\underset{\underset{R'''}{|}}{\overset{\overset{H}{|}}{C}}-CO\cdots$$

TABLE 5.3
Annual amounts (kg/ha/yr) of litter and nutrient elements for certain northwestern USA tree species (data of Tarrant et al., in Journal of Forestry, 1951, by permission, Society of American Foresters). Amount in kg/ha/yr (oven-dry weight basis)

Species	Litter	N	P	K	Ca	Mg
Red cedar	2,145	13.3	1.9	7.7	48.0	0.9
Douglas-fir (350 yr)	1,984	17.0	3.0	3.0	18.6	0.6
Douglas-fir (100 yr)	921	7.8	0.8	1.9	9.0	0.2
Western hemlock	1,049	8.1	1.1	2.0	6.3	0.7
Bigleaf maple	1,478	16.9	1.8	16.7	24.0	0.2
Red alder	1,314	30.9	1.7	13.0	10.8	0.1

have profound effects not only on the growth and activity of the organisms but also on the hydrolytic enzymes.

3 / Nitrogen is essential for plant growth and its major source for higher plants is organic matter itself. Much of the nitrogen in litter is present as proteins (Figure 5.3). These long-chain molecules are broken down to amino acids by enzymic hydrolysis of the peptide (-CO-NH-) linkages. The deamination or removal of nitrogen usually proceeds intracellularly in the micropopulation in contrast to the enzymic hydrolysis which is extracellular. The relative ease with which the soil population can transform organic nitrogen from litter and the amount available are often considered as the most critical controlling factors in determining the rates of decomposition. The amount of nitrogen in litter is usually determined in conjunction with other mineral elements. A complete review of nitrogen in forest soils is given by Wollum and Davey (1975).

4 / Many mineral elements, in addition to nitrogen, occur in forest litter. In forest soils, litter is a major source of nutrient elements for the vegetation and for this reason much attention has been directed to the nutrient content of plant litter. Several factors will affect the amount of nutrients. The species and concentration of elements in the foliage are two of the most important. Considerable differences can exist between species (Lutz and Chandler, 1946). Table 5.3 gives data for certain western North American species (Tarrant et al., 1951). The proportionately greater amount of calcium in red cedar litter and the very large quantity of nitrogen in red alder litter - a species that has symbiotic dinitrogen-fixing organisms in its roots - are evidence of major species differences. Scott (1955) noted that the lesser vegetation might contribute proportionately greater amounts of certain elements and his data (Table 5.4) show this for potassium. Lutz and Chandler (1946) concluded that, except for nitrogen, the annual return of major nutrients was greater in hardwood than in conifer litter. The

TABLE 5.4
Annual additions (kg/ha/yr) of litter, and selected nutrient elements contributed by the tree and lesser vegetation in white pine and mixed hardwood forests of the northeastern USA (data of Scott, 1955). Amounts in kg/ha/yr (oven-dry weight basis)

	White pine		Mixed hardwoods	
	Lesser vegetation	Tree	Lesser vegetation	Tree
Litter	259	1,399	303	1,762
N	3.8	9.1	4.3	13.3
P	0.39	1.82	0.48	2.61
K	3.3	5.9	3.3	8.6
Ca	2.5	9.4	2.5	14.0
Mg	1.3	4.8	1.3	7.1

greater amounts in hardwood leaves reflect generally higher concentrations of nutrients rather than larger amounts of litter. The nutrient concentrations in plant organs, especially leaves, are dependent not only on species but also on soil and general growing conditions, age, and stage of development of the vegetation. For these reasons, adoption of generalizations for specific situations should be avoided.

ENERGY TRANSFORMATIONS AND DECOMPOSITION OF LITTER

The organic matter which reaches the forest floor as litter represents stored energy. The decomposition of this material is a series of processes by which this energy is transformed. In certain forests the total annual litter may be less than the total net production of organic material (mainly foliage) which could be litter. The main causes for this difference would be consumption by herbivores and removal or loss due to human activities. Windstorms which blow over trees and logging with its resultant slash add large amounts of organic matter to the soil in a short time.

For the amount reaching the soil surface, there are five main consequences:

1 / There is a reduction in the original energy store – a decrease in the amount of litter after it reaches the soil, a result of its use by soil organisms.

2 / Concurrent with the reduction in the amount of energy, there is a conversion of some part into other organic forms. The body tissues of soil fauna consuming part of the litter and of microorganisms using the litter as energy substrate are examples of such conversion.

3 / There is a dissipation of energy in the form of heat and metabolic losses associated with the biological processes of decomposition.

Figure 5.4 Relationships between decomposition rate, litter deposition, and total organic content for an oak forest floor over one year. From W.A. Reiners and N.M. Reiners in *Journal of Ecology* (1970), by permission Blackwell Scientific Publications Ltd.

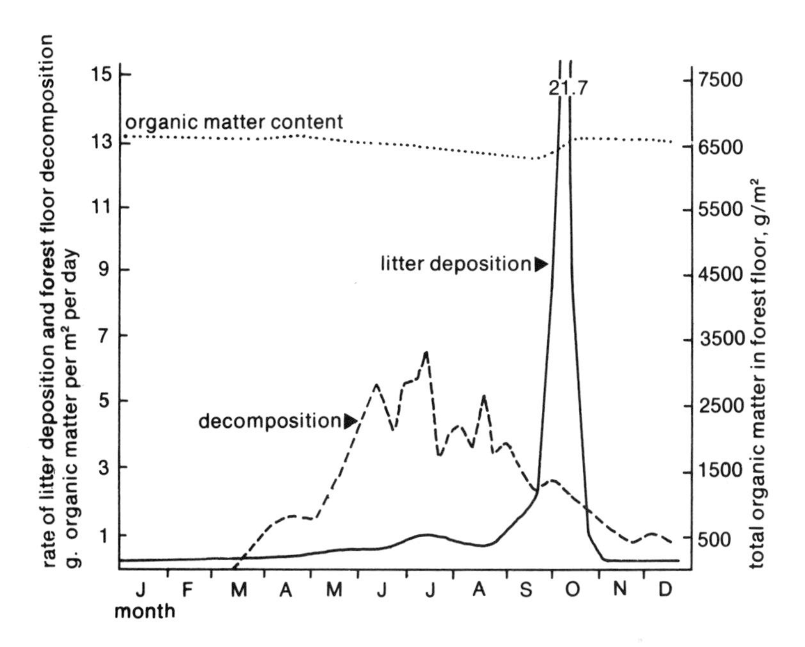

4 / The production of end products during the energy conversion - for example, carbon dioxide, ammonium and nitrate ions, and other inorganic compounds, as well as the relatively stable organic compounds associated with humus. Some of these products may move out of the soil system entirely; carbon dioxide may be lost to the atmosphere and nitrate ions may be leached downward to the ground water. A portion may be translocated to some other part of the soil and remain there in a more or less stable condition, as with the deposition of humic compounds at lower levels in the soil, or some part may be recycled into the soil-vegetation system.

5 / Litter and organic matter may be lost to the system by erosion or fire. Usually these causes are infrequent, but they may nevertheless exert an influence both in the short and the long term.

The relationship between decomposition rate, litter decomposition, and total organic matter content throughout one year for a forest of northern pin oak, white oak, paper birch, and red maple growing on a sandy upland in Minnesota is given by Reiners and Reiners (1970); see Figure 5.4. There is only a slight change

TABLE 5.5
Annual contributions of weight and energy by tree-shrub and herbaceous litter to forest floors of three Minnesota forests: oak, fen, and swamp (data of Reiners and Reiners, in Journal of Ecology, 1970, by permission, Blackwell Scientific Publications)

	Source	Oak*	Fen†	Swamp§
Litter	Trees	4.574	4.115	4.881
Weight	Herbs	0.097	0.299	0.112
kg × 10^3/ha/yr	Total	4.671	4.414	4.993
Energy	Trees	2208	1909	2374
kcal/m^2	Herbs	40	122	45
	Total	2248	2031	2419

Main tree species:
* Northern pin oak, white oak, paper birch, red maple.
† White cedar, black ash, northern pin oak, red maple, American elm.
§ White cedar, paper birch, black ash, American elm, yellow birch.

in the organic matter but the decomposition rate varies seasonally, as might be expected, in response to temperature and moisture regimes. Wallwork (1970) noted that 80-90 per cent of the energy bound up in litter and available to the soil community is captured by microfloral decomposers and their activities would reflect the changes in temperature and moisture. The annual contribution by the litter for three forest conditions is given in Table 5.5. On a weight basis, the annual litter fall represents 3.7, 4.5, and 3.1 per cent of the aboveground vegetation biomass and 52.4, 62.5, and 48.4 per cent of the net primary production of the oak, fen, and swamp respectively (Reiners, 1972). Rodin and Brazilevich (1967) compiled similar data for a wide variety of forests and these indicate generally that annual litter fall is greater than 50 per cent of the total net production. Ovington (1962) presents an energy budget for a Scots pine forest over a 50-year span (Figure 5.5). During this period, decomposition accounts for the greatest consumption of the primary production. Over the greater part of the life span of a forest the residual components of the primary production resulting from solar energy conversion are small compared to the quantities consumed by decomposition. Although these decomposition amounts might be termed losses, they represent the energy flow by means of which the cycling of materials is maintained in the forest-soil system.

Decomposition of organic matter depends on the sequence of feeding by soil fauna and flora (particularly the microflora). Associated with this is soil animal activity, which promotes the formation of soil aggregates. Often the first step in the decomposition process is related to the presence and availability of water-soluble compounds. Nykvist (1959 a,b; 1960, a,b; 1962) demonstrated in a series

Figure 5.5 Energy budget for plantation of Scots pine from time of planting. From J.D. Ovington in *Advances in Ecological Research* (1962)

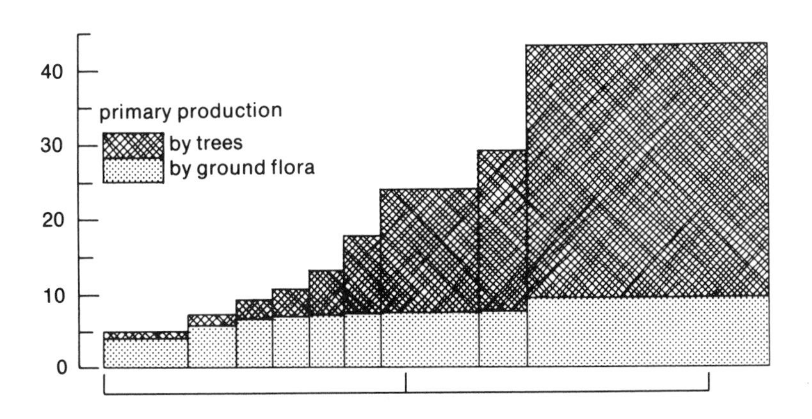

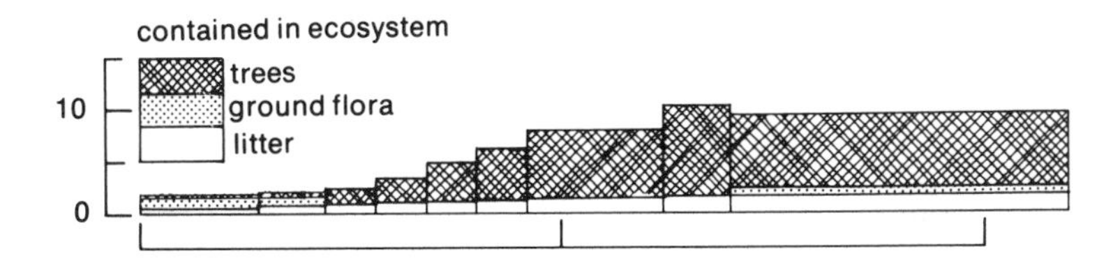

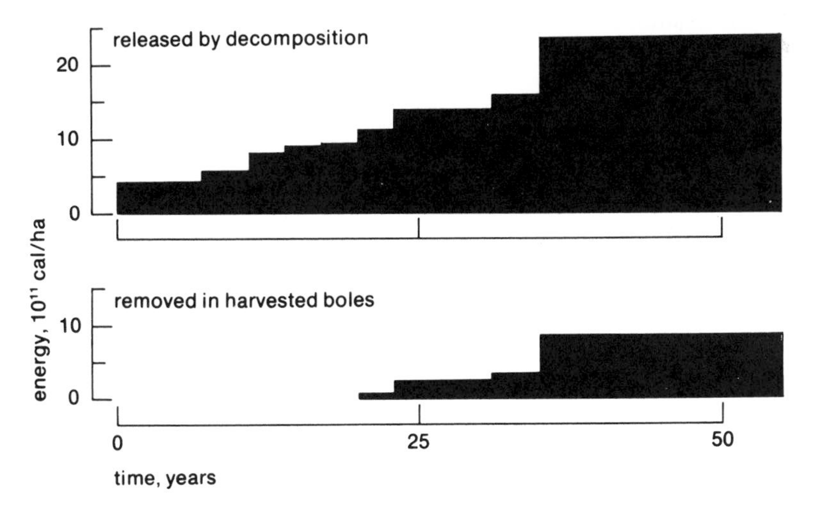

Figure 5.6 Decomposition of leaf discs by soil animals. From C.A. Edwards and G.W. Heath in *Soil Organisms* (1963), by permission North Holland Publishing Company

oak discs in 7 mm mesh bags △▬▬△ beech discs in 7 mm mesh bags ○▬▬○

oak in 0.5 mm bags △▪▪▪△ beech in 0.5 mm bags ○▪▪▪○

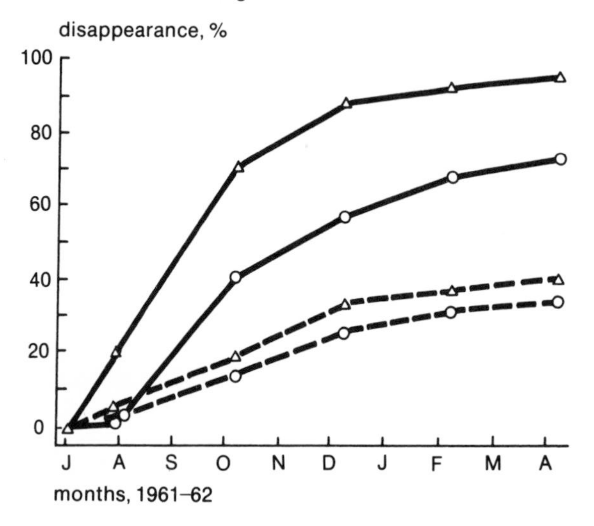

of studies that, while up to 22 per cent of the dry weight of fresh European ash foliage was readily leached in one day (the litter was only 3 to 4 hours old), less than one per cent was released from Scots pine and about one per cent from Norway spruce needles in the same period. European beech, birch, and alder released intermediate quantities of water-soluble materials in the first day. Decomposition rates were generally related to initial content of water-soluble substances.

The most useful technique for measuring rates of decomposition is to expose known quantities of litter to the soil population and measure the litter weight loss at succeeding intervals of time. An example of results obtained in this manner is shown in Figure 5.6. The influence of source of plant material is apparent and the influence of size of mesh is even more marked. Undoubtedly, the influence of size of mesh is primarily one of excluding soil animals which act as agents bringing about the comminution of litter. Witkamp and Olson (1963) followed the breakdown of confined and unconfined white oak litter in the southeastern USA and found that initially the rate of decomposition was similar in the bagged (1-mm mesh) and unbagged foliage, but that after June there was a sudden increase in the breakdown of unconfined leaves, caused by the fragmentation by physical (wind and raindrop action) and biological factors. They suggested that fragmentation by fauna is preceded by some initial microbial breakdown. Will (1967) concluded from a study of the breakdown of Monterey pine litter over a six-year period that 1-mm mesh bags which could be used to enclose the litter probably hampered the normal decomposition processes. The influence of mesh size, while important in temperate zones, may be of less consequence in tropical forests. Madge (1965) found that rate of disappearance was high from 1-mm mesh bags used to contain litter in Nigeria and concluded that the larger fauna such as earthworms were not a factor in litter consumption.

In general, decomposition rates may be expected to increase with temperature, providing there is adequate moisture and aeration. Mikola (1960), in a study of decomposition of Scots pine and European birch litter over a range of latitudes 60–68°N, concluded that, although birch leaves decomposed somewhat more rapidly than those of pine, the main controlling factor was the mean May-September temperature.

Daubenmire and Prusso (1963) studied the breakdown of foliage of 13 western North American species in an artificial environment and, while they found differences in decomposition in relation to species and temperature, they concluded with a caution about studies under artificial conditions. In the field, different species' litter may exhibit different rates of breakdown. To some degree this will reflect differences in preference by fauna such as has been shown for various earthworm species by Bornebusch and Holstener-Jorgensen (1953). Leaves of elderberry, buckthorn, alder, ash, birch, and elm were preferred over Norway spruce and larch; species such as black cherry, beech, and oak were intermediate. The microbial populations that develop in leaf litter can be quite different; Witkamp (1963) found for leaves of white and shumard oaks, American beech, red mulberry, and sugar maple that microbial counts over more than a year were positively correlated with litter breakdown. The type of soil material, particularly the nature of the forest floor, is important. Bocock (1964) found with litter from 26 species of woodland vegetation that it disappeared more rapidly on mull* than moder soils. He considered that this was largely due to the difference in populations of certain soil fauna - earthworms and millipedes being common on the mull soils. Bocock and Gilbert (1957) found that European birch and lime (Tilia spp.) leaves decomposed more rapidly on mull than on moder or peat soils, but that the rate of decomposition for oak did not vary with site.

The major result of decomposition is a reduction in the litter mass and a decrease in percentage concentration of carbon. The ultimate conversion of organic

* The terms *mull* and *moder* are defined in the following section on the forest floor.

material to inorganic compounds is termed *mineralization.* Commonly, the concentration of nitrogen remains stable, or increases or decreases at a lower rate than the reduction in carbon. Consequently the C:N ratio usually declines as decomposition progresses. Nitrogen is a key element in decomposition since the metabolism of organisms responsible for decomposition is dependent on the synthesis of proteins. Thus, if fresh litter is added to a soil, the buildup of an appropriate population of decomposers is contingent on adequate supplies of nutrients, especially nitrogen. For this reason and also because determinations of carbon and nitrogen in organic matter are well standardized, relatively simple, and inexpensive, the C:N ratio of the upper organic layers has been used as an index of organic matter decay rate. While the C:N ratio has been useful to some degree in agricultural soils, there should be caution in attempts to use it as a measure of decomposition rates per se in forest soils.

Witkamp (1963) found the C:N ratios of leaves of only limited value in characterizing their suitability for microbial attack and Kawada (1961) could find no clear relationship between C:N ratios of litter from a number of Japanese tree species and the type of humus produced. Increases in nitrogen contents for organic matter gathered from the surface soil layers may reflect appreciable nitrogen from other sources. For example, Bocock (1963) determined that nitrogen was added to oak litter from atmospheric precipitation, insect frass, and plant remains from the overstory canopy. Caterpillars, when abundant in the tree canopy, added up to 6.6 kg/ha nitrogen to the litter and up to 25 per cent of the nitrogen from atmospheric precipitation (equivalent to 1.3 kg/ha nitrogen) was absorbed by the litter over a 9-month period.

In general terms, the relationship between litter addition, decomposition, and storage (the amount of organic material that remains in a relatively stable condition in the soil) has been represented by Olson (1963) as shown in Figure 5.7. He considered that the ratio of annual litter production (L) to the amount accumulated on top of the mineral soil in a steady state (X_{ss}) provides estimates of the decomposition parameter (k). He determined k values for a range of forest conditions and found that they decreased from a maximum of about 4 in tropical forests to 0.01 for subalpine forests. Values for many temperate forests would be in the vicinity of 0.25. An advantage of this approach is that the decomposition rate (k) can be determined by measurements of the forest floor and annual litter fall. The disadvantages of this method lie in variation due to year-to-year differences in litterfall, the fact that organic matter moved down to mineral soil horizons is not taken into account, and finally the occurrence of sudden and major changes in both forest and soil because of such recurring factors as fire, wind, insect, and disease epidemics.

Figure 5.7 Schematic relationship between annual litter fall and forest floor accumulation. From J.S. Olson in *Ecology* (1963), by permission Duke University Press

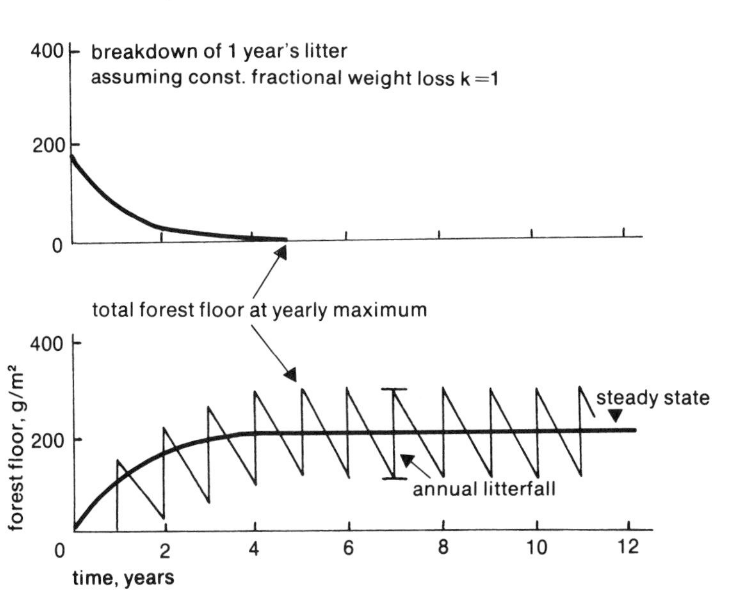

The forest floor

The forest floor and its biological processes attracted little, if any, attention from soil scientists until the mid-twentieth century. Foresters were virtually the only persons who attempted to describe, categorize, and develop concepts of what was happening in this most active part of the soil. Romell (1935) has described the early situation:

The important differences in the type of humus formation was (*sic*) early dimly conceived by nature-observing foresters. Hundeshagen (1830) although hesitating about taking a view so little scientific, distinguished two 'main conditions' with respect to the 'origin and effect of the humus.' He felt this distinction would be particularly important 'at least for practical application' in the same measure as 'the theory about these things is developed.' This remarkable forecast which seems completely forgotten in the later literature, came true through the keen nature observer, P.E. Müller (1887) who rediscovered Hundeshagen's two main

Figure 5.8 Diagrams representing mull, moder, and mor humus types

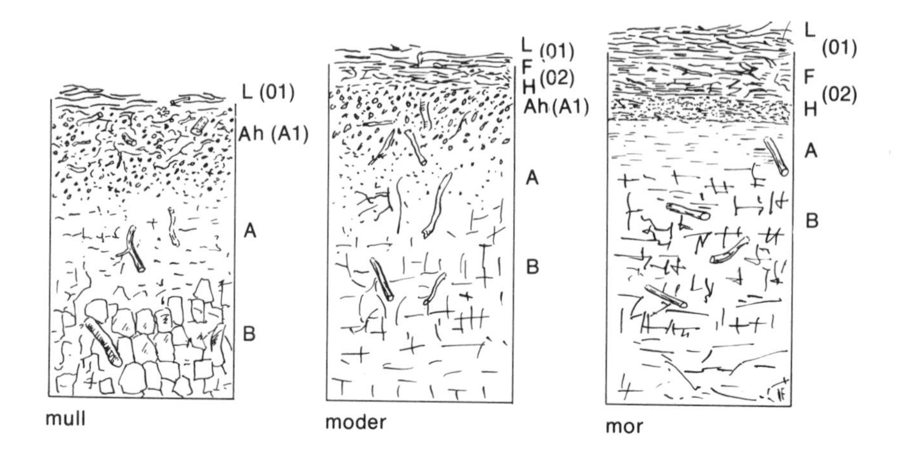

types of humus layer and was the first to apply consistently a point of view of soil biology to the problem. As testified by Hauch (1919), this work came as a revelation to Danish foresters, tapping around in the dark particularly with respect to the difficulties with beech regeneration. The work also laid the lasting ground for the biological forest-soil science.

P.E. Müller's description of two main types of forest humus forms - *mull* and *mor* (syn. raw humus) - has remained essentially the same to the present. The two types represent extremes and Müller was aware of intergrades between them. There have been numerous attempts to classify forest humus layers (Romell and Heiberg, 1931; Bornebusch and Heiberg, 1936; Wilde, 1946; and Hoover and Lunt, 1952. Hesselman (1926) suggested the use of L,F,H horizon symbols to describe the forest floor layers. Kubiena (1953) used the term *moder* to describe an intermediate form between mull and mor. Moder is essentially equivalent to the *duff mull* type (Wilde, 1946) but is a preferable term.

Drawings of the three main types - mor, moder, and mull are shown in Figure 5.8.

Mor or raw humus

The current litter (L) lies over a matted layer of partly decomposed material (F), in which the degree of decomposition increases with depth. Fungal mycelia (yellow and white strands) are often visible. The F layer grades into a lower horizon (H) of well-decomposed organic material, visibly unrecognizable as to plant origin. Normally the H horizon shows an abrupt transition to the underlying mineral soil horizon (A). The mor is generally acid.

Moder

The current litter (L) lies over a partly decomposed F layer. This F layer is not matted as in the mor, although if there is an abundance of fine roots it may appear so. It is transitional to an H layer which is usually quite thin <2-3 cm, and the lower portion usually shows a mechanical mixing of organic and mineral particles. The zone of mixing is often extensive enough that it can be distinguished as an A1 (Ah) horizon. Moder is usually acid.

Mull

Depending on the time of year, the current litter (L) may essentially be absent. In deciduous forests this would often occur towards the end of the growing season and just before major leaf fall. Although there may be individual pieces of organic matter such as twigs and small branches which are partially decomposed, they do not constitute an F layer. The uppermost soil horizon, unless a litter layer is present, is the A1 (Ah) which is an intimate mixture of well-humified organic matter and mineral soil. The horizon usually shows a well-developed granular structure. An H horizon is absent. Burrowing microfauna, especially earthworms, are commonly observed in this type.

These descriptions place emphasis on features of the horizon morphology which can be objectively determined. In many soils, disturbance, change in species composition or soil faunal populations may often result in apparent anomalies. To categorize the type is less important than to provide a clear factual description and measurement of its properties.

The forest floor - both amount and chemical composition - has received considerable attention, largely because it represents a residual pool of organic compounds and nutrients. The increase in forest floor amounts often follows gradients which affect forest growth. For example, in New Mexico, Wollum (1973) found that forest floor weights and nutrient element contents generally increased with increase in elevation and this reflected a gradient of increasing moisture. Differences in accumulation of certain elements between species are obvious; white fir accumulates proportionately more calcium than the other species. Wooldridge (1970), however, found no relationship between forest floors under ponderosa pine and elevation, aspect, or slope, but floors were heavier in stands growing on basalt compared with those on other soils. Gessel and Balci (1965) indicate the great differences in forest floor weights between immature coniferous forests and old growth stands in the Pacific northwest (Table 5.6). No effect of density on forest floor amounts was found in lodgepole pine stands

TABLE 5.6
Forest floor weights (oven-dry) and amounts of nutrients (kg/ha) for coniferous forests in Washington (from Gessel and Balci, in Forest-Soil Relationships in North America, 1965, by permission, Oregon State University Press)

Type of forest		Forest floor weight kg $\times 10^3$/ha	Nutrients – kg/ha		
			N	P	K
Old growth	Mor	157.9	2,040	142	127
	Duff Mull	103.4	1,393	115	103
Immature growth	Eastern Washington	28.5	327	29.1	42
	Western Washington	14.3	193	16.3	14.7

ranging from 1,730 to 35,089 stems per hectare and 34 to 99 years of age (Moir and Grier, 1969); the forest floor amounts ranged from 25.4 to 42.1 $\times 10^3$ kg/ha. Youngberg (1966) found a range of forest floor weights for Douglas-fir stands (immature) in Oregon - 23.9 $\times 10^3$ to 85.9 $\times 10^3$ kg/ha. Also working in the Pacific northwest, Williams and Dyrness (1967) reported forest floor weights under true fir-hemlock stands. The mean values and ranges are shown in Figure 5.9. Floor weights were least where the humus type was a fine mull and greatest where it was a thick duff mull (moder). The weights and nutrient contents for forest floors in aspen and birch stands in Alaska (Table 5.7) show the relatively large quantities which can accumulate in northern climates (Van Cleve and Noonan, 1971). In the southeastern US, Wells and Davey (1966) noted lower forest floor weights (approx. 9 $\times 10^3$ kg/ha) under hardwoods than under pine or pine-hardwood mixed stands (24-25 $\times 10^3$ kg/ha).

The accumulation of forest floor organic matter can be considerable under young plantations. Metz et al. (1970) reported amounts for 17-year-old plantations of four pine species in Virginia (Table 5.8). Tarrant and Miller (1963) noted for a 30-year-old plantation of Douglas-fir in Washington that the forest floor weight was 27.5 $\times 10^3$ kg/ha, but where red alder had been introduced into a part of the plantation the floor weight was 32.1 $\times 10^3$ kg/ha. Although these floor weights do not differ significantly, the nitrogen quantities - 158 kg/ha N for the Douglas-fir and 462 kg/ha N for the Douglas fir-alder - were quite different.

Figure 5.9 Forest floor weights under seven Abies-Tsuga types in the Cascade Range. The horizontal line is the mean for the type and the length of the bar represents the range. From C.B. Williams, Jr., and C.T. Dyrness (1967), by permission United States Department of Agriculture, Forest Service

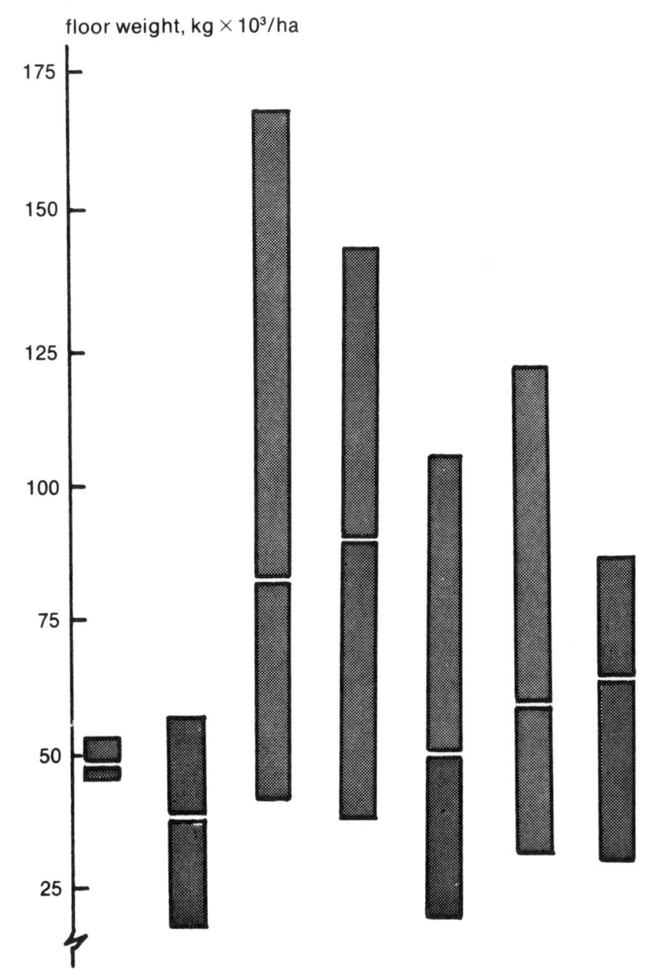

McFee and Stone (1965) studied the organic matter accumulation under yellow birch - red spruce forests in the Adirondack Mountains of New York (Table 5.9) and considered that the floor weights at ages greater than 300 years repre-

TABLE 5.7
Forest floor weights and amounts of nutrient elements (kg/ha) for aspen and paper birch stands in Alaska (data of Van Cleve and Noonan, in Soil Science Society of America Proceedings, 1971, by permission, Soil Science Society of America)

	Floor weight	Nutrients – kg/ha							
Forest	kg $\times 10^3$/ha	N	P	K	Ca	Mg	Zn	Mn	Fe
Birch	40.5	633	57.6	71.4	429	89.4	2.60	63.3	201
Aspen	42.0	651	61.0	84.2	641	108	4.38	33.6	223

TABLE 5.8
Amounts of forest floor and nutrient elements under plantations of 17 year-old pines (data of Metz et al., 1970)

		Nutrients – kg/ha				
Species	Floor weight* (kg $\times 10^3$/ha)	N	P	K	Ca	Mg
Loblolly pine	24.7	266	19.9	17.9	101	16.6
Shortleaf pine	15.2	177	12.9	14.1	85	12.8
Virginia pine	16.6	236	14.8	13.4	88	10.5
Eastern white pine	10.7	148	12.6	9.6	110	11.4

* Oven-dry weights were multiplied by (100 – % ash) and hence these data are on a volatile matter basis.

sented a steady-state accumulation. The amounts of certain nutrient elements contained in this forest floor are shown in Table 5.10. They considered that large variations in the weight of floor materials were caused by movement of litter by wind and its subsequent uneven accumulation, presence of old stumps and fallen logs which not only persist as large masses but also serve as catchment areas for drifting litter, and the sloughing of bark and limbs which results in greater accumulations about the bases of individual trees.

Mader and Lull (1968) found for white pine stands in Massachusetts that forest floor weights ranged from 11.4 to 96.2×10^3 kg/ha with a mean of 45.9×10^3 kg/ha. In soils where decomposition of litter fall is very rapid and soil fauna incorporate the humus with the mineral soil, a forest floor is only present as a transient phase and is essentially the litter accumulation of the season.

TABLE 5.9
Time of accumulation, mean thickness, weight, bulk density, and organic matter per cent for forest floors under yellow birch-red spruce stands in the Adirondack Mt., New York (data of McFee and Stone, in Soil Science Society of America Proceedings, 1965, by permission Soil Science Society of America)

Estimated age of organic matter accumulation (yrs)	Mean floor thickness (cm)	Mean floor weight ($\times 10^3$ kg/ha)	Mean bulk density (g/cm^3)	Mean organic matter (%)
90	8.1	130.9 ± 14.9*	0.16	64.0
135	14.0	180.3 ± 20.5	0.13	71.0
215	12.4	182.1 ± 17.5	0.15	78.0
290	11.9	185.7 ± 17.5	0.15	75.5
>300	12.4	196.6 ± 17.7	0.16	73.4
>300	20.0	265.1 ± 16.0	0.12	87.3
>325	17.3	265.8 ± 14.8	0.15	78.7

* 90% confidence intervals.

TABLE 5.10
Weight of forest floor and nutrient elements in undisturbed yellow birch-red spruce forests (data of McFee and Stone, in Soil Science Society of America Proceedings, 1965, by permission Soil Science Society of America)

Forest floor weight ($\times 10^3$ kg/ha)	Nutrients – kg/ha			
	N	P	K	Ca
242 ± 9.9*	3,200 ± 206	170 ± 11.2	102 ± 7.8	692 ± 98.6

* 90% confidence intervals.

6
Soil biology: organisms and processes

The array of living and dead plant material occurring in any soil harbours and supports populations of plants and animals ranging in size from the microscopic bacteria to the larger mammals such as rabbits and gophers. The complexity and interwoven nature of many of these populations reflect not only the varied food base on which they rely for energy but also the environmental differences in temperature and moisture, as well as the change in physical habitat occurring from the upper soil surface down to the mineral subsoil layers. The varied nature of the soil segregates populations and activities; in turn, the soil is subjected to the physical and chemical results of their actions. When we examine a particular horizon and note its properties and the prevalence of some particular organism such as earthworms in a mull soil or fungi in a mor, the question may arise as to whether the soil condition is there as a result of the organism or the organism because of the soil condition. Fortunately, such apparent 'chicken and egg' questions can often be resolved.

Perhaps some of the most striking examples to be found anywhere, of interrelated functions are those between certain microorganisms and the root systems of higher plants. The role of nodule bacteria in fixing nitrogen in leguminous agricultural plants has received well-deserved attention, but the widespread nodulation of some forest species and the symbiotic fungus-root associations of many trees far surpass in extent that of the few crop legumes. It is when these root-organism-soil interfaces are examined closely that the intricacy and dynamic nature of the forest-soil system become apparent.

Soil fauna

The animals which constitute the soil fauna are those that spend some part or all of their life cycle in the soil. Only for that stage of an animal's life which is spent in the soil is it considered part of the soil fauna. Those animals which do not spend all their life stages in the soil have been separated by Kevan (1962) into four classes: the *temporary* fauna, which spend a definite but temporary period underground, as do many larvae; the *periodic* fauna, those which leave and re-enter the soil irregularly; the *partial* fauna which have attributes of the previous two groups, are temporary, but periodic during their aerial phase - digger wasps, for example; *transient* fauna, those which spend only inactive stages in the soil - the pupae of insects such as the larch sawfly exemplify this form. With the exception of some of the mammals, the discussion of soil fauna will deal primarily with those animals which are *permanent* and spend all their life stages in the soil.

Several approaches may be taken to the classification of soil animals, apart from the primary taxonomic one. They may be classified in terms of size, habitat and activity (usually based on major food substrate). There are three main size classes (Wallwork, 1970).

macrofauna: animals with a body size greater than 1 cm. This includes earthworms, vertebrates, molluscs, and large arthropods.
mesofauna: body size is in the range of 1 cm to 0.2 mm and includes the mites (Acari) and springtails (Collembola), potworms, and larger nematodes. The lower limit is about the limit of viewing with an ordinary handlens. This group size has also been termed *meiofauna.*
microfauna: these are the organisms less than 0.2 mm in size and include the Protozoa as well as many of the smaller mites and nematodes.

Habitat preference by an organism, although important, is primarily a reflection of the degree to which a particular location provides the vital requirements of food, oxygen, water, and living space. These requirements will not only vary in the soil itself but be reflected also by migration of an organism from one place within the soil to another or by changes in the stage of development or seasonal activity of the organism itself. Kühnelt (1961) viewed animals in terms of their habitat relationships:

1 / Animals which inhabit the surface litter because they are too large to penetrate into the pore spaces of the underlying soil - snails are an example.

2 / Animals which can live in the pore space system of the soil (including the forest floor) as well as in fissures and root channels. The structure of the soil is a critical factor since these organisms have no capacity to burrow. Many are also sensitive to changes in air-water proportions in the soil pores; this includes not only those which cannot tolerate anaerobic conditions, when the pores are filled with water, but those small animals which can only exist in an active form in the films of water about soil particles and thus are sensitive to drying out of the soil.

3 / Burrowing animals such as earthworms and some ants which create their own space within the soil. They are often responsible for movement and mixing of soil materials, which is usually of considerable importance. Burrowing by mammals is more often only of local significance.

The activity of the soil fauna is related primarily to preferred food and movement. One group of animals exists in a carnivorous fashion by feeding on others. Bacteria and some protozoa are preyed upon by other protozoa. Mites are food for other arthropods, and earthworms are choice morsels for moles and predaceous snails. A second group of fauna feeds on live plants; for example, certain nematodes are parasites on roots of some plants. Other animals feed on fungi, algae, and bacteria. Many animals, particularly those in the forest floor, use dead plant material as a source of food. These saprophagous organisms can sometimes only consume material after it has undergone some previous microbial breakdown. Jacot (1939) described the initial breakdown of spruce litter as caused by a fungus (*Lophodermium piceae*) which reduces the internal structure so that the larvae of certain mites then proceed to eat the palisade tissue, depositing faeces which, in turn, provide a food base for midge larvae. The functionings of the animal and plant populations within the soil are thus interwoven and linked into very complex patterns and food webs.

Burges (1963) provides interesting observations of the breakdown of Scots pine foliage in England. Fungal infection of needles by *Coniosporium, Lophodermium, Fusicoccum*, and *Pullulana* (all fungi) begins 5 to 6 months before the needles die and fall as litter. These genera reach their peak development and then are succeeded by a wave of soil fungi. Some 13-15 months after litter fall, grazing mesofauna, including mites, remove much of the fungal spore production. It is now the turn of basidiomycetes whose mycelia become prominent. At the bottom of the F layer where it merges into the H horizon faecal pellets and exoskeletons of the mesofauna predominate and it is likely that chitin-attracted fungi are more evident.

When a soil profile is viewed in the field, few of the animals active in the soil are seen. The larger ones capable of relatively rapid movement leave as the soil is disturbed and the most conspicuous evidence of soil animals are the faeces, burrows, and channels, together with the particular state of litter deterioration that may be present. In order to obtain quantitative information about the soil animal population, it is necessary to use special techniques and apparatus.

THE MEASUREMENT OF SOIL ANIMALS

Methods of measurement of the animal population depend to a certain extent on the organism. Relatively large mobile animals such as small mammals and beetles can be caught in traps. Smaller organisms can be carefully removed by hand-

Figure 6.1 Diagram of a simple Tullgren funnel used to extract soil animals

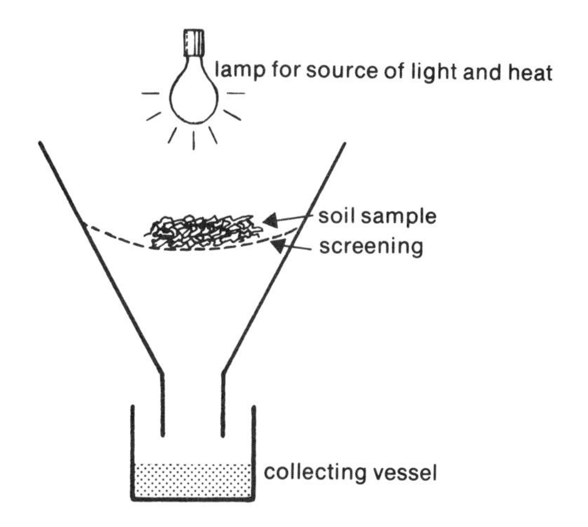

picking and sieving through samples of soil. Another group which includes the mites and springtails can best be obtained by subjecting a soil sample to heating and desiccation; as a result the organisms leave the sample and may be trapped. The basic apparatus used for this type of extraction is the Tullgren funnel (Figure 6.1). There are some animals which will not leave the soil when it is heated and desiccated. Many nematodes will encyst and can best be extracted by immersing the soil in water when they will swim out and may be removed. Apart from certain technological improvements, these common methods of extraction are the same as those used by Bornebusch in the late 1920s (Bornebusch, 1930). From these extractions it is possible to prepare estimates of the numbers and biomass of animals per unit volume of soil. Bornebusch recognized that these two parameters do not provide a measure of the intensity of organism activity and proceeded to measure the metabolic rates of many forest-soil organisms on the basis of their oxygen consumption under controlled conditions. When measures of feeding, assimilation rates, and age class distribution within the population are combined with data of numbers, biomass, and respiration rates, it is possible to obtain estimates of rates of energy flow for a soil population.

The energy budget for a soil population can be expressed by an equation

$$I = R + P + E$$

where I is the energy assimilated as food, R is the energy used in respiration, P is the energy bound up in body substance of the population, and E is the energy content of material eliminated from the animal population.

The estimation of any of the parameters of population and the determination of energy budgets is made difficult by variations in population size as well as by artifacts related to sampling. An example of a detailed study of one species is that of Wiegert (1970) who determined for the nest-building termite (*Nasutitermes costalis*) in a Puerto Rican rain forest that, although the species was prominent and numerically abundant in the insect fauna, the energy flow in the population was 5 to 6 kcal/m^2/yr which represented only a small fraction of the total input of energy and could be compared to the energy of annual leaf fall which was 2,200 kcal/m^2/yr. He concluded that the disparity was related to the limited amount of firm dead wood available to the termites as a food source.

Wallwork (1970) considered that the high biomass of certain animals, such as lumbricids, may be offset to a great extent by their relatively low metabolic rate and their contribution to energy release may be no greater than that of other smaller animals.

KINDS OF ANIMALS

It is not possible to deal with the subject of soil animals in detail here. There have been several excellent treatments since 1955 (Kevan, 1955; Kühnelt, 1961; Kevan, 1962; Murphy, 1962; Doeksen and van der Drift, 1963; Burges and Raw, 1967, and Wallwork, 1970).

Macrofauna

Vertebrates. These are very mobile and certain forms such as lizards, frogs, and snakes are generally found at the soil surface where they feed on insects and worms. Mammals such as rabbits, moles, and voles are notable for the burrows which they make; when vacated these frequently become filled with roots and are 'organic conduits' within the soil. In prairie-forest transition zone soils, gophers are frequently active in burrowing in the soil. The consumption of seed by small mammals may be quite remarkable, especially during periods of high population and can be significant for natural regeneration. Mice can be a detrimental factor in the early stages of forest growth by their girdling of stems.

Molluscs. Slugs and snails of various kinds are common in soils under deciduous forest, particularly where there is a relatively high level of calcium available. Food requirements are variable; some feed on live plants; others are saprophagous and consume litter while others are predatory and feed on earthworms. Newell (1967) suggested that the mucoids produced by snails and slugs may result in soil structure development either directly or by providing habitats for micropopulations of bacteria or actinomycetes.

Earthworms. In relation to soil more attention has been given to this group of animals than to any other since Darwin described their activities in 1881, drawing attention not only to their effects in consuming litter but also to the manner in which they buried old Roman ruins. There are a number of species belonging to several genera:

Lumbricus terrestris - the common earthworm, common in Europe and other continents. It burrows quite deeply, 1 to 2 metres and feeds on surface litter, but is less important than other species in actively mixing the soil materials (Bornebusch and Holstener-Jørgensen, 1953). The long vertical burrows are often lined with clay films. It is found in coarse mull soils under deciduous forests, but is common in pastures and gardens.

Lumbricus rubellus - a reddish worm which occurs over a wide range of forest soils under deciduous and conifer species. Bornebusch (1930) found it abundant in beech mor. It is considered to occupy a shallower zone in the soil than the larger *L. terrestris* and is typically found just beneath the upper litter.

Lumbricus castaneus - in Denmark it has been associated with mulls under deciduous forest but is more commonly found in old field soils.

Lumbricus festivus - is not so abundant but Bornebusch (1930) considered that it could be found in a wide range of conditions similar to *L. rubellus.* Langmaid (1964) noted this species in the colonization of an acid podzol (Spodosol) in New Brunswick.

Allolobophora longa - about the same size and often confused with *L. terrestris*. Bornebusch and Holstener-Jørgensen (1953) considered it was more active than *L. terrestris* in mixing organic and mineral soil beneath the surface. Kühnelt (1961) stated that from May or June to September and October these earthworms withdraw to a depth of 20 to 40 cm and remain inactive in roundish chambers.

Allolobophora calaginosa - an earthworm active in forest soils in Europe and North America. It is numerous in coarse mull soils in eastern North America and common under deciduous forest in Wisconsin (Nielson and Hole, 1964). Stegman (1960) found it under hardwoods, white and red pine, and white cedar in central New York.

Octolasium spp. - there are a number of species of this genus in forest soils, of which the most common is *O. lacteum*. This species has been found in soils under conifers and hardwoods. Eaton and Chandler (1942) found it only occasionally in northeastern USA forests.

Dendrobaena octaedra - a small earthworm found primarily in surface organic layers, usually in the lower litter, L and F horizons. Consequently, it is found more frequently in moder and mor humus types than in mull soils.

The activities of earthworms in the soil are therefore very much dependent on the species which occur. *Allolobophora calaginosa* and *Allolobophora longa* are

Figure 6.2 Changes in cover and exposure of soil surface, thickness and weight of forest floor during the period of 15 months in a soil characterized by earthworm activity. From G.A. Nielsen and F.D. Hole in *Soil Science Society of America Proceedings* (1964), by permission Soil Science Society of America

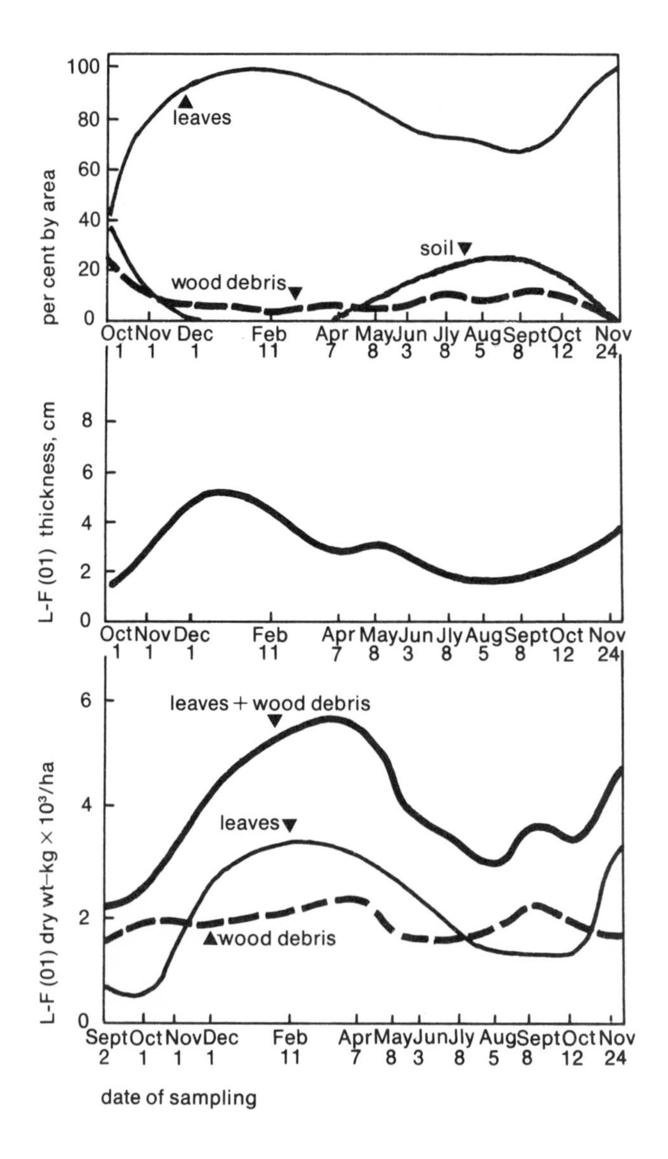

capable of mixing organic material with the mineral soil as does *L. terrestris* to a lesser extent. Nielson and Hole (1964) studied the activities of *L. terrestris* and *A. caliginosa* under deciduous species including oak and black cherry in Wisconsin. They found that when they reduced the amount of litter experimentally, the number of *L. terrestris* middens decreased and that by doubling the litter the number of middens virtually doubled. This indicates that, in their situation, amount of substrate (litter) was a factor limiting earthworm activity. Figure 6.2 illustrates the changes in quantities of forest floor materials and the amount and type of surface soil cover in a soil with an active earthworm population. Powers and Bollen (1935) suggested that earthworms are 'colloid mills' and it is generally considered that the development of mull horizons are largely a result of certain earthworm activity. Observations of the colonization of profiles by certain earthworms and the development of a mull within 3 years (Langmaid, 1964) is evidence of this. Also Nielson and Hole (1964) found that removal of the A1 (Ah) horizon did not suppress activity of *L. terrestris* and *A. calaginosa.* Certain earthworm activity can undoubtedly facilitate the development of granular structure and may, indeed, be the primary cause of it.

Lunt and Jacobson (1944) analysed earthworm casts and found them to contain higher total nitrogen, organic carbon, exchangeable and total calcium than the surrounding soil in a hardwood forest. In addition to physical and chemical changes which may result from earthworms, Satchell (1967) suggests that it is their effect in conditioning plant remains for microbial decomposition that is the most important action of Lumbricidae in the ecosystem.

Millipedes. Millipedes comprise a large number of species occurring widely in forest soils. They feed mainly on dead and decaying organic material and may be found predominantly in the surface organic layers. Some species will descend into lower mineral and soil layers during winter frosts (Kevan, 1962).

Termites. In subtropical and tropical forests termites move soil materials and mix organic material with it. The presence of those species constructing mounds is obvious and Nye (1955) has described such forms in West Africa. The main effect of mound-building termites is that with time a gravel-free upper layer is created which may move down slope as a soil creep (Nye, 1954).

Mesofauna

Ants. While common in a wide range of forest soils, these animals have not received much study. Lyford (1963) concluded that the A horizon of many brown podzolic soils in New England consists of material moved by ants to the surface from the B horizon and the rate of buildup may be 2.5 cm over a 250-year period. The ants Lyford studied had nests which penetrated the soil to depths of

35 to 40 cm. Salem and Hole (1968) studied the effects of another species of ant which builds nests in openings bordering woodlands, in contrast to those of Lyford's study whose nests were under the forest canopy. They found that ant nests would extend up to 1.6 m below the surface. Ants were effective in reducing the bulk density of the surface soil materials from 1.5 to 0.8 g/cm^3 and increased the contents of available nutrients compared with the surrounding soil.

Mites (Acari). Mites are one of the most numerous animals in most forest soils. Kevan (1962) said they are among the most important of all soil animals. There are several hundred families of mites represented in the soil fauna (Wallwork, 1970) and they are particularly prevalent in the organic matter of the forest floor. Bornebusch (1930) found mites in Danish forest soils to number 1000–10,000 per m^2. Mites that feed primarily on dead plant material are undoubtedly important in the breakdown of the forest floor, but in addition certain mites may move organic material vertically in the profile.

Springtails (*Collembola*). Springtails, like mites, abound in most forest soils. Their name is derived from the fact that many of them possess a springing organ which is folded forward under the abdomen and when this is suddenly released, the animal springs into the air. Although commonly found in the upper soil layers, some will penetrate into mineral soil, especially if it is loose. Poole (1961) found in Douglas-fir plantations that the number of springtails averaged 46,700 per m^2. Eaton and Chandler (1942) estimated the number of springtails in a coarse mull at 8,019 per m^2 compared with 14,370 per m^2 in a mor. Generally, springtails are more prolific in moder and mor humus types than in mulls. This may be related to the feeding preference that springtails have for fungi and fungal spores. The collembola have a wide range in food materials - living and dead plant material, faeces, bacteria, and algae.

Potworms (*Enchytraeids*). There are a number of species of these small white worms in forest soils. They are thought to feed primarily on dead organic matter, but some may feed on other material including the faeces of small soil fauna. Potworms also ingest mineral soil particles and O'Connor (1957) concluded that, under a Douglas-fir plantation with a moder soil in North Wales, potworms were important in mixing mineral soil and plant debris. The number of worms ranged from a low of 42,000 per m^2 in late winter to an early summer maximum of 250,000 per m^2.

Nematodes. There are many species of these small round worms, and those which have received most attention are parasitic in crop plants. Wallwork (1970) considers that the feeding activity of soil nematodes does not significantly affect the decomposition of organic matter. Many nematodes are predaceous on other organisms and, in turn, are a source of food for other members of the soil population.

Microfauna

Protozoa. Apart from bacteria, protozoa are representative of the smallest soil animals. There are three main types of protozoa in soil. Flagellates are prevalent in soils with abundant organic matter. Rhizopods, typified by amoeba are variable in their food requirements: some are predatory on bacteria and other organisms, while others absorb from decomposing organic substrates. Ciliates subsist primarily on soil bacteria. Stout and Heal (1967) quoting data from Heal (unpublished) plotted the vertical distribution of a group of rhizopod protozoa (Testacea) in an acid brown forest soil under oak in northern England and showed pronounced differences in species composition for L,F,H and A1 (Ah) horizons. Out of 20 species studied, only two occurred in the A2 (Ae) horizon and, then, sparsely.

Soil microflora

Four groups of organisms constitute the soil microflora: bacteria, actinomycetes, fungi, and algae. As with soil animals the microflora population can be considered as: kind, amount, and activities. It is convenient to separate the *autotrophs* which obtain their carbon from carbon dioxide for incorporation into their body tissue from the *heterotrophs* which obtain their carbon from complex organic materials. Algae are distinguished from bacteria, actinomycetes, and fungi in that they are autotrophs which contain chlorphyll pigments and are capable of photosynthesis. They are termed photoautotrophs to distinguish them from chemautotrophs which obtain energy by oxidation of inorganic compounds of nitrogen, iron, and sulphur. Autotrophs require an external source of energy in order to synthesize carbon from carbon dioxide and it is considered that this is more energy-consuming than the heterotrophic assimilation of carbon from organic sources. The soil micropopulation may also be considered as *aerobes* which require the presence of free oxygen for their activity or *anaerobes* which grow in the absence of free oxygen. Separation on this basis is often not clearcut; many forms are *facultative*, that is, although they may exhibit a preference for one form of aeration, they can exist in either. In well-aerated soils, films and interstices of water provide anaerobic microsites.

THE ESTIMATION OF MICROFLORA POPULATIONS AND THEIR ACTIVITIES

There are two main procedures used in studying the numbers of organisms. One is based on direct microscopic examination and the other involves the culture and plating of suspensions from the soil.

Microscopic methods
Various techniques may be used but most involve staining the organisms. A suspension of soil in fixative solution may be spread on a clean slide, dried, stained, and examined under the microscope. Sometimes the soil is suspended in agar gel and then subsequently stained and dried. The contact slide method involves inserting a clean glass cover slide into a slit made in the soil with a knife. The slide is then left there for several days or weeks. It becomes coated with soil solution and particles. After the slide is taken from the soil, the particles are gently removed and the preparation is fixed and stained so that any organisms on the slide can be examined under a microscope. Kubiena (1938) has used a special microscope for direct examination of microorganisms in the soil but its use has been limited.

Dilution and plating
A known amount of soil is shaken with a volume of water and from this a series of dilutions are made of the order of $1:10^2$, $1:10^3$,..., $1:10^8$. Aliquots are then taken from each dilution and placed on agar nutrient media in dishes. These are then incubated and the colonies that develop are counted. It is assumed that each colony develops from a single organism, and from the colony counts in relation to the dilutions and the original amount of soil it is possible to estimate the number of organisms in the soil.

There are several weaknesses inherent in both procedures. Microscopic techniques involve taking very small samples from the soil and the variability in numbers can be extremely large within a few centimetres' distance in the soil and also from hour to hour and day to day. The microorganisms which are found on smooth glass surfaces of slides buried in the soil will obviously not include those for whom this site is unsuitable. The agar nutrient cultures will allow the preferential development of those organisms for which the particular nutrient composition is most appropriate. This form of selectivity is used by investigators to segregate organisms in a soil suspension by providing different substrates on which they may develop preferentially.

The activity of the microflora can be measured in the simplest manner by assessing the disappearance of the organic matter or some portion of it when exposed to a microbial population. Frequently pure substrates such as proteins or various carbohydrates may be used. A second technique is to measure the amount of a substance that may be liberated from a soil and which can be related to a specific set of organisms. For example, the amount of nitrate nitrogen may be used to reflect the activity of nitrifying bacteria. The measurement of carbon dioxide production and/or oxygen consumption is also used as a measure of general activity within the soil population.

TABLE 6.1
Plate counts of microorganisms in soils under three forest covers in central Ontario (data of Chase and Baker, reproduced by permission of the National Research Council of Canada from the Canadian Journal of Microbiology, 1954)

Tree cover	Horizon	Organic matter (%)	Plate counts – $n \times 10^5$/g (oven dry soil)		
			Bacteria	Actino-mycetes	Fungi
Sugar maple	F-H-A1	23.0	4.2	2.8	2.6
	B2	9.6	1.0	3.5	1.7
	B3g	5.2	0.7	1.0	0.1
Eastern hemlock	F-H	65.0	3.5	4.6	13.1
	A1-A2	1.4	0.7	3.4	1.0
	B2	6.4	0.7	0.4	1.0
Eastern white pine	F-H	19.0	2.6	7.4	11.8
	A1-A2	3.7	1.0	0.7	1.5
	B2	2.8	1.4	2.3	0.1

Assessment of soil micropopulations is often undertaken in relation to specific treatments of soil either in the field or in the laboratory under controlled conditions. Under natural conditions there are often fairly abrupt changes in substrate supply which may be seasonal, as with litterfall. Winogradsky considered that soil populations could be separated into two broad groups: *autochthonous* or native microorganisms which are characteristic for a particular soil and the *zymogenous* organisms which, although present, only develop under the influence of specific soil treatments or change in properties.

The distribution of the four main types of microorganisms varies. Algae are concentrated at the surface. The other three – bacteria, actinomycetes, and fungi – will vary somewhat, but generally predominate in the zones of organic matter, as illustrated in Table 6.1. The functions and particularly the transformations of organic and inorganic material by these organisms in forest soils will be considered in the following sections.

BACTERIA

Bacteria may be classified on the basis of their form-rods, cocci or spore-forming bacilli, and on staining responses such as gram positive or gram negative. Separation of bacteria on the basis of their physiological functions and in particular their source of carbon and energy is more appropriate in a consideration of the soil population.

Heterotrophic bacteria
The principal factor limiting bacterial growth is scarcity of food or a lack of a suitable and available energy supply (Clark, 1967). The main substrates are simple compounds - sugars and starches which undergo rapid decomposition. Complex carbohydrates such as cellulose and the hemicelluloses are hydrolysed by enzymes produced by the bacteria extracellularly, as are the proteinaceous materials. Aerobic bacteria which decompose cellulose are active mainly when the soil reaction is greater than pH 6.0. Alexander (1961) considers that fungi, rather than bacteria, are mainly responsible for cellulose decomposition in forest floor materials. Certain anaerobic species of *Clostridium* can decompose cellulose. Many kinds of bacteria, both anaerobic and aerobic, can decompose hemicelluloses and this is probably one reason why these disappear relatively rapidly in the soil, compared to cellulose. Protein decomposition is accomplished by a number of anaerobic and aerobic bacteria. There are several chemical reactions by which proteins are broken down, and intermediate compounds are produced. A common end-product is ammonium nitrogen (NH_4^+). This mineralization of organic nitrogen to an inorganic form is important, partly because it provides a source of nitrogen to other microflora and partly because it is a source of nitrogen for many higher plants. The absorption of nitrogen and other nutrient elements by microorganisms and incorporation into their protoplasm exemplifies *immobilization* of these nutrients.

The overall action of heterotrophic bacteria on the breakdown of complex plant tissue in forest soils is difficult to assess. It is known that they have some effect in 'softening up' litter for faunal consumption, and bacterial counts are usually high on faecal material of soil animals such as earthworms (Went, 1963), but in acid forest soils they probably play a minor role in decomposition compared to fungi. The function of bacteria, as gut flora of soil fauna, in facilitating decomposition of ingested plant remains is probably of some significance.

Dinitrogen-fixing bacteria
Asymbiotic. These are heterotrophic organisms distinguished by their use of gaseous nitrogen from the atmosphere as their nitrogen source. A number of genera are aerobic - *Azotobacter, Achromobacter*, and *Biejerinckia* are representative and of these *Azotobacter* has received the greatest attention. While important in arable soils, it does not appear to be a component of forest soils, even when the pH values are suitable. *Beijerinckia* will grow well in acid soils but is apparently found only in tropical soils. The main genus of anaerobic bacteria fixing nitrogen is *Clostridium.* Species of this genus are more acid-tolerant than *Azotobacter* and may be a common inhabitant of many forest soils. Corke (1958) reported most probable numbers (MPN) per gram of soil in the humus layers to

TABLE 6.2
Amounts of nitrogen fixed in mull layers from under sugar maple and mor layers under American beech (data of Knowles, in Soil Science Society of America Proceedings, 1965, by permission, Soil Science Society of America)

		Annual N fixation - kg/ha/yr	
Tree species	Soil layer	Aerobic	Anaerobic
Sugar maple	Mull A1 (+ 1% glucose)	5.0	55.2
	Mull A1 (no glucose)	–	44.0
American beech	Mor L	–	0.8
	Mor F & H	–	0.9

be 19,000-92,500 and pH ranges of 3.8-5.5 for *Clostridium* with no detectable *Azotobacter*. Table 6.2 illustrates the amounts of nitrogen which Knowles (1965) estimated were fixed in mull horizons under sugar maple and mor under American beech. It will be noted that anaerobic fixation dominates, but in the mor layers it is only a small amount compared to the mull soil. The addition of glucose and the subsequent increases in fixed nitrogen indicates that carbohydrate supply may be a limiting factor.

Symbiotic. The isolation of a bacteria responsible for nitrogen fixation as a symbiont was done by Beijerinck in 1888, although the significance of certain legume crops in the nitrogen budget of the soil was well known. The symbiosis between bacteria *Rhizobium* spp. and many legumes has been extensively investigated, particularly for agricultural crop species. Many trees are members of the Leguminosae, particularly in the tropical and subtropical forests. In temperate regions of North America, black locust is a tree legume which enhances soil nitrogen (Ike and Stone, 1958) and this fact, that black locust has a nodulated root system with dinitrogen-fixing *Rhizobia*, has been utilized in the afforestation of mineral soils with little or no organic nitrogen content.

In many forest areas of the world the major symbiotic fixation of nitrogen occurs with non-leguminous species. The nodules of these species show many similarities to those of legumes but the microorganisms responsible are thought to be actinomycetes (Bond, 1970). There are 13 genera in which nitrogen fixation either has been proven or is likely to occur: *Alnus, Arctostaphylos, Casuarina, Ceanothus, Cercocarpus, Coriaria, Discaria, Dryas, Eleaganus, Hippophae, Myrica, Purshia,* and *Shepherdia.* Youngberg and Wollum (1970) include *Artemisia* and *Opuntia* in their list of nodulated non-legumes as well as most of the above.

There are several ways in which the presence of nodulated higher plants can be significant in forest soil systems.

1 / They can colonize exposed mineral soil surfaces and become a major factor in the buildup of nitrogen and organic matter. Crocker and Major (1955) estimated that the species of *Dryas, Shepherdia,* and *Alnus* fixed 62 kg/ha/yr nitrogen in colonizing glacial moraines in Alaska. Even when the amounts fixed are not large as with *Myrica asplenifolia* in the boreal forests of eastern North America, they may be significant because they provide a milieu suitable for accelerated successional development following natural or man-made disturbances. For these reasons, nodulated species are often chosen for land reclamation of mine spoils, blow sand, and erosional surfaces.

2 / Many nodulated species grow in association with other trees under natural conditions or they may be planted together. *Alnus* spp. are common in many forest regions and their presence is generally associated with improved growth of the associated species (Tarrant and Trappe, 1971). Mean annual accretions to soil nitrogen under five species of *Alnus* ranged from 12 to 300 kg/ha/yr. The accumulation of nitrogen by *Ceanothus* in Oregon was studied by Zavitkoski and Newton (1968) who concluded that it may fix nitrogen in amounts ranging from 0 to 20 kg/ha/yr. The relatively short period during which the major accumulation of nitrogen in the litter takes place is illustrated in Figure 6.3.

3 / Certain shrubs, *Purshia* and *Ceanothus* for example, may provide browse for deer and the high nitrogen levels of the nodulated species undoubtedly reflect higher protein levels. The effect may be indirect; Davidson (1970) reported that white-tailed deer browsing was greater on pine growing near black locust.

Autotrophic bacteria

These microorganisms derive their carbon from carbon dioxide and their energy by oxidation of inorganic compounds. They may be considered on the basis of the compounds which they utilize:

Nitrogen. Ammonium nitrogen is produced as a result of many decomposition processes. Its conversion involves several genera of which *Nitrosomonas, Nitrobacter, Nitrosocystis,* and *Nitrosogloea* are the most commonly recognized. Conversion of ammonium nitrogen to nitrate nitrogen is termed *nitrification* and takes place in two steps. The first is the oxidation of the ammonium form to nitrite:

$$2NH_4^+ + 3O_2 \rightarrow 2NO_2^- + 4H^+ + H_2O + \text{energy}.$$

Nitrosomonas spp. are the bacteria which are primarily responsible for this transformation. The second step is the oxidation of nitrite to nitrate:

$$2NO_2^+ + O_2 \rightarrow 2NO_3^- + \text{energy}.$$

Figure 6.3 Dry weight of snowbrush litter and nitrogen content of the litter in stands of various ages. From J. Zavitkovski and M. Newton in *Ecology* (1968), by permission Duke University Press

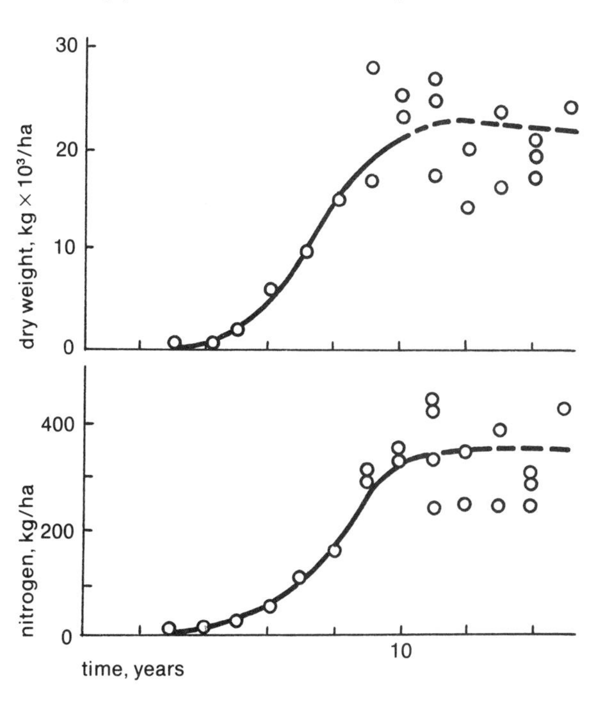

Nitrobacter spp. are involved in this oxidation process. Accumulation of nitrites in the soil can prove toxic to many higher plants, but such occurrences are uncommon in natural soils. The ammonium cation (NH_4^+) can be retained as an exchangeable ion by the clay-organic colloids. The nitrate form (NO_3^-) may be readily leached from soil. This difference may be of importance where higher plants are preferential in their nitrogen source or where leaching may be common.

Nitrification is most active in soils that are circum-neutral in reaction; favourable moisture and temperature conditions must also occur if the organisms are to be active, and there must also be a supply of ammonium substrate. Acid forest soils (pH less than 5.5) usually show slight evidence of nitrification. Chase and Baker (1954) found little nitrification in acid forest soils under sugar maple, eastern hemlock, and eastern white pine. They concluded that this was related to

TABLE 6.3
Most probable numbers (MPN) of nitrifying bacteria in an acid forest soil under sugar maple-American beech, eight years after fertilizer treatment (from Corke, in Proceedings, First North American Forest Soils Conference, 1958, by permission, Michigan Agricultural Experiment Station)

	Most probable numbers /g (oven dry wt)	
Treatment	Nitrosomonas (oxidize NH_4 - N)	Nitrobacter (oxidize NO_2 - N)
Control	260	240
Phosphate	300	250
Lime	11,600	5,300
Lime-phosphate	32,900	30,600

acid conditions and a low population of nitrifiers. Corke (1958) working in the same area with soils that had received lime or lime plus phosphate treatments found that there was a dramatic increase in nitrifiers over an eight year period (Table 6.3). It seems likely that nitrification is a common transformation process in mulls, even if acid, but of little consequence in moder and mor types (Bernier and Roberge, 1962).

In contrast to nitrification, there are bacteria which utilize nitrate nitrogen as a source of oxygen rather than the gaseous form. Certain species of *Bacillus, Pseudomonas, Achromobacter*, and other genera can reduce nitrate nitrogen to gaseous nitrogen (N_2) and nitrous oxide (N_2O). *Denitrification* is most likely in anaerobic situations with a source of nitrate nitrogen and circum-neutral reaction. Alexander (1961) states that the denitrifying population becomes large enough to be significant only at pH values greater than 5.5.

Volatilization losses of nitrogen from the soil can also take place as a result of chemical processes; ammonia may volatilize at pH values above 7.0 and nitrite may decompose to nitric oxide (NO) in acid soils.

Sulphur. Several species of bacteria utilize inorganic forms of sulphur. Elementary sulphur can be oxidized to sulphuric acid by certain common bacteria. Flowers of sulphur are added to acidify soil in nursery management. Other bacteria can utilize the sulphates which may result from decomposition of organic compounds in the plant and animal remains. Hesse (1957) attributed the occurrence of sulphates in forest subsoils in east Africa to microbiological transformations beginning with organic sulphur compounds in the tree litter. Certain anaerobic bacteria reduce sulphates to hydrogen sulphide (H_2S); these bacteria are not active in soils with pH values less than 5.5. Alexander (1961) considers that in neutral-alkaline soils (pH 7–8) under anaerobic conditions they are responsible for the corrosion of iron and steel because the sulphides they produce react with the iron to produce ferrous sulphide.

Iron and manganese. The iron bacteria oxidize ferrous iron to the ferric state with the production of ferric hydroxide. Conversely, under anaerobic conditions other bacteria can reduce ferric hydroxide with the production of soluble ferrous compounds. It is probable that the changes in iron form and colour associated with impeded drainage and gley conditions may reflect the activities of iron bacteria to a large degree. Oxidation and reduction of manganese compounds are known to occur in a somewhat analogous manner to the changes for iron.

ACTINOMYCETES

These microorganisms occupy an intermediate position between the bacteria and fungi. They are found in forest soils (Table 6.1) although they are generally more abundant in grassland or arable soils that are not acid. They are heterotrophic and can decompose cellulose and hemicelluloses as well as more readily attacked simple carbohydrates. Certain actinomycetes can decompose chitin. It is thought that actinomycetes are the symbiont responsible for dinitrogen fixation in the nodules of many non-leguminous plants.

FUNGI

Fungi are heterotrophs, predominantly aerobic and for many forest soils they are the major factors in the decomposition of many complex organic compounds. Typically a fungus exists as a filamentous mycelial network of individual strands or hyphae in the soil. Often the mycelia are coloured and readily seen, particularly in L and F horizons of moder and mor humus types and the fungal occupation of the soil can be intense. Burges (1963) determined that in the H layer of a podzol (Spodosol) under Scots pine there were 5.56 m of hyphae per cubic centimetre of soil and the corresponding value for the A2 and B2 horizons were 3.78 and 0.37 m/cm^3 respectively. Although the greater abundance of fungi in acid soils, compared to bacteria and actinomycetes, is usually attributed to their being more acid tolerant, it could also be related partly to moisture relations in the upper soil organic layers; these frequently dry out and provide less suitable niches for sustained bacterial development.

A large body of information about fungi in forests comes from forest pathology studies of parasitic fungi. The term 'brown rot' refers to fungal decomposition primarily of cellulose, leaving the lignin material behind. 'White rot' is a condition where lignin is decomposed but cellulose remains. Garrett (1963) has pointed out that in the colonization of substrate there is frequently a fungal succession. Initially, those fungi capable of utilizing the sugars and simple carbon compounds will colonize first – sometimes before the plant material is dead and has become part of the litter (Burges, 1963). They are followed by the cellulose decomposers and lastly by fungi which attack the most resistant compounds and lignins. This sequence may not hold for all soils; for example, Handley (1954) suggested that a lignin-protein complex could coat cellulose and protect it from attack. He and others found that lignin may disappear before cellulose in the decomposition of some litter.

Mycorrhizal fungi

During the latter half of the nineteenth century, the close association between fungal hyphae and the roots of certain trees was observed. Frank in 1885 coined the term *mycorrhiza* (meaning fungus root) to describe this root formation. Since then there has been knowledge of an increasing number of higher plants which have mycorrhizas. One of the most notable groups is the Orchidaceae, but there are many other plants including agricultural crop species. Rayner (1927) and Hatch (1937) present accounts of the early work on mycorrhizas and Harley (1969) gives a more recent comprehensive review.

Mycorrhizas are generally distinguished as being one of two kinds, *ectotrophic* or *endotrophic*. The ectotrophic have a mycelial sheath about the root and hyphae which penetrate intercellularly, that is between the cortical cells (Figure 6.4). The pattern of interconnecting hyphae between cortical cells is termed the *Hartig net*. This type of mycorrhiza is common on forest trees in temperate regions and is the form which will be discussed primarily. Sometimes hyphae may also penetrate intracellularly and then the term *ectendotrophic* is used to describe the mycorrhiza form. Endotrophic mycorrhizas do not normally have an external sheath and the hyphae are found consistently within the plant cells. Often large vescicles are produced inside or outside the root.

Infection of tree seedlings by mycorrhizal-forming fungi takes place in the first growing season after the primary leaves have been produced. The infection may be from other mycorrhizal roots or from spores. Not all roots of a tree will be infected and the form of the infected roots will vary. Diffuse, pyramidal, coralloid, and nodular are terms which describe certain of the forms. In the genus *Pinus* the form is consistently a dichotomous branching. Slankis (1958) showed

Figure 6.4 (1) Mycorrhizal roots on 20 month old white spruce seedling, 10 months after inoculation, X4. (2) Non-mycorrhizal roots of white spruce seedling, X4. (3) Cross-section of mycorrhizal root from (1) showing hyphal mantle and Hartig net, X400. (4) Cross-section of non-mycorrhizal root from (2), X400. Photo courtesy of R.D. Whitney and Canadian Forestry Service, from R.D. Whitney in *Forest Science* (1965), by permission Society of American Foresters

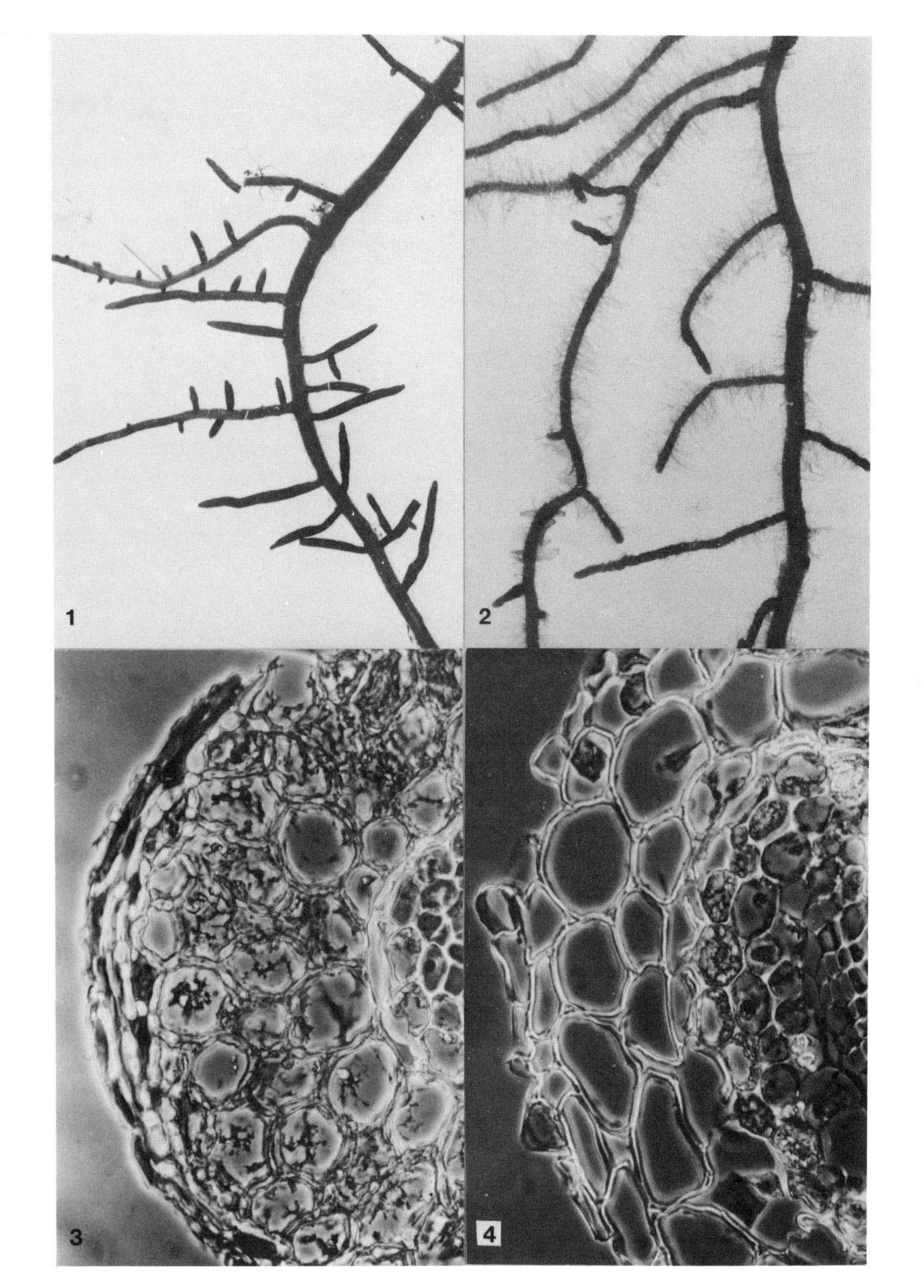
1
2
3
4

that fungal auxins could induce the morphological changes associated with mycorrhizal root forms without the root being infected. The infection of roots is considered to be seasonal with the life of a mycorrhizal root extending up to a period of a year. Interest in determining the reasons for mycorrhizal infection and the possible physiological nature of the symbiosis was stimulated by the experiences of foresters in Australia, Great Britain, and the prairie region of North America. They found either with the introduction of exotics in certain soils or in the afforestation of lands such as prairie and heaths that seedlings without mycorrhizal roots showed less survival and grew at a lower rate than those trees which were infected. Hatch (1937) in a classic study concluded that in pine the ectotrophic mycorrhiza increased the absorbing surface area of the root because: (a) infected roots lived longer than the uninfected; (b) infected roots were larger in diameter; (c) the increased branching habit of infected roots produced a larger surface area; (d) the hyphae of the external sheath formed a larger absorbing surface; and (e) the fungus delayed suberization of the endodermis and cortex. The seedling size and absorption of mineral nutrients were greater for infected than for uninfected seedlings grown in prairie soil. He also observed that less mycorrhizal development occurred when seedlings were grown in fertilized (especially with nitrogen) soils. Bjorkman (1942) studied the response of one-year Scots pine seedlings to combinations of increasing nitrogen and light supplies. He found that the degree of mycorrhizal infection increased with increase in light. Addition of nitrogen decreased the degree of mycorrhizal development at a given light intensity, but resulted in an overall increase in seedling size. He concluded that degree of infection was related to the carbohydrate status of the root tissues and that high levels of nitrogen resulted in increased protein synthesis and a reduced carbohydrate status. The carbohydrate status of the plant is not the only factor to control infection; the balance in nutritional supply is probably equally important. Certainly in nurseries conifer seedlings grown in fertilized soils show extensive mycorrhizal development.

Considerable attention has been given to the nutritional aspects of mycorrhizal symbiosis. The fungus receives simple carbohydrates if not other substances from the tree. Carbohydrates derived from the tree may move distances of up to 12 cm from the mycorrhiza in external hyphae (Reid and Woods, 1969). Melin and his co-workers in the 1940s and 1950s demonstrated that phosphorus, nitrogen, calcium, and sodium can be transported to pine seedlings via the fungus mycelium. Harley and his co-workers found in studies of excised, infected, and uninfected mycorrhizas of European beech that rate of absorption of phosphorus was enhanced in the infected roots. Stone (1950) found increased absorption of phosphorus for Monterey pine seedlings with extensive mycelial development. It appears that the mycelium of the mycorrhizal fungus not only absorbs a nutrient such as phosphorus but may accumulate it and subsequently release the stored phosphorus to the host tree.

An aspect of mycorrhizal infection that is receiving some attention relates to the capacity of the fungus to exert a control on other microorganisms at the root surface. Marx and Davey (1969a,b) provided evidence that pine seedlings with mycorrhizal roots were less susceptible to attack by *Phytophthora cinnamomi* than those with uninfected roots.

In the large areas of forest soils supporting indigenous vegetation, the presence of mycorrhizal roots is most likely a natural, common phenomenon. Until recently most concern for the presence of mycorrhizal roots has been felt where exotic tree species are introduced or where trees are established on lands which had not previously supported such vegetation. It is now apparent that the use of fungicides or other soil sterilants, particularly in forest tree nurseries, may result in the removal of mycorrhizal fungi. Such losses are more likely with endomycorrhizal than ectomycorrhizal fungi, since the latter have spores which are wind-disseminated and hence permit more rapid soil recolonization. The presence of mycorrhizal roots on seedlings used to vegetate spoil banks and wastelands is particularly important and in these situations it is often the specific fungus that is critical. Marx and Bryan (1975) cite the importance of the fungus *Pisolithus tinctorius* as an ectomycorrhizal symbiont in the vegetating of spoil lands.

ALGAE

Algae occur almost exclusively in the surface soil layers where light and moisture supply are adequate. They are usually subdivided into green, blue-green, yellow-green, and diatoms. Certain blue-green algae are capable of dinitrogen fixation, but it is generally less than that of the non-symbiotic dinitrogen-fixing bacteria. Algae are found below the soil surface at depths to which no light can penetrate and Alexander (1961) considers these as heterotrophic variants which utilize the oxidation of organic carbon rather than light as a source of energy.

The major importance of algae, particularly those that fix nitrogen, is to be found in exposed mineral surfaces which they can colonize and in which they can initiate the accumulation of organic carbon and nitrogen, a first stage in soil development.

The rhizosphere

The *rhizosphere* is the zone of soil at a root surface and immediately surrounding it where the microbial population is altered both quantitatively and qualitatively by the presence of a root. The symbiotic relationship shown by mycorrhizas is a particular example of a rhizosphere effect. Studies of the rhizosphere population employ the conventional techniques used in microbiology. The numbers of organisms such as bacteria are expressed as a ratio of those in the rhizosphere to the number in the soil, unaffected by roots - the R:S ratio. Typically ratios may be of the order of 10 to 20 or more (Katznelson, 1965). The rhizosphere effect may involve one or more of several relationships:

1 / The increased number of organisms, especially bacteria may be a response to substrate materials exuded from the root while alive or from normal fine-root mortality. Carbohydrates, amino acids, and vitamins are known to be derived from live roots and there are other compounds which may also be released. This often results in a zone richer in substrates immediately about the root than in the surrounding soil.

2 / The increased metabolic activity associated with a greater population of microorganisms can affect the solubility of nutrients and the rate of soil mineral weathering caused by higher carbon dioxide concentrations and organic acids from the micropopulation.

3 / The presence of bacteria involved in nitrogen transformations may result in a difference in nitrogen supply to the higher plant. Alexander (1961) notes that the members of one of the groups that characteristically have high R:S ratios are those responsible for ammonification - the hydrolysis of complex organic nitrogen forms and resultant production of ammonium nitrogen. This would be of major benefit to those higher plants such as many forest tree species which not only can use ammonia nitrogen but also may absorb it preferentially over nitrate nitrogen. Dinitrogen-fixing bacteria such as *Azotobacter* may in certain circumstances be a significant feature of the rhizosphere.

4 / The rhizosphere population can produce growth-controlling compounds such as auxins which may affect the form of root development. This is true for certain mycorrhizal fungi (Slankis, 1958).

5 / The rhizosphere may be a factor in host-pathogen relationships. A pathogen may be stimulated in some way by exudates of the host. One of the clearest examples of this is the movement of zoospores of *Phytophthora cinnamomi* to roots of avocado in response to an exudate produced by the root (Zentmyer, 1961). Alternatively, the rhizosphere may include organisms in its population which may inhibit pathogen development as shown by certain mycorrhizal fungi for *Phytophthora cinnamomi* in pine seedling roots, already mentioned.

6 / The rhizosphere population of microfauna may, in turn, result in changes in the population of soil fauna such as mites and nematodes and their presence may affect the higher plant adversely or otherwise.

7 / The presence of a rhizosphere population related to the rooting habit of the higher plant species ensures a pattern of soil population development which can have considerable long-term effects on the soil. The rate of weathering of

soil minerals and development of structure are two processes that can be affected markedly by the soil micropopulation.

Knowledge of the rhizosphere population of forest tree species other than mycorrhizal and nodule associations is scant. Ivarson and Katznelson (1960) studied rhizosphere populations on roots of yellow birch seedlings grown for 28 weeks in soil layers brought into the greenhouse. They found that the R:S ratios for microorganisms were smaller than for most agricultural crop plants, but that ammonifying and methylene-blue reducing organisms were predominant in the rhizosphere; there was a greater rhizosphere effect in the B horizon than in the upper A horizon, although the total numbers of microorganisms were less. Neal et al. (1964) found different rhizosphere populations of bacteria and *Streptomyces* in rhizospheres of Douglas-fir roots with three different kinds of mycorrhizas. In a more general investigation Jurgensen and Davey (1971) found no rhizosphere effect in the presence of anaerobic non-symbiotic dinitrogen fixers with roots of yellow birch or red oak.

7
Soil chemistry

Soil chemistry is concerned with the nature, chemical composition, properties, and reactions of soils. The chemical composition of a soil will reflect to a large degree the geologic origin of the soil materials. A soil developed in limestone will be characterized by high levels of calcium and perhaps magnesium. Often the mineralogical state of the geologic materials may be altered by physical weathering and transportation forces such as ice, wind, or water. Tills deposited by ice show a mixing of minerals and rocks, whereas there is a sorting of particles both by size and density when they are moved by wind and water. In the soil itself, weathering processes occur and it is these processes and their products which are the focus of attention in soil chemistry.

The measurement of soil chemical properties employs techniques and procedures from the science of chemistry. Such methodology is necessarily arbitrary and selective, but the results obtained are used to characterize many of the chemical properties of soil and often soils are distinguished by the intensity of one or more of these properties. Since the soil is a dynamic system and the measurement of soil chemical properties is largely selective, caution should be exercised in ascribing distinction in chemical attributes to soils without considering the methods of investigation and the degree of variation involved. There is probably no greater pitfall for the student of soil chemistry than the assumption that by measurement of a chemical property to the *n*th decimal place, using sophisticated apparatus, he has defined the soil.

The expression of soil chemical properties is not only arbitrary but it commonly employs a value established at some specific time. For example, the amount of phosphorus in two particular horizons may be compared and that horizon with the greater amount can be considered richer in phosphorus. Yet in terms of the uptake of phosphorus by a plant's roots, the more critical factor may be the rate at which the phosphorus is available to them, not the absolute amount present at any time. Increasingly, therefore, soil chemical properties are considered not as static attributes but in terms of rate and flow. The method of expression of the properties should be the one most significant from a consideration of the soil and not necessarily the conventional measure of the chemist.

Weathering of soil materials

PHYSICAL

Weathering requires the presence of water and, depending on the specific reaction, the rate of weathering is temperature-dependent to a greater or lesser degree. Many weathering processes take place on surfaces and the larger the surface area available to react, the greater the amount of weathering. Physical weathering brings about an increase in surface area. Often the mode of geological formation is a factor in the increase in surface area by creating particles of smaller size - *comminution.* Wind-, ice-, and water-transported materials are abraided or ground to produce smaller particles. Within the soil, larger particles may be further broken down by the *freeze-thaw* action of water in pores and crevices. If these openings are occupied by water in the liquid state and the water is then frozen, it will expand as ice and under the large pressures thus developed the particle will be fractured. At the soil surface, particularly where coarse rock fragments are exposed to large diurnal changes in temperature, differentials in mineral thermal expansion may result in surface *exfoliation* of rocks. The intensity of such weathering will vary with the mineralogical composition, particularly in size and colour of crystals. In many areas, fires, natural or man-caused, can result in stresses brought about by rapid temperature changes in rocks at the surface, thus causing further exfoliation.

Tree and other plant roots will often grow into small openings in rocks and as they grow they expand and exert pressure against the surrounding walls. This pressure may ultimately result in a fracture of the material. Windthrow of forest trees often results in the upheaval of soil fragments held by roots and these may be broken in the process and exposed to further physical weathering at the surface.

CHEMICAL

The minerals and compounds of soil particles subjected to physical weathering are not chemically altered, but chemical weathering processes result in changes both in the composition and amount of materials as well as physical changes to the minerals. Chemical weathering usually reduces the density of particles, making them more porous; frequently it brings about increases in volume. In simple terms, chemical weathering is a twofold process involving: (1) removal of elements - either lost completely from a particular soil system or translocated from one place to another; (2) transformations in which one compound is changed by weathering into another. It is convenient to separate individual chemical reactions for purposes of discussion but it should be realized that, although one or more may be dominant in a specific soil or at a particular time of year, they should not be viewed in isolation from the soil itself.

Solution and hydrolysis

When a fresh mineral surface is exposed, any compounds that are readily soluble in water will go into solution. Halite (NaCl) and carnallite ($KMgCl.6H_2O$) are examples. As solution takes place, there is usually a progressive loss of the soluble materials. In some soils (not usually forested soils) the solution may move upwards to the surface in response to water removal by evapotranspiration, and soluble materials such as sodium and magnesium sulphate may be precipitated. The cycle of solution and precipitation may be repeated.

Hydrolysis refers to the reaction between ions of water and a mineral. An example is the hydrolysis of a primary silicate such as potassium feldspar:

$$KAlSi_3O_8 + H_2O \rightleftharpoons HAlSi_3O_8 + KOH.$$

The $HAlSi_3O_8$ is unstable and decomposes to form $Al_2O_3.3H_2O$ (gibbsite) and H_2SiO_3 (silicic acid).

Hydrolysis releases basic elements (potassium in this example) and complex molecules of aluminium and silicon which may recrystallize into secondary silicate minerals usually to form clay minerals. The release of basic ions including a number of nutrient elements and the formation of clay minerals are of profound significance in soils because of their effects on other chemical and physical properties. The intensity of hydrolysis increases with increase in temperature in the presence of moisture; thus soil materials in tropic and subtropical areas may be expected to exhibit such weathering to a greater degree than soils in more temperate latitudes. It should be kept in mind that for many soils with a high clay mineral content, the clays may not be derived by weathering *in situ* from primary minerals, but have a geological origin as in lacustrine deposits.

Hydration

Water has an ability to form additive compounds with other substances. Crystals of gypsum ($CaSO_4.2H_2O$) are an example of this reaction. Iron oxide as hematite (Fe_2O_3) can combine with water to form limonite ($FeO(OH).nH_2O + Fe_2O_3.nH_2O$). Hydration often results in secondary materials softer than the primary ones from which they were derived and there is usually an increase in mineral volume. Colour changes may also occur; hematite is dark red whereas limonite is yellowish-brown. The reverse process to hydration is dehydration and it may occur where drying out of the soil may be extreme, as at the surface or about large pores which are open to the surface.

Carbonation

Water which moves into soil will contain some dissolved carbon dioxide. As it passes through the surface horizons where biological activity may be intense, it will pick up greater amounts of CO_2 and this carbonic acid solution will react with a number of minerals in the soil to produce carbonates and bicarbonates. An example is the weathering of limestone (calcite).

$$\underset{\text{insoluble}}{CaCO_3} + H_2O + CO_2 \rightleftharpoons \underset{\text{soluble}}{Ca(HCO_3)_2}.$$

This reaction is dependent on the partial pressure of carbon dioxide (pCO_2) and therefore, as the solution containing the soluble bicarbonate moves through the soil to a zone where the carbon dioxide concentration is lower, there will be a reversal and insoluble carbonates may be desposited. In a number of forest soils this deposition occurs at the lower boundary of the solum. Conversely, especially on bare soils when evaporation losses from the surface are large, the soluble bicarbonate may move to the surface where it is precipitated as carbonate nodules.

Carbonation will also cause weathering of silicate minerals, although Sticker and Bach (1966) consider it plays a minor role compared to other weathering processes.

Oxidation and reduction

Many cations present in compounds in the soil may readily lose or gain electrons and thus be oxidized or reduced. As a result, the lattice structure of a mineral becomes unstable and renders it susceptible to other weathering reactions such as hydrolysis. The ability or tendency of the soil solution to supply electrons is measured by its *redox potential.* This potential is primarily governed by the amount of dissolved oxygen and carbon dioxide in the soil solution and the degree to which it is acidic or basic. The oxidation-reduction of iron involving the ferrous (Fe^{2+}) to the ferric (Fe^{3+}) states is common in many soils. In soils where saturation conditions exist on a seasonal basis, oxidation-reduction will be a major weathering process.

Chelation

Many organic compounds have acid, ketone, or alcohol groups which can function as electron donors. Such organic molecules have the ability to react with metal ions and retain them in their structure. Elements such as iron, aluminium, copper, and others may be held and moved in the soil solution in this form. The stability of the chelating molecule varies. Certain organic compounds occurring in the foliage of many species of trees have been idientified as polyphenols and are effective in mobilizing iron (Bloomfield, 1957). Sticker and Bach (1966), basing their conclusions on work by other investigators, consider that chelation can be responsible for the decomposition of silicate minerals.

The intensity and duration of weathering processes in soils will depend upon many factors: moisture regimes, temperature regimes, the mineral composition of the soil materials and their specific surface areas, the nature of the vegetation and process of organic decomposition which will determine the kinds and amounts of organic acids and chelating compounds, and time. The greater the intensity of weathering and the longer the period of time over which it has occurred, the larger the proportion of secondary minerals and the greater the proportion of minerals highly resistant to weathering will be. The thickness of solum normally increases with intensity of weathering. Erosional forces which result in removal and deposition, movement of soil materials by fauna and drainage changes alter the soil and modify not only the weathering processes themselves but also their results as evidenced by the soil profile.

Clay minerals – products of weathering

The four most common elements in the earth's crust are oxygen (46.5%), silicon (27.6%), aluminium (8.1%), and iron (5.1%) (Jackson, 1964). The most abundant primary minerals are the silicates, which have been shown by X-ray studies to have a basic unit of structure. Oxygen atoms are large and occur at the four corners of a tetrahedron with a silicon atom inside (Figure 7.1). One of the simplest silicates is the mineral olivine $(Mg,Fe)_2SiO_4$, which consists of silica tetrahedra, with each alternate tetrahedron inverted; on weathering it yields magnesium and iron and a clay mineral, nontronite. The most abundant silicates are the feldspars. These are aluminosilicates in which the silicon-oxygen tetrahedra are joined together by sharing of oxygen atoms and where some silicon atoms have been replaced by aluminium. The imbalance in electrical charge resulting from this substitution is offset by the introduction of other smaller ions. Anorthite ($CaAl_2Si_2O_8$), albite ($NaAlSi_3O_8$), and orthoclase ($KAlSi_3O_8$) are examples of feldspars which yield calcium, sodium, and potassium ions, upon weathering.

A group of silicates, of which the micas are an example, are termed layer or *phyllosilicates.* These minerals consist of sheets of silicon-oxygen tetrahedra attached to octahedral sheets of oxygen, hydroxyl, and aluminium, iron, or magnesium. The sheets may occur in pairs, tetrahedra plus octahedra. The crystal is then referred to as 1:1. Or they may be in the form of a tetrahedra plus octahedra plus tetrahedra sandwich, in which case the crystal is a 2:1 arrangement as shown in Figure 7.2. In the formation of clay minerals of this layer type, cationic replacement is common and there can be considerable variation in both the species of ions and their abundance. When the replacement is internal as, for example, when an aluminium atom replaces a silicon in the crystal, it is termed *isomorphous replacement.* These differences will determine to a large degree the properties of the mineral. The most common types of clay minerals are:

Figure 7.1 Diagram of a silicon-oxygen tetrahedron

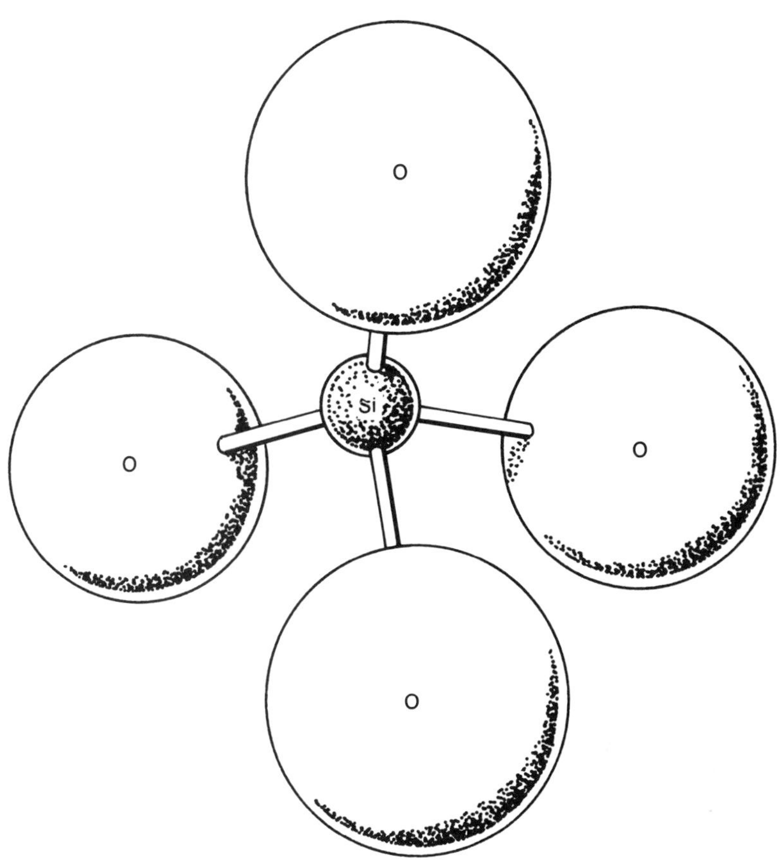

Montmorillonite. These are 2:1 minerals capable of considerable lattice expansion. Water and ions can move in and out of the lattice interlayer space so that the clay exhibits considerable volumetric changes – swelling or shrinking as it wets or dries. Magnesium and iron are found in the lattice frequently replacing aluminium and giving rise to a negative charge in the layer.

Illites. These are micaceous clay minerals of the 2:1 type. A proportion of the silicon atoms in the tetrahedra are replaced by aluminium, and potassium occupies interlayer positions. The interlayer space is much less than in montmorillonite because of closer bondings between lattices and the mineral does not allow water to enter and leave in the same manner as montmorillonite and hence does not show similar volumetric changes.

Figure 7.2 Diagram of a 2 : 1 type clay mineral. Large circles represent oxygen atoms, small circles silicon, and dotted circles aluminium atoms

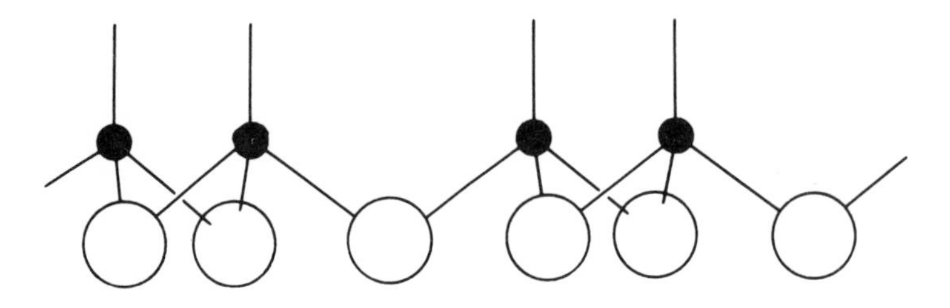

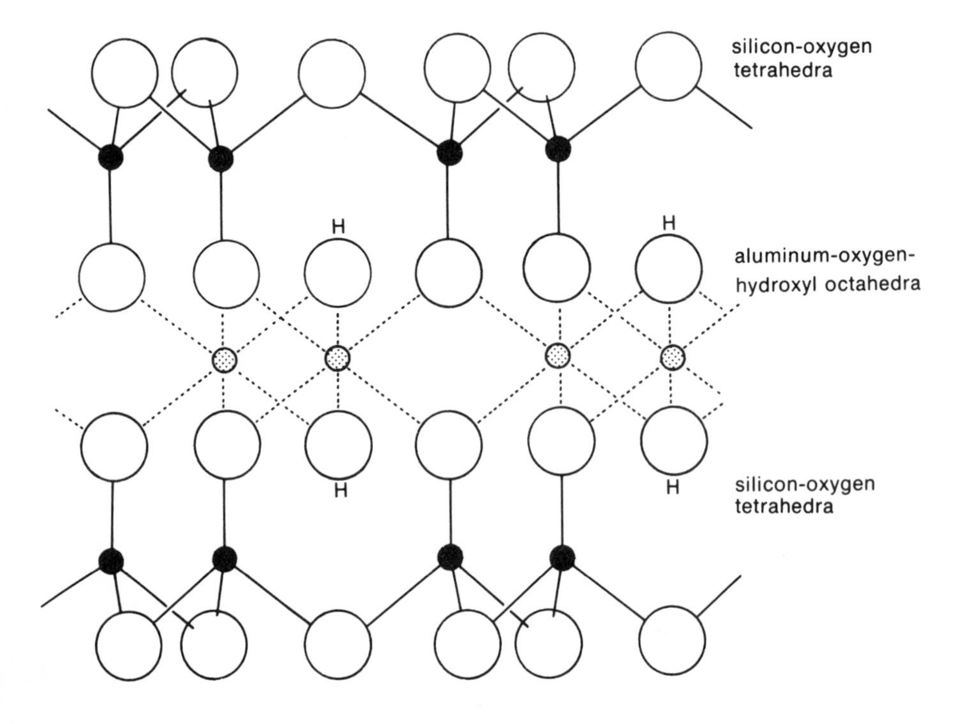

Vermiculite and chlorite. These are both 2:1 type minerals and each has an expanding lattice, although not to the same degree as montmorillonite. Typically, vermiculite has magnesium in the octahedral sheet where it replaces aluminium

and also has magnesium in the interlayer space. Chlorite has both magnesium and aluminium in the interlayer space.

Kaolinite and halloysite. These are clay minerals with 1:1 lattices. Halloysite is similar to kaolinite except that water molecules are held in the interlayer space. This water can be given up as the clay readily dehydrates. Kaolinite is a mineral in which there is little isomorphous replacement. Kaolinite therefore carries little electrical charge. The bonding between layers in kaolinite is strong and prevents water and other ions from entering the interlayer space. It does not shrink and swell with changes in moisture content. Kaolinite comprises particles in the upper range of clay sizes.

Allophane. This term describes amorphous, clay-size particles containing aluminium, silicon, oxygen, hydrogen, and other elements, primarily magnesium and iron. Allophane may result from weathering of soil minerals but it is particularly characteristic in soils developed in volcanic ash deposits.

Hydrous oxides. Gibbsite ($Al(OH)_3$) is representative of a number of hydrous aluminium oxides and occurs as a residual product from intensive soil weathering. Bauxite is mainly aluminium oxide.

Hematite (Fe_2O_3) and Goethite ($Fe_2O_3.H_2O$) are two commonly occurring oxides of iron. They may be present as amorphous coatings around larger soil particles or as crystals. They impart the red and reddish-brown colours typical of many soils.

The occurrence of the clay minerals, which have been described briefly, varies greatly and in many soils the clay separate will contain a number of different clay minerals. Kaolinite and hydrous aluminium oxides tend to predominate in soils that have undergone long periods of intense weathering and, although more common in tropical and subtropical zones, they are by no means restricted to them. Illites and montmorillonite occur more frequently in soils developed in temperate regions.

Cation exchange capacity

Both clay minerals and organic materials, particularly those of colloidal nature, possess a net negative charge. There are several origins of this charge: (1) In clay minerals, isomorphous replacement gives rise to electrical imbalance and, where a magnesium (Mg^{2+}) or iron (Fe^{2+}) replaces an aluminum (Al^{3+}) or an aluminum, a silicon (Si^{3+}), a net negative charge will occur. (2) At the edges of clay minerals the layers are often irregular and ionic bonds occur which are unsatisfied; these can be either negative or positive and therefore may increase or decrease the overall net negative charge of the particle. (3) Hydroxyl groups in clay minerals and carboxyl (COOH) and other groups in organic materials may dissociate with the production of hydrogen ions (H^+). This gives rise to negatively charged sites on the mineral or organic particle.

The total of negatively charged sites on specific soil particles or of a sample of soil including inorganic and organic components represents its ability to hold positively charged ions (cations). Quantitatively, it is the *cation-exchange capacity* (CEC) of the soil or soil component and it is conventionally expressed in milliequivalents per 100 grams of material (meq/100g). One milliequivalent is the amount of an element or compound that will combine with or replace one milligram of hydrogen. The value is obtained by dividing the atomic weight of an element in milligrams by its valency. For example, 1 meq calcium (Ca^{+2}) = 40.08/2 = 20.04 mg, 1 meq potassium (K^+) = 39.10/1 = 39.10 mg, and 1 meq ammonium (NH_4^+) = (14 + 4)/1 = 18 mg. For minerals a distinction is often made between the capacity which is a result of isomorphous replacement - *the permanent charge* - and that which depends on ionization of hydrogen from hydroxyl groups. This latter charge will vary with the concentration of hydrogen ions in the soil solution surrounding the clay mineral and is therefore termed a *temporary charge.*

The cation exchange capacities of soil clay minerals and organic material exhibit a wide range. Representative values (meq/100 g) would be: kaolinites 5-15; chlorite and illite 60-180; organic 50-250. The convention of expressing results on a weight basis can be particularly misleading where appreciable capacity is related to organic matter. The bulk density of organic matter may be one fifth to one tenth that of mineral soil and therefore the exchange capacity should be expressed on a volume rather than a weight basis.

The species of ions held or adsorbed at exchange sites depends on a number of factors. The force of attraction is generally greater with increase in charge; thus $H^+ < NH_4^+ = K^+ < Ca^{2+} < Mg^{2+} < Al^{3+}$. However, ions in solution may be hydrated and, since oxygen is a very large ion, the closeness with which a hydrated ion may be held will determine the strength with which it is held. Al^{3+} and Mg^{2+} are large hydrated ions compared to K^+ and NH_4^+ and will be held less closely than the latter two. The density of charge on the clay or organic particle will affect the strength by which an ion is retained. The degree to which the cation composition will be altered depends on the concentration of ions in the soil solution and, if there is a high concentration of an ion that otherwise has a low capacity for adsorption, it will displace other ions. Any ion such as Ca^{2+} which may be strongly retained is responsible for flocculation of clay mineral colloids and this increases aggregation in soils; Na^+, on the other hand, is noted for its ability to disperse clay colloids. Often the sum of the exchangeable cations other than hydrogen and aluminium is expressed as a percentage of the total cation exchange capacity and is termed the *base-saturation per cent.*

MEASUREMENT OF CATION EXCHANGE CAPACITY

The usual procedure is to place a sample of soil in a column where it is leached with a solution of $1N$ neutral ammonium acetate. The leachate will contain the cations displaced by the NH_4^+ and these may be determined quantitatively by appropriate chemical analyses. The soil in which the exchange sites are now occupied by NH_4^+ is first washed with alcohol to remove any excess ammonium acetate and then treated with either dilute hydrochloric acid or sodium chloride. The second leachate contains the displaced NH_4^+. The amount of ammonium nitrogen is determined quantitatively, usually by the Kjeldahl procedure, and hence the cation exchange capacity of the soil. In soils that contain free calcium carbonate the ammonium acetate will dissolve some of the calcium and consequently the amount in the first leachate will represent more than exchange Ca^{2+}. Other leaching solutions such as barium chloride-triethanolamine have been used for these soils.

ANION EXCHANGE

Within the colloidal system in the soil there exist positively charged exchange sites at which anions such as Cl^-, NO_3^-, $H_2PO_4^-$, and PO_4^{3-} as well as others may be adsorbed. Many factors affect the degree of adsorption; generally it is greater as acidity increases.

Soil reaction - pH

Soil reaction refers to the 'active' hydrogen ions in the soil. It is measured in units of pH, where the pH value, numerically, is the logarithm of the reciprocal of the hydrogen ion concentration in the soil solution:

$$\mathrm{pH} = \log(1/[H_a^+]),$$

where $[H_a^+]$ is the concentration of active hydrogen ions. In an aqueous solution the product of the concentration of H^+ and OH^- ions is 10^{-14}. Thus, when the concentrations are equal and the solution is neutral, the pH is 7. If hydrogen ions predominate, the solution is acid and pH values will be <7; conversely if hydroxyl ions predominate, the pH values will be >7 and it will be alkaline.

The pH of a soil is closely related to the nature of the soil colloids, both clay minerals and organic. The dissociation of hydroxyl (OH) and carboxyl (COOH) groups from these colloids releases hydrogen ions into the soil solution. Organic acids are another source of hydrogen ions as well as aluminium which may hydrolize to yield hydrogen:

$$Al^{3+} + H_2O \rightarrow Al(OH)^{2+} + H^+,$$

$$Al(OH)^{2+} + H_2O \rightarrow Al(OH)_2^+ + H^+.$$

Figure 7.3 Seasonal changes in pH values for F and H layers of iron-humus podzol measured in three different suspensions. From P.E. Vézina in *Ecology* (1965), by permission Duke University Press

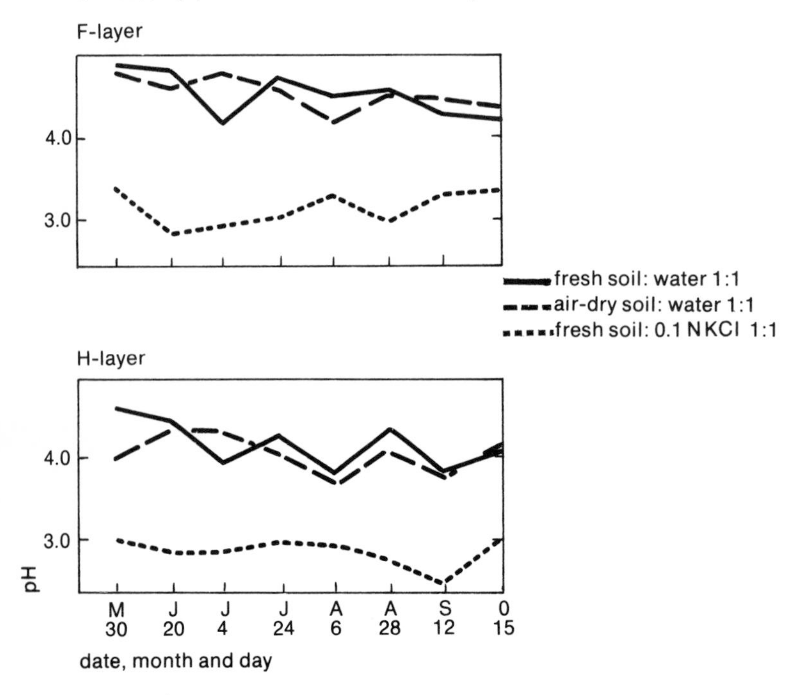

Thus aluminium is as much a characteristic ion of very acid clay mineral soils as is hydrogen.

Hydrogen and other ions in the soil solution exist in equilibrium with the cations which are adsorbed about the surfaces of the soil colloids. This swarm or cloud of cations is referred to as a double layer. The thickness of this double layer will vary; cations which are adsorbed more tightly (divalent as compared to monovalent ions) will result in a thinner double layer. If the concentrations of cations in the soil solution are increased, this will have the effect of compressing or reducing the thickness of the double layer. The thinner the double layer, the easier it is for hydrogen (or other ions) to move into the soil solution and the lower will be the observed pH of the soil solution. The movement of hydrogen and aluminium ions from a tightly held position on the soil colloid surface out into the soil solution illustrates the property of *exchange* or *reserve acidity*. Thus, when they are tightly bound, they do not affect directly the pH of the soil solution.

Figure 7.4 Buffer curves for a poorly buffered soil (A) and a soil buffered against acidity (B)

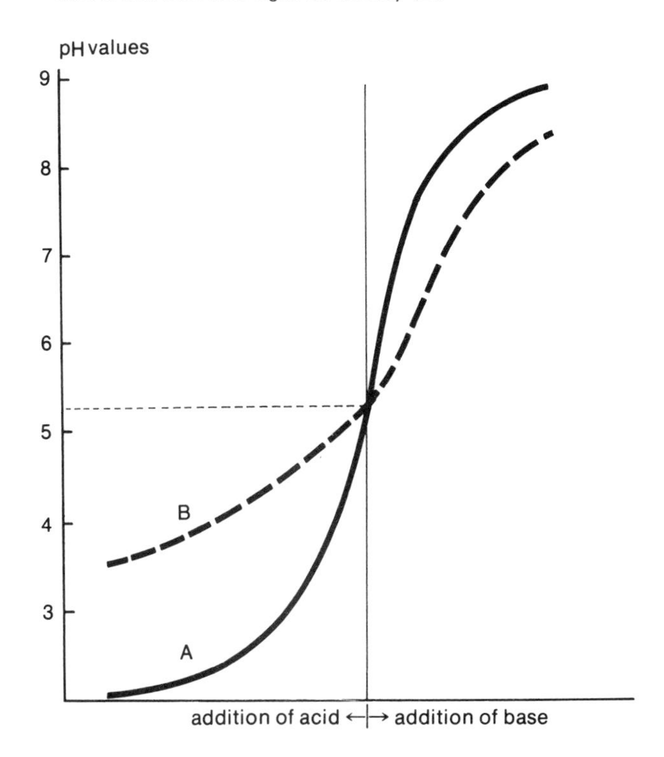

Seasonal variation and effect of the type of solution on soil pH values are illustrated for organic layers in Figure 7.3. It will be seen that variations of nearly half a unit occur during the season and the effect of using O.1*N* KCl solution in the suspension results in lower pH readings and these are less variable throughout the season than are the values for suspension in water.

The movement of hydrogen ions between the double layer and the soil solution is dependent on changes in ionic concentration of the solution. If a base is added to the solution, hydrogen ions will move to the solution from the colloidal complex and consequently minimize the change in pH value. The ability of a soil to resist changes in pH value is termed *buffer capacity.* Soils with little or no organic and clay mineral components will have low buffer capacities and increase in amounts of these components will result in larger capacities. Figure 7.4 illustrates buffer curves for two soils, one of which is poorly buffered; the other is buffered on the acidic but not the basic side. In forest soils the greater buffer capacity will provide more uniform ionic conditions and this may be important

Figure 7.5 Variation in pH values with depth in a Petawawa soil. From P.J. Rennie in *Journal of Soil Science* (1966), Oxford University Press, by permission of The Clarendon Press, Oxford

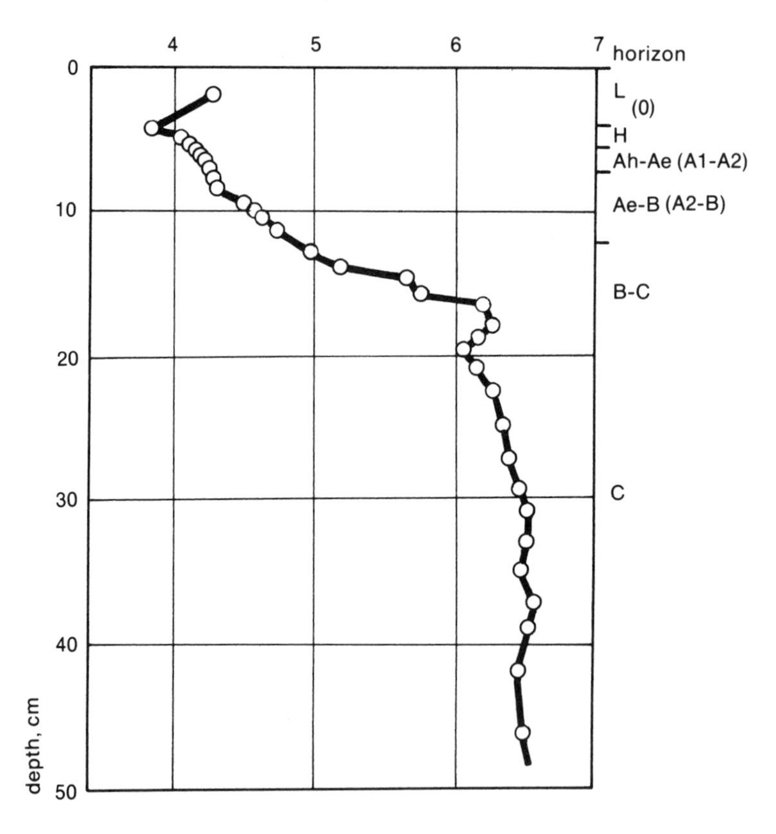

for many members of the soil microorganism population which may be sensitive to changes in soil pH values. The activity of organisms in the soil results in carbon dioxide production and this will result in increased acidity of the soil solution.

Typically, forest soils show considerable heterogeneity both laterally and vertically in relation to horizon differences in the profile. The distribution of organic matter for example is non-uniform and this can influence soil pH values. Figure 7.5 shows the vertical sequence of pH values for a soil profile in which differences in pH values, even within one horizon, occur. For many forest soils it is unrealistic to refer to *the* 'pH of the soil'; not only may it change seasonally, but in many instances there will be marked differences with depth.

MEASUREMENT OF SOIL REACTION

Colorimetric and electrometric methods are both used, but because of its relatively low cost, portability, and sturdiness, electrometric equipment is becoming more common than the use of dyes which are the basis for colorimetric use. With dyes it is possible to determine pH values to the nearest 0.2 unit; using pH meters, much finer differences can be recorded. Although the precision with which the pH value can be determined can be great, especially with a pH meter, the significance in relation to the soil may be questionable. The size of the individual soil sample and the number of samples may affect pH values obtained. This was found to be important for organic layers by Van Groenewoud (1961). He suggested that the measured pH of large composite samples should be used as a mean pH, instead of the calculated average of small samples and that maximum and minimum pH be used instead of calculated standard deviation. It should be noted that since the values of pH are logarithmic, a difference of one unit as between pH 5 and pH 6 for example, represents a tenfold change in hydrogen ion concentration.

Significance of pH values in forest soils

The influence that pH values may have in regulating the soil organism population has already been mentioned. Certain microorganisms, such as nitrifying bacteria, can only function within a range of pH values. As a result, the specific activity or transformation of materials associated with such an organism will be present only when soil pH values are suitable. Nitrification would not normally be expected to occur in a soil of pH 4 for this reason. Pathogenic organisms such as certain of the damping-off fungi which attack conifer seedlings in forest tree nurseries can be controlled to some extent by regulation of soil pH. The form and solubility of many chemical compounds occurring in forest soils is often pH-dependent. As a soil becomes more acid, elements such as manganese, copper, and zinc become more mobile in ionic form. The solubility of phosphorus in soil is particularly pH-dependent. The pH of soil in the rooting zone can thus control to some degree the amount and relative availability of nutrient elements or those such as aluminium which at low pH values can occur in toxic quantities.

Soil elements

A very large number of elements are present in forest soils and in this discussion only those which are of major importance in plant nutrition or those which are used to characterize forest soils and reflect soil properties of significance will be considered. The expression of the quantities of an element or compound in soils varies with country, region, and investigator. For those whose primary field of interest and education is not chemistry, this variation is most exasperating and at times confusing. It has often rendered comparisons between results difficult and thereby reduced the value of much past investigation.

Sampling. Soil used for chemical analysis is either taken from the soil as bulk samples or it is taken as a volumetric sample which enables the results to be expressed on a soil volume, in addition to a weight basis. It is increasingly common to take both a bulk and volume sample.

Pretreatment. The value of a property may vary depending upon whether the soil is analysed when fresh and moist or is air-dried, the most common form of pretreatment. It has already been noted that pH values will be different for air-dried and fresh samples (Figure 7.3). Hesse (1971) has summarized the chemical effects for a number of elements. Generally they are variable and depend largely on the kind of soil and method of analysis employed.

Extraction. For most elements the total amount of the element in the sample is seldom determined. Such determinations are time-consuming, costly, and except for certain purposes provide little useful data that can be related to higher plants or soil processes. Most analyses use one or more methods of extraction which either remove an element in a specific form or have been found appropriate in the use of the data for some particular purpose. For example, water-soluble phosphorus can be distinguished from phosphorus that may be extracted by a solution of sodium bicarbonate or dilute sulphuric acid. The amount of an element extracted by one solution may correlate well with growth responses of a plant species, whereas the amounts removed by another solution may not. The chemical procedure for extraction may itself change the form in which an element or compound occurs, thereby creating artifacts. It is most important that the extraction procedure used should be specified. Too often the amounts of nutrient elements are stated as 'available,' a term that by itself means little, if anything.

Determination. Methods for rapid, accurate, and precise qualitative and quantitative analysis have developed rapidly since the 1950s. Emission and absorption spectrophotometers, spectrographs, paper and gas chromatographs have become common instruments in laboratories. Many of these instruments have solved or simplified some of the previous problems. The presence of interfering elements in certain analyses not only necessitated their removal but introduced possible errors. Certain forms of absorption photometry obviate this problem to a large extent. The accuracy of data in final form is limited by the least accurate step in the analytical procedure. Failure to recognize this is often compounded by an inability to use only significant digits. The precision chosen for the original measurement is influenced greatly by the variation in the population sampled and the size of the sample. In much soil sampling the variation is large and the sample taken small; hence high precision is often not worthwhile.

TABLE 7.1
Commonly used units of expression of analytical data and element – oxide conversions

Concentrations	Element	X factor	Oxide	X factor	Element
Parts per 100; 1×10^2; per cent - %	P	2.292 =	P_2O_5	0.436 =	P
Parts per 100,000; 1×10^5; mg/100g	K	1.205 =	K_2O	0.830 =	K
Parts per 1,000,000; 1×10^6	Ca	1.399 =	CaO	0.714 =	Ca
Parts per 2,000,000; 2×10^6	Mg	1.658 =	MgO	0.600 =	Ca

NOTE: For cations, to convert ppm to meq/100g:

$$\frac{\text{ppm}}{10 \times (\text{equivalent weight of cation})} = \text{meq/100g}$$

$$\text{equivalent weight} = \frac{\text{atomic weight (in mg)}}{\text{valency}}$$

Expression of results. With few exceptions, results of soil chemical analyses are usually expressed as a concentration on a soil oven-dry weight basis. The oven dry-weight is based on drying in an oven at 100–110°c to constant weight. The units of concentration often vary with the element. Frequently carbon and nitrogen are expressed in per cent and phosphorus or sulphur are expressed as parts per million (ppm) or as milligrams per 100 grams of soil (mg/100g). Sometimes the concentrations are parts per two million (pp2m). The reason for this is that in north temperate agriculture it is convenient to assume that one acre-furrow-slice (the volume of soil with a surface area of one acre and a depth of 6-7 inches) of mineral soil weighs 2×10^6 pounds. Thus a statement that the concentration of calcium in the soil is 50 pp2m is equivalent to stating that it is 50 pounds per acre-furrow-slice. This mode of expression has no place in forest soils. The concentration of cations is normally in milliequivalents per 100 grams of soil (meq/100g) and is by convention termed 'exchangeable.' In much of the previous literature elements were given in the oxide form, for example, P_2O_5, K_2O, CaO and MgO. This practice is not recommended. Table 7.1 summarizes the conversion units for concentrations and oxides. The value of including bulk density determinations together with other analytical data has already been emphasized since it enables the values to be considered on a soil volume rather than a weight basis. The inclusion of estimates of accuracy should also be routine.

There are a number of books dealing with analysis of soils and plants (Association of Official Agricultural Chemists, 1955; Black (ed.), 1965; Chapman and Pratt, 1961; Hesse, 1971; Jackson, 1958; Metson, 1961; Richards (ed.), 1954; Wilde et al., 1972).

CARBON

Carbon occurs in soil in one or more of three forms: inorganic carbon as in carbon dioxide and carbonates, the organic carbon of litter (decomposed plant and animal remains), and living organisms and as elemental carbon, the most common form in forest soils being charcoal.

The total carbon in a soil sample is seldom determined; more usually only a determination for organic carbon is made after removal of carbonates. The procedure for organic carbon involves measurement of the amount of carbon dioxide evolved after treatment. Oxidizable carbon determinations are carried out using chromic acid oxidation. Instead of measuring the CO_2 evolved, some procedures add a known excess of chromic acid and determine the residual acid by titration with a reducing agent (ferrous sulphate). By difference the amount of carbon oxidized by the chromic acid can be calculated. This type of determination is one of the most common. In the Walkley-Black procedure no external heat is applied but in the Schollenberger method, heat is applied. In these techniques little or no carbon in elemental form is measured. Jackson (1958) stresses that in the presentation of results distinction should be made between organic carbon and oxidizable carbon.

Another type of procedure used to estimate organic matter involves ignition at moderate or high temperatures and expressing the results as a weight loss (per cent) - *Loss-on-Ignition* (LOI). For coarse-textured soils with little to no clay content and no carbonates the LOI values may approximate the amount of organic matter, but for most soils the presence of clay and carbonates renders LOI determinations of little value as estimates of organic matter.

Determinations of organic or oxidizable carbon are frequently converted to organic matter by multiplying by the factor 1.724. The use of this factor assumes a consistent proportion of carbon in all organic matter forms, an assumption that is highly questionable.

The determination of the complex forms in which carbon is present in soils is beyond the scope of the present consideration. Essentially the process involves fractionation, extraction, and separation of the compounds and their identification. Apart from determinations for specific compounds in the soil humus, one of the common initial fractionations uses alkali and acid treatments:

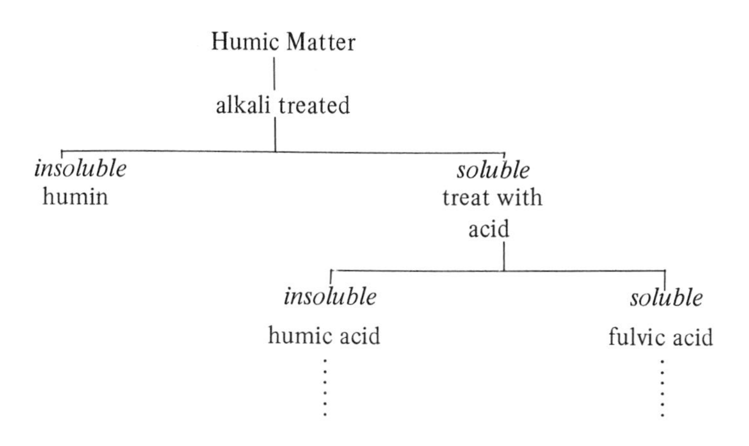

NITROGEN

Most nitrogen in soils is in organic form. The inorganic forms of ammonium and nitrate have received particular attention from soil chemists, but nitrites and gaseous oxides (N_2O and NO) can occur and gaseous nitrogen (N_2) is a major component of the soil atmosphere.

One of the most common determinations is for 'total nitrogen' and two procedures may be used. The Dumas method involves the combustion of the soil sample with copper oxide in a stream of carbon dioxide; the nitrogen from organic compounds is released as gaseous nitrogen and measured manometrically. In the more commonly used Kjeldahl procedure, complex organic nitrogen forms are hydrolysed by acid digestion to ammonium form. The ammonia is released by treatment with alkali and steam distillation, absorbed in boric acid, and determined quantitatively by titration with a standard dilute acid. The Kjeldahl method normally includes only organic and ammonium forms, unless a modification in the procedure is made.

The ammonium (NH_4^+) nitrogen in soils occurs chiefly as an exchangeable cation on the clay-humus colloids. It is extracted usually by leaching with sodium chloride (NaCl) and replacing the exchangeable NH_4^+ with Na^+. Sometimes potassium chloride may be used in place of sodium chloride. The amount of ammonium is determined as in the Kjeldahl process or colorimetrically using Nessler's reagent. Jackson (1958) has emphasized that exchangeable ammonium determinations should be made quickly after taking field samples because of possible oxidation of ammonium to nitrate and nitrite if the samples are left in a warm moist state.

Nitrate (NO_3^-) nitrogen is extracted by using either a potassium chloride or a copper sulphate solution in the same manner as that by which the ammonium

Figure 7.6 Effect of inoculation with garden soil and added ammonium-N on nitrification of limed forest soils. From F.E. Chase and G. Baker in *Canadian Journal of Microbiology* (1954), by permission of the National Research Council of Canada

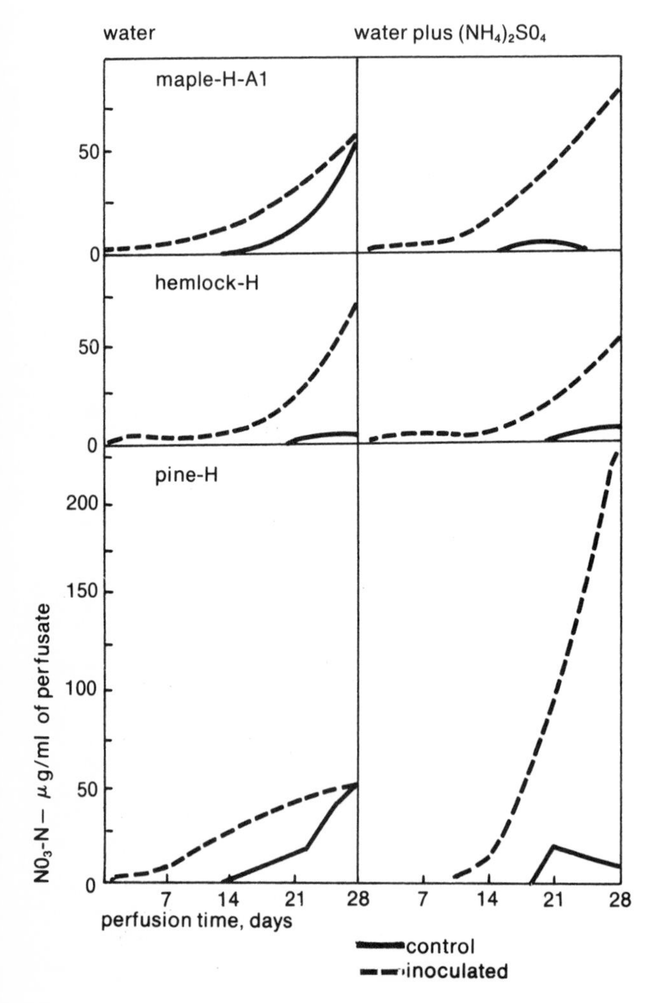

nitrogen was removed. The amount of nitrate is determined colorimetrically after treatment with phenoldisulphonic acid. Alternatively the nitrate may be reduced to ammonium by treatment with Devarda's alloy and alkali. The ammonium is then determined as in the Kjeldahl procedure. This alternative procedure

Figure 7.7 Sum of ammonia and nitrate-N in flasks after various periods of incubation, expressed as a percentage of total initial nitrogen - Garpenberg soil, F-H and B horizons of podzol. From C.O. Tamm and A. Pettersson (1969), by permission Royal College of Forestry, Sweden

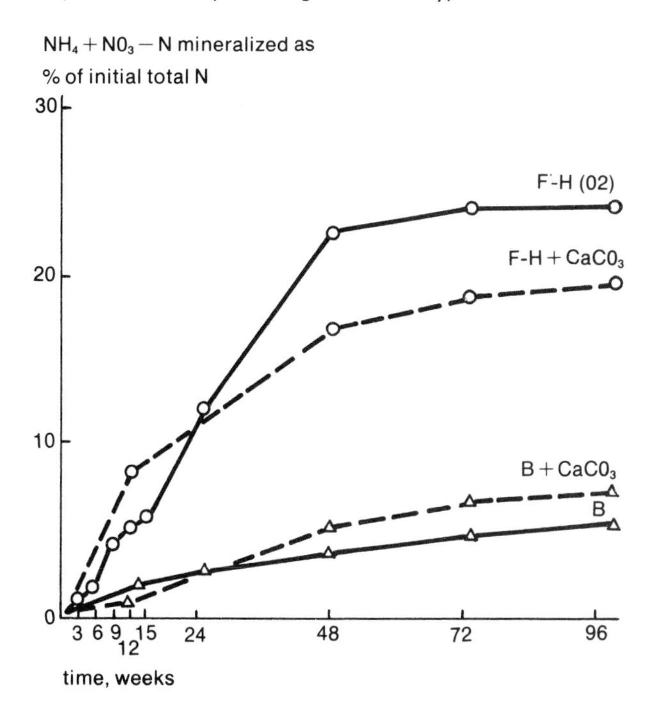

is the one used if nitrate is to be included in the 'total nitrogen' determination by the Kjeldahl method.

The measurement of rates of nitrogen mineralization (i.e. the ammonium and/or nitrate production) in soils is accomplished by incubating a fresh sample and then periodically sampling and determining the quantities of nitrate and ammonium nitrogen present. A sophisticated technique is that of the perfusion apparatus (Lees and Quastel, 1946) whereby samples may be drawn off for analysis without any disturbance of the sample itself. Often in mineralization studies an inoculation with nitrifying soil may be made to determine if lack of nitrification results from absence of nitrifying bacteria. Addition of ammonium may also be made to indicate if nitrate production is limited by lack of substrate. Results using such procedure in a perfusion apparatus are illustrated in Figure 7.6. The stimulation resulting from inoculation is apparent but initially there is a one to two week period of establishment of the nitrifying population. A supplementary source of ammonium results in increased rates only for the pine humus samples. Most nitrogen mineralization studies are undertaken for periods of time - a few days or several weeks, but particularly for certain forest soils the length of the incubation period may be important. Tamm and Pettersson (1969) studied mineralization for periods of more than one year and found that, although rates of mineralization might be stimulated initially in some soils by addition of lime, after a period of several weeks the amount mineralized was greater in some unlimed soils but not in others (Figure 7.7).

Ammonium and nitrate nitrogen are the sources of nitrogen for higher plants, except for those species which host symbiotic and dinitrogen-fixing organisms in their roots. Ammonium ions may occur on readily exchangeable sites in the soil cation exchange complex or they may be 'fixed' in interlayer positions in clay minerals where they are released slowly. Nitrate ions, if present, are in the soil solution and are consequently readily available and easily moved. They are, therefore, much more susceptible to leaching loss than ammonium nitrogen.

Studies of species' preferences for form of nitrogen are somewhat conflicting and deal mainly with conifers. Leyton (1952) found Sitka spruce grew well with either. Swan (1960) had some indication that jack pine seedlings grew better with an ammonium than a nitrate nitrogen source, but the result was inconclusive because of associated changes in pH values. A more comprehensive and controlled set of experiments by McFee and Stone (1968) demonstrated that Monterey pine and white spruce both grew larger with ammonium source than with nitrate source nitrogen under a range of soil-root conditions.

PHOSPHORUS

The basic sources of phosphorus in forest soils are the phosphorus-containing minerals, primarily apatite-$Ca_{10}(PO_4)_6F_2$. As mineral phosphates weather, they can yield three phosphorus ions - $H_2PO_4^-$, HPO_4^{2-}, and PO_4^{3-}. Phosphorus may be absorbed by plants, mainly as $H_2PO_4^-$ and HPO_4^{2-}, and incorporated into organic compounds which may be returned in litter to the soil, where mineralization can release phosphorus for reabsorption by roots. Much of the phosphorus from the primary mineral source forms new compounds within the soil, especially with calcium, iron, and aluminium. Some portion of the phosphorus anions may be adsorbed by clay minerals in the soil. The amount of phosphorus in the soil solution which is available to plants at any one time is usually a very small proportion of the total soil phosphorus.

The form of phosphorus in the soil is highly dependent on soil pH values. In acid soils, with pH values less than 5.0, the $H_2PO_4^-$ ion reacts with iron and aluminium to form insoluble compounds. It has been mentioned that with increasing acidity, aluminium ions are more prevalent and the reaction may be

Figure 7.8 Relationship between 'available' phosphorus, Bray and Kurtz – 0.03 N NH_4F plus 0.025 N HCl extraction and a soil with adjusted acidity. From R. Waito and R.C. Gregory (1973)

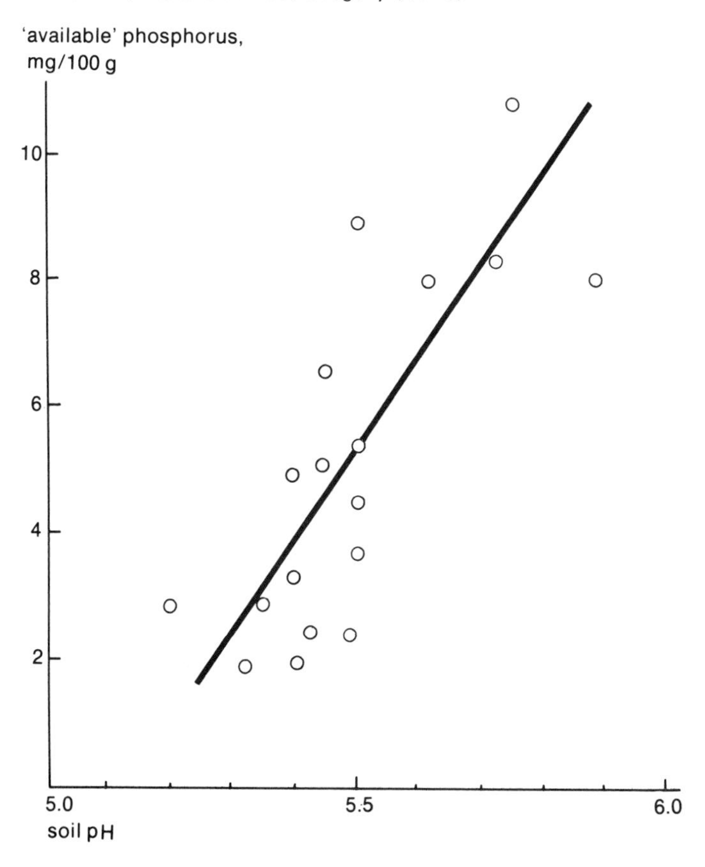

$$Al^{3+} + H_2PO_4^- + 2H_2O \rightleftharpoons \underset{\text{insoluble}}{Al(OH)_2\,H_2PO_4} + 2H^+$$

At soil pH values greater than 6, insoluble calcium phosphates are formed. Effect of changes in soil pH over a limited range on amounts of 'available' phosphorus is indicated in Figure 7.8. In soils characterized by free calcium carbonate and high pH values and those which are acid with high contents of active iron and aluminium the 'available' level in the soil solution may be minimal. It is considered that maximum 'availability' of phosphorus occurs at a soil pH value of about 6.5. Since knowledge of the exact nature of the soluble phosphorus in the soil solution is very limited, Larsen (1967) has pointed out that it would be an oversimplification to assume that even the bulk of soluble phosphorus is present as the ions, $H_2PO_4^-$ and HPO_4^{2-}. For this reason it is preferable to refer to 'available' phosphorus only in relation to specific extractants. In addition to the changes in form of phosphorus already mentioned, it is known that certain organic compounds can chelate phosphorus particularly as iron and aluminium phosphates and thus increase the availability of this element.

Phosphorus in organic forms derived from plant and animal tissue probably provides an important source of phosphorus in the uppermost forest soil horizons. Although the organic chemistry of soil phosphorus has been studied, the nature of much of the phosphorus is unknown (Black, 1968). The largest known group of soil organic phosphates are the *inositol hexaphosphates* based on a six-carbon ring:

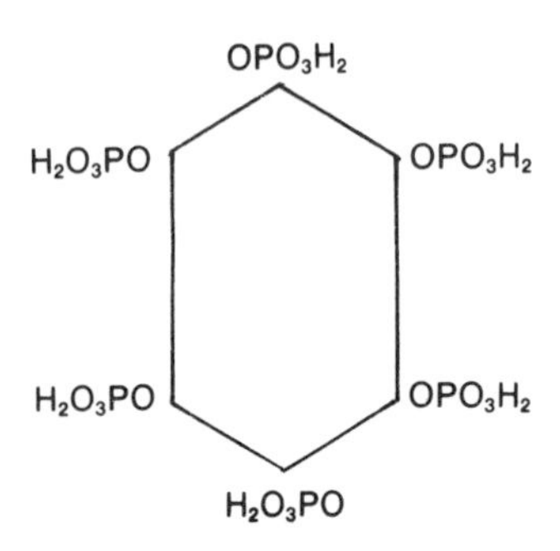

and occurring most likely as salts of calcium, magnesium, iron, and aluminium. Other organic forms, such as phospholipids, have been found in soils.

In general, phosphorus in the soil may be viewed as consisting of several pools, the larger being the insoluble or fixed inorganic phosphorus in the soil and the organically fixed phosphorus in the living components of the soil and vegetation which it supports. An intermediate size pool of phosphorus in dead organic material is in various stages of decomposition and mineralization. The smallest pool is that in the soil solution (Figure 7.9). The degree to which plants may take up organic phosphorus directly, although documented for certain agricultural crop plants, is not well known for forest species.

The determination of soil phosphorus is done colorimetrically by treating the solution in which the phosphorus has been extracted from the soil with reagents which complex with the phosphorus to form a colour, the intensity of which varies with the concentration of phosphorus in the solution. One colorimetric procedure uses a combination of phosphate and molybdate ions which on reduction gives a blue colour; another uses vanado-molybdate which gives a yellow colour with phosphorus.

Figure 7.9 Schematic representation of pools of phosphorus in soil.

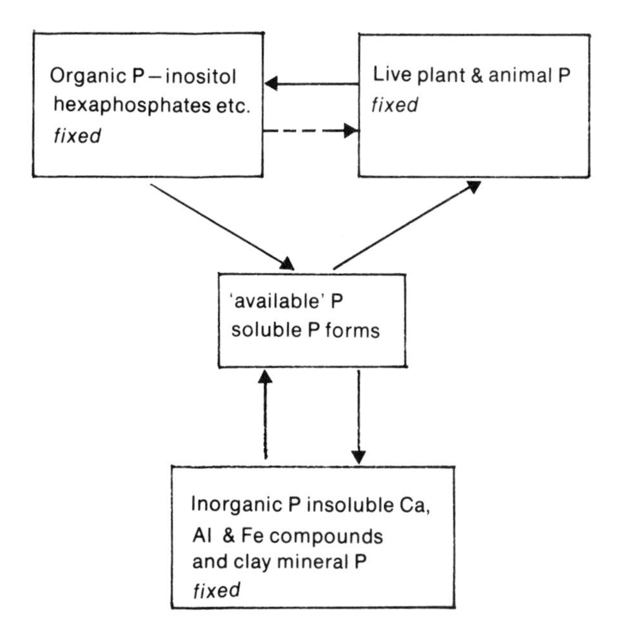

The extraction of soil phosphorus may be accomplished using many different solutions. The most common have been presented by Jackson (1958).

1 / Total phosphorus. Two of the most commonly used procedures involve either a fusion with sodium carbonate with subsequent solution in acid or digestion with perchloric acid.

2 / Extraction with moderately alkaline solutions. Olsen's procedure employs a sodium bicarbonate solution.

3 / Dilute acid extractions of various kinds have been used to provide correlations with agricultural crop response, particularly on acid soils, but are generally not used on alkaline soils.

4 / Ammonium fluoride and dilute acid are the basis of the Bray and Kurtz solutions. These are considered to include some phosphorus held by trivalent iron and aluminium ions. This extracting solution has been used successfully in assessment of availability of phosphorus to conifer seedlings in forest tree nurseries in eastern Canada.

5 / Anion exchange resins have been used as an extractant.

Much of the knowledge of soil phosphorus comes from studies in which radioactive phosphorus (^{32}P) is added to soil, and plant growth response and phosphorus uptake are monitored. The fraction of the soil phosphorus which is isotopically exchangeable is referred to as *labile* phosphorus. Kilian and Lumbe (1972), using a wide range of soils, applied ^{32}P in fertilizer and monitored the growth and uptake of P in Norway spruce, Scots pine, black alder, and rye and demonstrated that the amount of 'available' phosphorus varies not only with soil but also with the plant species grown. It is clearly erroneous to consider that a soil has an absolute phosphorus-supplying capacity.

In mineral soils the presence of most of the soil phosphorus in fixed or insoluble forms ensures that leaching losses in the soil solution are usually minimal. In organic soils, however, fertilizer phosphorus may move relatively easily.

POTASSIUM

Unlike nitrogen and phosphorus, all potassium in forest soils is in mineral form; it is an element that is not known to enter into the structure of organic compounds. It occurs in a wide range of soil minerals, which upon weathering may release potassium ions. These ions may be present in the soil solution or as exchangeable cations on clay or organic colloids and are readily available for absorption. Some potassium ions become 'fixed' in the interlayer spaces of 2:1 clay minerals and in this position are not readily available. In a soil, the various states in which potassium occurs exist in equilibrium with each other:

Mineral K $\rightleftharpoons$ K^+ in solution $\rightleftharpoons$ exchangeable K^+ $\rightleftharpoons$ K^+ in fixed lattice of clay mineral.

As a result of these equilibria, addition of potassium ions into the soil solution will increase the amount of exchangeable and fixed potassium. Absorption by plants from the soil solution or leaching losses will tend to bring potassium out of the fixed form.

In the organic horizons of forest soils or soils with high organic contents the level of exchangeable potassium is usually considerably greater than in the mineral soil horizons and 'fixing' does not occur. Losses of potassium, particularly fertilizer potassium may be large in coarse-textured soils if they are acid (Krause, 1965).

The most frequently used extraction of soil exchangeable potassium is by leaching with 1*N* neutral ammonium acetate. The potassium in the leachate is then usually determined by flame photometry.

CALCIUM AND MAGNESIUM

Calcium is present in primary minerals, as an ion Ca^{2+} in solution, held in exchangeable form, or in secondary minerals, one of the most common being calcium carbonate. Common primary minerals which are a source of calcium upon weathering are, in order of decreasing ability to release calcium (Burger, 1969):

calcite, dolomite, apatite, bytownite, diopside, grossularite, labradorite, augite, horneblende, oligoclase, and albite.

The presence or absence of calcium carbonate in a soil is usually one of the first qualitative determinations made either in the field or the laboratory. A small soil sample is treated with dilute acid and, if free carbonates are present, it will effervesce.

$$CaCO_3 + 2\,HCl \rightarrow CaCl_2 + H_2O + CO_2\uparrow$$

The usual determination for calcium and magnesium is by atomic absorption photometry. The extraction from the soil for exchangeable calcium and magnesium is a 1*N* neutral ammonium acetate solution, unless free carbonates are present, when an alternative method is employed using barium-triethanolamine.

Magnesium occurs together with calcium in the mineral dolomite, but in most soils the principal sources of magnesium may be subject to leaching losses, although in well-drained soils these are probably minimal. Movement of these and other ions in the ground water may be considerable and quantitatively reflects the nature of the geological deposits. Troedsson (1952) studied the chemical composition of ground water over a three-year period and found that the concentrations were almost constant with only small seasonal differences. He calculated that where the ground water was flowing down a 1:6.5 slope at a rate of 0.5 m/hr the estimated annual content (loss) of nutrient elements was, in kg/ha: Ca = 39.4; Mg = 17.1; K = 5.3, and Na = 36.8.

SULPHUR

Sulphur is present in soil minerals as sulphides which on weathering yield sulphate (SO_4^{2-}) ions. It is the sulphate form that is absorbed by plants. Both the aerobic oxidation of sulphides to sulphates and the reduction of sulphates to sulphides under anaerobic conditions involve autotrophic bacteria. The breakdown of organically-bound sulphur in litter decomposition is also microbiological.

Determination of total sulphur or sulphate content of forest soils is not common. Although the sulphate anion can be retained in an exchangeable form, the significance of this form is not thought to be great in terms of plant supply.

IRON AND ALUMINIUM

These elements are common in most soils. They occur in primary silicates, clay minerals, and as secondary hydrous oxides. One of the characteristics of soil development is a tendency for these two elements to occur in greater concentrations in parts of the soil as weathering progresses. Both, but particularly iron, can be mobilized by certain organic compounds (chelates) produced by forest vegetation. The form of iron in the soil is dependent on pH and oxidation-reduction conditions as mentioned in the discussion on weathering.

The amount of iron may be extracted depends upon the solution used. Two commonly used extraction solutions are dithionite-citrate and ammonium oxalate. The iron determined by these extractions is referred to as 'free iron oxide.' Other solutions have been used to differentiate iron and aluminium of different 'ages' of weathering. For example, Ball and Beaumont (1972) used extractions of oxalic acid and pyrophosphate to distinguish between 'fresh plus aged' and 'fresh' iron plus aluminium, respectively.

MANGANESE

Manganese dioxide (MnO_2) is probably the most common primary form in which manganese occurs in the soil. Upon chemical or biochemical weathering, manganese may be released and is probably absorbed as the manganous (Mn^{2+}) ion by plants. The amount of divalent manganese in a soil is dependent on pH values. In acid soils it is relatively abundant, but as acidity decreases it occurs more in the form of insoluble oxides often associated with iron oxides.

The usual determination for manganese is in leachates for exchangeable cations using 1*N* neutral ammonium acetate.

TRACE ELEMENTS

The metals boron, copper, zinc, and molybdenum are essential nutrients. Others, like nickel and mercury, normally occur in small quantities, but may be present in larger quantities in mine spoils and areas previously denuded by atmospheric contaminants.

Generally low soil contents of these elements are found in acid, leached, highly weathered soils. To varying degrees organic soils may have small amounts of boron and copper. For higher plants which have root nodules where symbiotic dinitrogen fixation occurs it is known that molybdenum requirements are higher than for non-nodulated species.

Oxidation-reduction

Although often considered as an addition or removal of oxygen, oxidation and reduction should be considered in terms of movement of electrons.

$$\text{reduced} \rightleftharpoons \text{oxidized} + \text{electrons}.$$

The most common factor affecting oxidation-reduction status of a soil is the soil water content, reducing conditions becoming prevalent as a soil is progressively waterlogged. Organic components from the vegetation play a part in reduction under anaerobic conditions.

One of the usual methods used to assess the oxidation-reduction state is potentiometric. A platinum electrode is placed in the soil and will either lose electrons to an oxidizing agent in the medium and hence gain a positive charge or gain electrons in the presence of reducing conditons. Normally the same potentiometer used to measure soil pH when equipped with a platinum electrode and calomel half-cell can, after standardization, be used to measure the oxidation-reduction potential E_n. The potential is measured in volts and is dependent not only on the oxidation-reduction balance but also on the nature of ions (Hesse, 1971) and it may be considered:

$$E_n = E_o + \frac{0.0002T}{n} \log \frac{a_{ox}}{a_{red}}$$

where E_o is the standard oxidation-reduction potential of the system, a_{ox} and a_{red} are the respective activities, usually expressed as concentrations of ions, T is the absolute temperature, and n represents the number of electrons. Oxidation-reduction potentials are pH dependent and, owing to the complex nature of soils, the interpretation of measurement is often difficult. Lafond (1950) measured potentials of several humus types and found that mull types tended to have positive potentials whereas mor types showed negative potentials.

8
Soil fertility

Soil fertility is defined as 'the status of a soil with respect to the amount and availability to plants of elements necessary for plant growth' (Soil Science Society of America, 1973). The definition implies that amount of growth or yield is a variable, dependent on the level of soil fertility but that many other factors such as type of plant and growing conditions may also significantly affect growth. Jacks (1956) distinguished three stages in the evolution of man's use of soil: (1) the shifting cultivation stage when human activity has only an ephemeral effect on the soil; (2) soil-exhausting agriculture associated with population increases and permanent settlement; (3) a soil-conserving or fertility producing stage associated with an urbanized society. The same three stages can be found in man's use and subsequent management of forests. Even in natural forests, where man may be only an observer, soil fertility is not a completely stable factor but changes with stage of forest succession and with soil profile development. Occurrence of fire and extensive wind throw can result in sudden dramatic changes in soil fertility. A soil, particularly one with the heterogeneity of many forest soils, cannot be considered to have a unique single, static level of soil fertility.

Since the assessment of soil fertility is only possible in terms of a plant's growth, characteristics of the species and of the plant population, such as density, are of importance. A soil may be quite capable of supplying essential elements for plant X but not for plant Y. The amount and availability of elements to support plants at one density may be considerably different if the density is increased tenfold.

Other soil properties may alter the level at which one or more nutrient elements are available to a plant. Soil reaction (pH) values can affect the form and solubility of several nutrient elements. Moisture and aeration levels can affect the uptake of nutrients, if not their availability, to various degrees.

The study of soil fertility has, therefore, been undertaken largely by using plant or crop growth as the measure (the plant is a *phytometer*) and then relating other measures of soil or plant to it. Of all soil properties, fertility is the one with which man is most involved; it is the property that can be readily changed by man in his exploitation or management of the land.

Essential elements

Soil fertility involves the capacity of soils to supply elements necessary for plant growth. An *essential element* is one that is necessary for the plant to complete its normal growth processes and development. Symptoms of inadequacy in the supply of an essential element are corrected only by supply of the element.

There are twenty elements which are considered to be essential for plant growth. These are: *carbon* (C), *hydrogen* (H), and *oxygen* (O), which constitute the bulk of the dry matter of a plant and are obtained from carbon dioxide and water; *nitrogen* (N), *phosphorus* (P), *potassium* (K), *calcium* (Ca), *magnesium* (Mg), *sulphur* (S) are often referred to as major or *macronutrients* since the amounts in which they are absorbed are considerably greater than those of *iron* (Fe), *manganese* (Mn), *boron* (B), *copper* (Cu), *molybdenum* (Mo), *zinc* (Zn), *cobalt* (Co), *chlorine* (Cl), *sodium* (Na), *vanadium* (V), and *silicon* (Si) which are termed trace or *micronutrients.* Of the micronutrients, not all are required by all plants and this is particularly true for Co,Cl,Na, and V. Many other elements may be found in plants and large differences in concentrations may occur between species when analyses of similar organs are compared. Young and Guinn (1966) found that concentrations of aluminium in foliage of balsam fir and eastern hemlock were nearly fourfold the concentration in red spruce needles; manganese values were much greater in red spruce, balsam fir, and eastern hemlock (1150–1500 ppm) compared with those in eastern white pine needles (375 ppm).

Measurement of soil fertility

The assessment of soil fertility usually involves measurements of one or more of the following: elements in the plant, elements in the soil, and growth of the plant or some portion of it. Some of the features of each of these will be considered before discussing the relationships between them which are used for diagnosis and prediction.

ELEMENTS IN PLANTS

The earliest attempts to assess differences between forest trees were undertaken in the late nineteenth century by European foresters who reasoned that, by

TABLE 8.1
Estimates of nutrient demands, based on 100 years' rotation. Values are for total standing crop at 100 years plus contents of thinnings, litter was excluded (data compiled by Rennie, 1955 from European sources)

	Nutrient elements – kg/ha		
Crop	Ca	K	P
Pines	502	225	52
Other conifers	1,082	578	101
Hardwoods	2,172	556	124
Agricultural crop	2,422	7,413	1,063

measuring the chemical constituents contained in forest vegetation, they would gain information about the nutrient removal by a forest and that the relative fertility of soils could be established when species growing on different soils were compared. It is now recognized that the determination of the amount of elements in a standing forest biomass is necessary, particularly when the cycling of nutrient elements is considered, but it does not always reflect level of soil fertility. Even if a species of the same age and density grows on two different soils, the possibility that the soils differ only in respect of their supply of one or other of the essential elements is extremely remote. Any differences in the stage of forest development or density will similarly render such simple comparisons of doubtful value in assessing fertility. One problem that does not exist in plant analyses is that of availability. If it is in the plant, the element must have been available.

Frequently, the content of an element in a tree or stand is treated as if it were synonymous with uptake, but this is usually not so. *Uptake* refers to the amount absorbed by a plant, whereas content at any time measures only the net amount which is the difference between uptake and losses due to leaching and/or exudation from the plant and amounts contained in dead tissues – leaves, twigs and branches, bark, fruit, flowers, and roots. In seedlings the content will more closely approximate uptake, but as a tree grows older the disparity, which will not be the same for all elements, will become larger because of litterfall and leaching. An example of the type of estimates that may be obtained is given in Table 8.1. The nutrient demands vary not only with the type of crop but also with the element. Demands for calcium by hardwood species are twice those of other conifers, yet their demands for potassium are similar. The large demands by agricultural crops illustrate why, when abandoned farmlands are reforested,

TABLE 8.2
Estimates of nutrient removed by two-year-old nursery seedlings at time of harvest. Foliage for basswood and white ash not included, all values for seedling densities of $269/m^2$ (data of Armson and Sadreika, 1974)

	Nutrient content – kg/ha				
Species	N	P	K	Ca	Mg
Eastern white pine	67	10	23	16	5
Black spruce	37	3	22	8	2
Basswood	329	60	213	202	58
White ash	187	52	118	76	23

infertility is frequently encountered in the establishment of the trees. This is especially true when the soils are coarse-textured and have only small organic contents. The nutrient contents of seedlings removed from tree nursery soils are illustrated in Table 8.2 and both species differences and elemental differences are apparent. The phosphorus removal by black spruce is one third that of the eastern white pine, but potassium removals are similar. It is interesting to compare Tables 8.1 and 8.2; over a 100-year period the nutrient demand for potassium by hardwoods is 556 kg/ha (Table 8.1) but a similar amount would be removed by basswood seedlings in a little over two crops (4 to 5 years).

ELEMENTS IN THE SOIL

Measurements of fertility based on determination of amounts of nutrient elements in the soil suffer from several weaknesses. The question of the number of soil samples necessary to provide an estimate of desired accuracy is important and for many elements it is commonly found that the variation is high, particularly for small numbers of samples. Further, the heterogeneity of many forest soils both vertically in relation to horizon development and laterally intensifies the difficulties of sampling. The variability in root distribution further complicates the sampling problem. This may be contrasted with the situation of arable crops in agriculture, where for many species the root zone is primarily that of the plow layer which is given homogeneity by cultivation practices. For forest soils the greater the uniformity of soil and root distribution the more likely that a chemical determination may provide a measure of the soil fertility.

A second difficulty in determining soil nutrient levels is related to the method of extraction of a particular element. For cations, determinations of exchangeable amounts are used most commonly as being equivalent to amounts 'available'

TABLE 8.3
Means and ranges of phosphorus in pitch pine (10 trees) and soils in which they grew. Root surface sorption zone is the volume of soil within 1 cm of any pine root surface. Root system sorption zone is the segment of a sphere with a mean radius of rootspread and depth of rooting (from Voigt, in Soil Science Society of America Proceedings, 1966, by permission, Soil Science Society of America)

	Phosphorus (mg)	
	Mean	Range
Total P in tree	154	36 - 507
Annual uptake	19	5 - 56
Amount in root surface sorption zone		
H_2O - soluble	2	0.5 - 10
HOAc - NaO Ac soluble	23	4 - 102
H_2SO_4 soluble	31	5 - 137
NH_4F - HCl soluble	35	6 - 154
Amount in root system sorption zone		
H_2O - soluble	148	18 - 546
HOAC - NaO Ac soluble	1.61×10^3	$200 - 5.95 \times 10^3$
H_2SO_4 soluble	2.17×10^3	$267 - 8.00 \times 10^3$
NH_4F - HCl soluble	2.43×10^3	$300 - 8.98 \times 10^3$

to a plant. The determination of phosphorus can be made using a wide array of methods which will provide different results, depending on soil and plant. An illustration of the difference in amounts of phosphorus which may be extracted is shown in Table 8.3. In this example the amount of phosphorus in the soil has been related to two measures of soil volume - the root surface sorption zone and the root system sorption zone. Normally this is not done and most frequently analytical results are expressed as concentrations on a soil weight basis which, as Leaf and Madgwick (1960) have stressed, is unsatisfactory; values should be based on soil volume.

PLANT GROWTH

The fertility of the soil is ultimately expressed in the growth of the plants it supports. A fundamental measure of growth is the dry matter produced, either on an individual plant basis or as amounts per unit area of land surface. Where trees or forests are being managed for some particular objective, the measurement of growth usually reflects the objective. In a forest tree nursery individual seedling size is important; in a seed orchard the yield of cones or seed would be used; for wildlife purposes the amount of browse would likely be the important measurement. When wood production is the major objective, volume or weight of stemwood will most likely be the growth parameter. The choice of parameter may be of importance where one measure of a plant may respond differently either in amount or time of response. For example, Leyton (1954) observed growth responses to changes in fertility were reflected first by an increase in current foliage size and by height in the subsequent growing season.

PLANT-SOIL RELATIONSHIPS

The basic relationships between the plant and soil supply of nutrients serve to define much of our understanding of soil fertility. One of the first attempts to quantify the relationship between plant growth and fertility was made by J. Von Liebig (1803–73) whose Law of Minimum stated that growth was limited by the absence or deficiency of one constituent (element), all others being present in adequate supply. As the factor in minimum increased or decreased in supply then so would the growth of the plant. Two assumptions are implicit in the Law of Minimum; one is that only one element can be limiting at a time and the second that the relationship of increase in growth with increase in nutrient supply is essentially linear.

The general realtionship between growth and nutrient supply is shown in Figure 8.1. It will be seen that, although it is a complex curve, portions of it are capable of more simple mathematical expression. A nutrient *deficiency* exists when growth increases in response to increase in supply of a limiting nutrient. The general shape of the curve in the deficiency range is sigmoidal and can be separated into three components. The first is at the lowest level of supply and is an exponential type of response, $y = ax^n$. It would not be expected to occur in most forest soils unless exceptional conditions of low supply existed. In the middle part of the deficiency range, the relationship is essentially linear, $y = a + bx$ and, as will be noted later, this frequently represents the relationship between plant growth and plant nutrient level. The third and upper part of the curve was expressed mathematically by E.A. Mitscherlich and is one of diminishing returns $dy/dx = (A - y)C$ where dy is the increase in yield which occurs with an increase dx in nutrient supply. A is the maximum growth and C is a constant. At the upper part of the response curve where there is little change in growth with increase in nutrient supply, the fertility is considered *optimum*. Further increase in nutrient supply can result in a reduction in plant growth and a *toxic* condition prevails. Sometimes the reduction in growth can result not from a poisoning of the plant metabolically but from a reduction in supply of another essential ele-

Figure 8.1 Diagrams illustrating the general relationships between plant growth and nutrient supply. The overall response curve (centre) may be viewed as comprising four separate curves: (i) $y = a.x^n$, (ii) $y = a + bx$, (iii) $dy/dx = (A - y)\,C$ (Mitscherlich equation), (iv) $y = a + bx + cx^2$

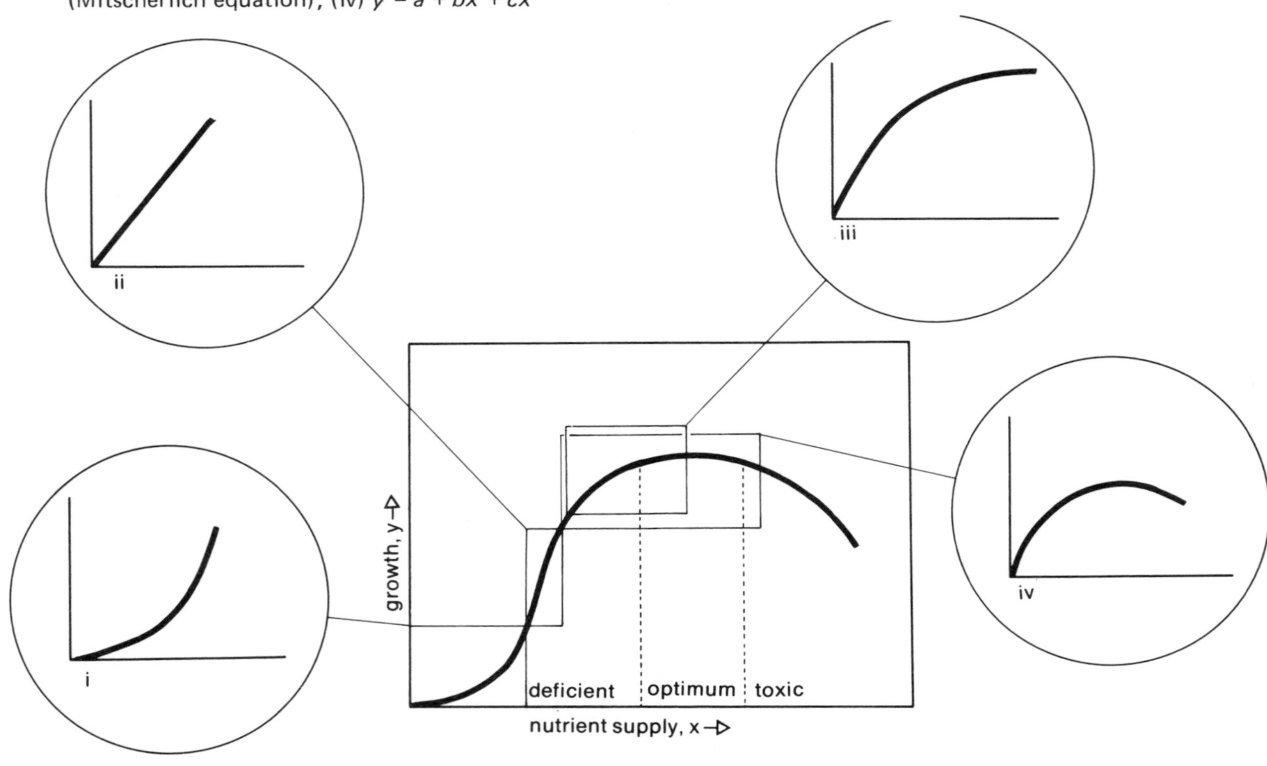

Figure 8.2 Growth response (mean seedling dry weight) by one-year-old white spruce seedlings to different levels of nitrogen and phosphorus. From K.A. Armson and V. Sadreika, *Forest Tree Nursery Soil Management and Related Practices* (1974)

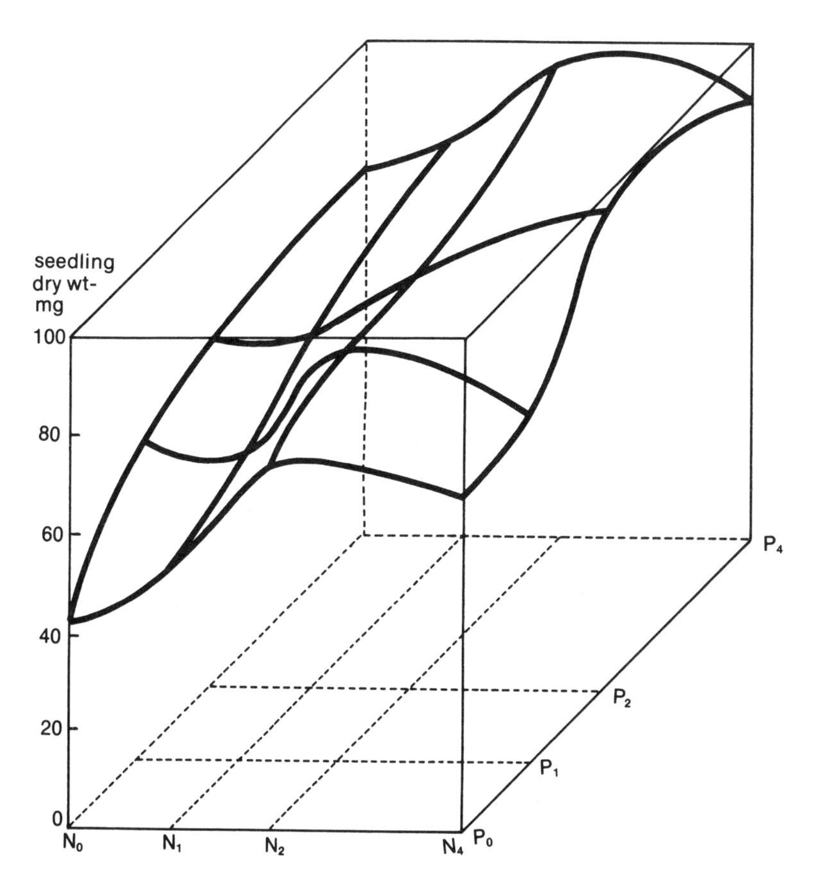

ment. Additions of potassium can result in reductions in uptake of calcium and magnesium and under certain conditions excessive supply of nitrogen as ammonium (NH_4^+) can reduce the uptake of potassium. These effects are examples of ion *antagonism.* The reduced growth associated with such antagonism is more properly considered an induced deficiency rather than a toxicity. In many fertility studies the range of fertility encompasses deficiency, optimum, and toxic regions and a curve of the form $y = a + bx + cx^m + dx^n$ ... may be fitted.

It is seldom that two soils differ in fertility only in terms of one nutrient element, and in the manipulation of soil fertility frequently more than one element may be added to the soil. When the growth responses to different levels of two nutrients are graphically portrayed, the result is a response surface. An example of a response surface is shown in Figure 8.2. It can be seen that the response to phosphorus at the lowest level of nitrogen (N_0) is essentially of the Mitscherlich type, whereas at both the low (P_0) and highest (P_4) phosphorus levels the response curves to nitrogen are of the complex type shown in Figure 8.1.

When the growth response of a plant to additions of two or more nutrients added in combination is significantly greater or less than the sum of responses obtained when the nutrients are applied separately, there is an *interaction.* This is illustrated in Table 8.4. When no potassium was added, the increase in growth due to nitrogen (0.06g) and phosphorus (0.08g) alone was relatively large, but the response to them when applied together (0.41-0.18 = 0.23) exceeds the sum (0.06 + 0.08 = 0.14g) of the individual effects of the elements applied singly. The amount of response (0.23-0.14 = 0.09g) due to the combined effect of these two elements is a measure of their interaction. When potassium is applied, the growth response to nitrogen and phosphorus and their interaction are slightly less.

Growth response to changing fertility can be altered by a number of different factors; one of these is the species; another is the density of the stand. An illustration of the effect that density may have is shown in Figure 8.3 for white spruce seedlings. The difference in response to a given fertility level by plants at different density often reflects quite simply the increased demand by a larger number of organisms, but it may also reflect such things as mutual shading of foliage and differences in water relations which determine a plant's metabolism and consequently its nutrition. Interaction between soil moisture and fertility is generally acknowledged, but the specific relationships are difficult to quantify, especially under natural conditions. Wright (1957) stated that a heavy thinning of poor growing Corsican pines delayed drying out of the soil and resulted in a temporary increase of nitrogen, potassium, and magnesium levels in the foliage of these trees. The general pattern is for growth to increase to a maximum with increase in soil moisture supply and then decline as a restricted aeration and increased leaching losses of nutrient elements occur; Bengtson and Voigt (1962) found this pattern of response for slash pine. Variations in the interaction response can occur. Klemmedson and Ferguson (1969) in a study of the response to nitrogen by bitterbrush at low and high moisture supply found that, when the bitterbrush was grown in a surface soil, an increase in nitrogen resulted in decreased growth, but the decrease was diminished at the higher moisture supply. The growth of bitterbrush increased with nitrogen supply when it was grown

TABLE 8.4
Growth responses (seedling dry weight – grams) of 240 white spruce seedlings from a fertilizer trial at Swastika Nursery, Ontario (from Armson and Sadreika, 1974)

Nutrient treatment	– N – P	+ N – P	– N + P	+ N + P
– K	0.18	0.24	0.26	0.41
+ K	0.17	0.22	0.22	0.34

	Response to N		Response to P		
	– P	+ P	– N	+ N	NP interaction
– K	0.06	0.15	0.08	0.17	0.09
+ K	0.05	0.12	0.05	0.12	0.07

Figure 8.3 Increase in total dry weight of two-year-old white spruce seedlings at varying seedbed densities to three soil fertility levels: I, lowest fertility; II, intermediate fertility; III, highest fertility. From K.A. Armson (1968)

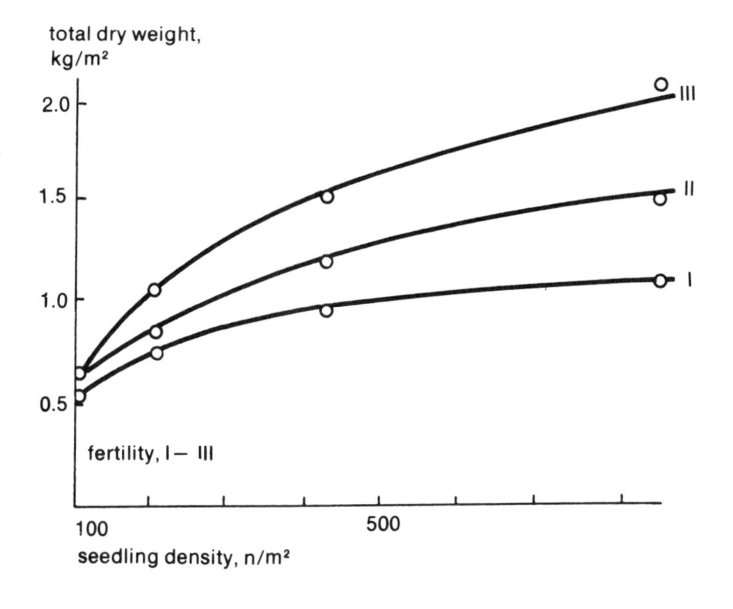

Figure 8.4 Plant growth response related to soil nutrient supply (—) and plant nutrient level (---). Adapted from H.L. Mitchell (1939)

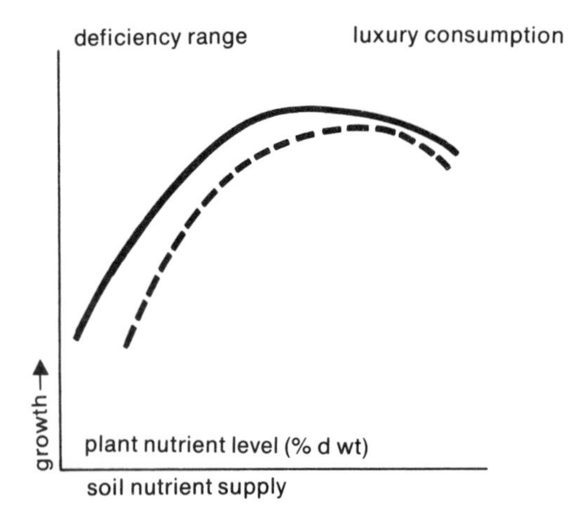

on deeper subsoil and then the growth response was enhanced at the higher moisture supply.

Considerable attention has been devoted to the relationships between plant growth and the level of nutrients in the soil and in the plant. For trees the most frequently sampled tissue is the foliage and element levels are usually expressed in concentrations per leaf dry matter. Figure 8.4 illustrates the general nature of the relationships. The parallel in change of plant nutrient concentration and growth with increase in nutrient supply is a commonly observed phenomenon and serves as a model for the diagnostic use of plant tissue analysis to detect nutrient deficiencies. There is a zone of plant nutrient increase in concentration not associated with plant growth increase; this increase in content is referred to as *luxury consumption.*

When there is an increase in the supply of a deficient nutrient and a rapid increase in plant growth, the concentrations of other elements in the plant may decrease. This does not necessarily reflect a decrease in the amount absorbed, but only that the rate of growth exceeds the rate of absorption of the non-limiting element. The decrease in concentration with increase in growth is a *dilution effect* (Figure 8.5). In certain instances, usually at low levels of nutrient supply, the effect of a small increase in supply of the deficient nutrient may result in a proportionately greater rate of growth than the increase in rate of absorption. The result is a decrease in the concentration of the deficient nutrient which is then followed by an increase in concentration. This change in concentration by a limiting element is termed the Steenbjerg effect (Steenbjerg, 1954).

Figure 8.5 Changes in nutrient concentrations with plant size for a limiting nutrient (nitrogen) and a non-limiting nutrient (potassium). Adapted from K.A. Armson (1968)

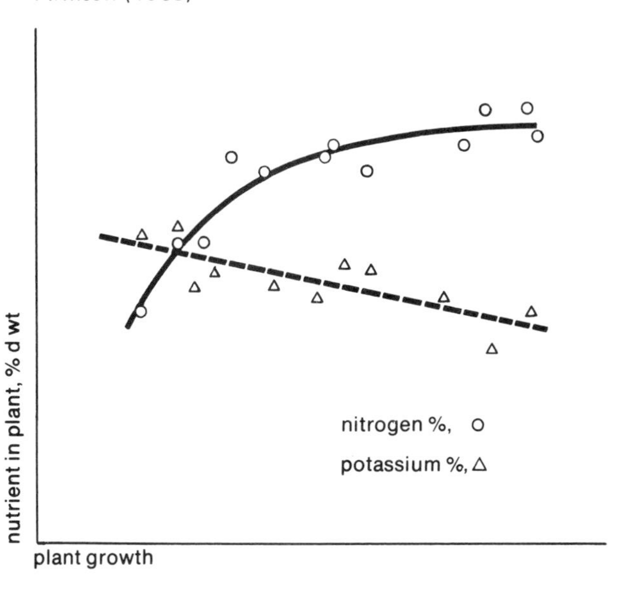

Diagnostic and predictive techniques

Richards and Bevege (1971) distinguished between the use of plant analysis for diagnostic and predictive purposes. While this may be a suitable division at times, more frequently the technique employed is used to determine not only what element(s) may be deficient – the diagnosis – but the degree of deficiency and expected growth increases – prediction. Foliage colour symptoms and comparative chemical analyses are essentially diagnostic techniques and have little if any predictive value. Morrison (1974) has reviewed the information available on the nutrition of conifers and summarized its use for diagnostic interpretation. Assessment of soil fertility is often a complex process in which the symptoms reflecting fertility differences and nutrient deficiencies may relate to a causal factor which is not soil fertility. For example, if a significant proportion of absorbing roots of a plant are killed by either a parasite or a sudden change in a soil property such as soil moisture or aeration at a time when the plant is in an active growing stage, the reduction in absorption of essential elements may cause not only a decrease in their levels within the plant but visual symptoms, usually in the foliage. Often the visual symptoms and foliage analysis at this stage reflect low fertility in terms of nitrogen because it is absorbed in larger quantities than most other mineral elements and its uptake more closely parallels the rate of growth.

The rate of growth and stage of development of the plants can profoundly influence the results obtained by soil and plant analysis. Vegetation which becomes established naturally on soils of low fertility normally shows a concomitant low rate of growth. Although chemical analysis of the soil may show a low 'available' level of one or more elements, the plant may show neither visual symptoms of deficiency nor abnormal levels of nutrients. The amount of growth, however, will be less than comparable vegetation on a soil of higher fertility. As the vegetation develops it increases in size and this is commonly accompanied by an increase in total demand for nutrients; for example, as a tree crown increases, so usually does the foliage mass and its nutrient contents. When the stage of development is reached where the rate of supply of nutrients by the soil is not equal to the required uptake rate by the vegetation, then a deficiency exists, and usually it will be reflected by both visual symptoms and tissue analysis. The plant population is seldom uniform so that the symptoms may show only on a few plants within a stand and then in relation to differences in soil fertility related to soil heterogeneity. Under natural conditions normal mortality reduces the total nutrient demand. A certain amount of the nutrient capital is returned to the soil with the dead organic material. The net result is that the residual vegetation may develop without any further sign of nutrient deficiency. The same general pattern of development will occur on soils of different fertility, but as the nutrient supplying capacity increases to the level where it may no longer be limiting, other factors such as soil moisture supply may become critical.

Much of our knowledge of forest soil fertility has been derived from studies in which artificially regenerated vegetation has shown abnormal growth, frequently with foliage discoloration. Plants set out in operational regeneration are usually grown and cultured in nurseries under relatively optimum conditions of soil fertility; as individual plants they typically have high relative growth rates. When they are out-planted into soils with low fertility, the occurrence of deficiency symptoms is to be expected. The nutrient deficiency is accentuated by the fact that in establishing the new plant there is inevitably some destruction of the root system and an even greater inability of the plant to absorb nutrients. As the growth of the plant adjusts to the fertility of the soil, the initial symptoms will moderate and often disappear.

VISUAL SYMPTOMS

Foliage discoloration and growth abnormality are often the first outward signs of deficiency observed in plants. It is important to recognize that, usually, overall growth reduction resulting from nutrient deficiencies takes place *before* visual symptoms appear. Thus, with intensive forest management the emphasis should be on anticipating possible deficiencies rather than on waiting to treat them after they occur.

Symptoms are not to be equated with causes. As has been mentioned, the visual sign of a nutrient deficiency may reflect a mortality of roots caused by some other factor. The form of symptom can vary greatly with both species and growing condition. Further, the nature of the nutrient element involved and its mobility within a plant can influence the nature of the symptom. Some of the more general symptoms of deficiencies are the following:

Nitrogen. Reduction in foliage size and a general yellowing or *chlorosis* are the usual symptoms. In evergreens there is usually a reduction in the amount of foliage and older foliage is carried for fewer years; the result is that crowns are thinner and branching is finer than for non-deficient plants. Immediate response to an increase in nitrogen supply is the greening of yellow foliage.

Phosphorus. Foliage becomes purplish red to reddish brown, particularly the newly developing leaves associated with apical meristems. There can be considerable variation in colour; sometimes the foliage also exhibits chlorosis. Root development of phosphorus deficient plants is usually more sparse than normal.

Potassium. In conifers chlorosis shows at the distal end (apex) of the needle and develops towards the base. As more of the needle becomes chlorotic, death (necrosis) of the apical tissue results in a red-brown coloration. The boundary between the yellow and green portions of the foliage is usually distinct. In pines and spruces, older foliage is shed earlier than normal and this shedding often occurs at the time of new shoot extension or shortly thereafter. Uppermost, youngest foliage shows less developed chlorosis or is green in comparison to older and lower foliage. The crowns of a potassium deficient tree are thin compared to a healthy crown. At times dull blue-green or reddish to purple colours have been noted in foliage of potassium deficient plants. In broadleaved species the chlorosis is typically marginal and sometimes more highly developed at the leaf apices. The chlorosis may be more yellow-white than yellow and sometimes rather than chlorosis the margins are red to reddish brown.

Calcium Symptoms on conifers in the field are rare. From experimental studies with conifers, the prominent visual symptoms of deficiency are a browning and necrosis at apical meristems while the lower foliage remains green. Root mortality is common. In broadleaved trees the youngest tissues are affected first and newly developing leaves are chlorotic and small. The pattern of yellowing is variable in its progression and may be succeeded by red-brown areas of dead tissue.

Magnesium. For conifers the typical pattern is for a chlorosis (often golden yellow) to develop first in the distal parts of the needles and then increase progressively along the leaf. Sometimes there is an alternate banding of yellow and green sections of the leaf. The yellowing of the foliage becomes most pronounced in the latter part of the growing season on the current foliage; necrosis and reddish brown coloration follow the chlorosis. In broadleaved trees the chlorosis is interveinal and in compound leaves has been noted first in lowermost leaflets.

Sulphur. The younger foliage of conifers becomes chlorotic and this later becomes necrotic tissue and turns red-brown. In broadleaved species the leaves are pale yellow-green. Sulphur deficiencies are unlikely in the vicinity of urban and industrial zones.

Iron. Chlorosis resulting from iron deficiency in both conifers and broadleaved trees appears in the upper younger foliage. In needles the yellowing first appears at the base of the foliage and in broadleaved trees it is interveinal. Iron chlorosis is most common on alkaline soils with free carbonates in the rooting zone and is often termed 'lime-induced chlorosis.' In many plants iron chlorosis results not from the inability of the soil to supply iron or the plant to absorb it but from the plant's inability to translocate and utilize iron that is absorbed.

Manganese. Chlorosis usually develops at the foliage margins first and is interveinal although a band of green may occur on either side of main veins. In some broadleaved species the leaves or leaflets appear wrinkled. In conifers manganese deficiency shows a more general yellowing of the new needles.

Zinc. Deficiency of zinc produces a chlorosis or bronze colour in the upper younger foliage. Older foliage is shed and there is a reduced growth or stunting of shoot growth resulting in a 'tufted' or 'rosette' appearance of foliage. Dieback of top shoots is common.

Boron. Lack of boron produces dwarfed and deformed leaves with chlorosis and occurrence of brown spots in many broadleaved species. Terminal buds and associated tissues are reduced and are frequently killed. In conifers, shoot 'dieback' or deformation is common; associated with restricted shoot elongation is the bunching of needles to produce 'rosettes.' Older foliage often appears normal.

Copper. Deficiency in young conifer seedlings shows as a needle tip burn and needles may be wavy or twisted. In older conifers the shoots often droop and branches are recurved. Terminal buds may die. In broadleaved trees chlorosis is often variable; sometimes a 'cupping' or distortion of the leaf is apparent. Dark, necrotic areas are often prevalent. Copper deficiency has typically occurred in acid, highly weathered, coarse-textured soils and in some peats.

Molybdenum. Little is known of molybdenum deficiencies. It is considered to be important in those woody species with nodulated roots and dinitrogen-fixing symbionts because of its known role in the fixation of nitrogen. In hardwoods molybdenum deficiency may appear as marginal chlorosis or scorching.

Visual symptoms vary from plant to plant and with soil and other growing conditions. There are a number of detailed reviews of visual symptoms. Chapman

(1966) and Sprague (1964) both present descriptions for agricultural and orchard species and Leaf (1968), Morrison (1974), and Stone (1968) review deficiency symptoms for forest trees.

COMPARATIVE CHEMICAL ANALYSES

Samples from trees or stands considered to exhibit satisfactory growth together with samples from the soils in which they are growing may be analysed and the levels of nutrients in these samples are used as standards to compare with analyses from other vegetation and soils. At best, comparative analyses serve to provide ranges of values and where the comparisons are between individuals of the same species, similar age, growing in the same location on the same soil materials, the data may be of value (Madgwick, 1964), especially where nutritional differences may be related to some aspect of soil use (Armson, 1959; Heiberg and Loewenstein, 1958). The greatest unreliability of comparative values may be expected when data are extrapolated from one location to another or from a tree or stand at one age or stage of development to another of differing age and development.

MATHEMATICAL RELATIONSHIPS

Visual symptoms and comparisons of results of plant and/or soil analyses are used as guides for diagnosis. The curvilinear relationships which have been illustrated (Figure 8.1) can be utilized for both diagnostic and predictive purposes. Sampling problems associated with soil heterogeneity and root distribution have already been mentioned. It is customary in plant analysis to sample only a portion of a tree or stand. Leaf (1973) has reviewed the effects of the type and position of sample tissue and its age and time of sampling; a comprehensive account of plant tissue analysis has been given by Goodall and Gregory (1947). There are three forms of relationship commonly used: plant growth - plant nutrient, plant growth - soil nutrient, and plant nutrient - soil nutrient. In determining these relationships the use of experimental fertilizer additions is most common.

Plant growth - plant nutrient relations

The curvilinear nature of the change in plant growth with internal level of a limiting nutrient (Figure 8.4) is the basis for establishing not only the deficient element but quantitatively the expected increase in growth if the plant nutrient level is increased. Sometimes, over the lower range a linear relationship may apply. The technique may be applied to a population of varying growth as was done by Leyton and Armson (1955), who determined for a particular Scots pine stand that tree height could be related primarily to foliage levels of nitrogen and potassium:

$$Y = 11.79\ \mathrm{N\%} + 5.70\ \mathrm{K\%} - 10.18,$$

where Y is the height in centimetres. This technique can also be applied to trees or stands which have received some form of fertility amendment and the relationship can be broadened to include other measures of the stand in addition to nutrient levels. Bevege and Richards (1972) used such a procedure to establish both artificial levels of phosphorus for loblolly and slash pine and a growth prediction:

$$\log B = a_1P + a_2P^2 + a_3 \frac{\mathrm{BA.SI}}{\mathrm{Age}} + a_4 \frac{\mathrm{SI}}{\mathrm{BA.Age}},$$

where B is the average annual basal area increment preceding or subsequent to foliar sampling, P is the percentage of phosphorus in foliage, BA is the basal area at time of sampling, SI is the site index (dominant height at 25 years), age is the age of stand, and a_1, a_2, a_3, a_4 are coefficients of the equation. Plant growth – plant nutrient relationships of these forms can only be expected to have validity for uniform plant populations growing on similar soils.

Plant growth – soil nutrient relations

Attempts to relate plant growth to soil nutrient supply have generally been less productive in terms of both diagnostic and predictive uses. Pritchett (1968) used both foliar analysis and soil analysis for phosphorus and pointed out that soil analysis may be used to predict fertilizer needs prior to stand establishment whereas plant tissue analysis is restricted to vegetation already established. Figure 8.6 shows the predicted volume increases for slash pine three to eight years after application of two different rates of phosphatic fertilizer in relation to soil phosphorus. It can be seen that when soil phosphorus is greater than 1 ppm, little differential in growth increase exists between the two phosphorus treatments, but as soil phosphorus decreases below 1 ppm the differential increases markedly. The equations for growth increase Y (volume %) in terms of soil extractable P (X) were

$$Y = 965.1 - 776.7X + 320.9X^2$$

for 19.5 kg/ha P, and

$$Y = 1463.4 - 1397.5X + 320.9X^2$$

for 39 kg/ha P.

Plant nutrient – soil nutrient relations

A relationship between level of plant nutrient and some measure of a soil nutrient is common but does not necessarily indicate deficiency, although it does usually reflect supplying ability of the soil. Strong correlations between soil

Figure 8.6 Predicted volume increases for slash pine 3 to 8 years after application of 19.5 and 39 kg/ha phosphorus from superphosphate in relation to ammonium acetate (pH 4.8) extractable P in surface of non-fertilized flatwoods sands in the southeastern United States. From W.L. Pritchett in *Forest Fertilization - Theory and Practice* (1968), by permission

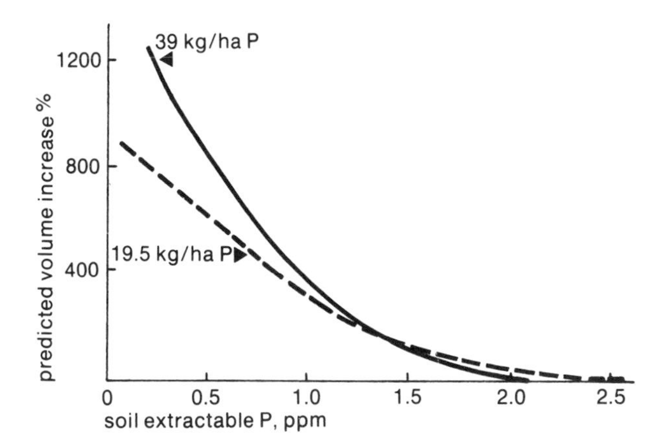

exchangeable potassium and potassium concentrations in the foliage of a variety of woody and herbaceous species in New York were suggested by Walker (1955) as a useful technique to delineate areas most suitable for establishment of a species with known response to soil potassium levels. Another approach to the use of plant-soil nutrient relationships is illustrated by Mitchell and Chandler's (1939) study of nitrogen and deciduous tree growth. Their approach is useful in rating the nutrient-supplying ability of several soils. It is particularly appropriate with an element such as nitrogen for which soil chemical analyses seldom correlate well with growth. They applied nitrogen fertilizer at four different levels on several soils and measured both the growth response and foliar nitrogen levels. The relationships between foliar nitrogen of red oak and fertilizer nitrogen applied on three different soils is shown in Figure 8.7 a,b. First the foliar nitrogen for red oak on soil *N* was plotted against fertilizer nitrogen and a curvilinear response was obtained (Figure 8.7a). The foliar levels for red oak from another soil (*ON*) which received another series of smaller amounts of fertilizer nitrogen were then plotted. It will be seen that these lie on the left-hand side of the curve and the foliar nitrogen (1.66 per cent) of the control plot relates to a fertilizer level equivalent to 17 kg/ha less than for the control of the *N* soil. Consequently, if the *X*-axis is moved to the left an amount equal to 17 kg/ha so that the control *ON* soil is now zero, the nitrogen-supplying ability of the *N* soil is 17 kg/ha.

Figure 8.7 (a) Foliar nitrogen values for red oak growing on two soils (N and ON) plotted against rates of fertilizer nitrogen application. (b) Foliar nitrogen values plotted against adjusted X-axis in which control plot for ON soil is arbitrarily rated 0. Values for AN soil are included. Adapted from H.L. Mitchell and R.F. Chandler, Jr. *The nitrogen nutrition and growth of certain deciduous trees of northeastern United States* (1939), by permission Harvard Black Rock Forest

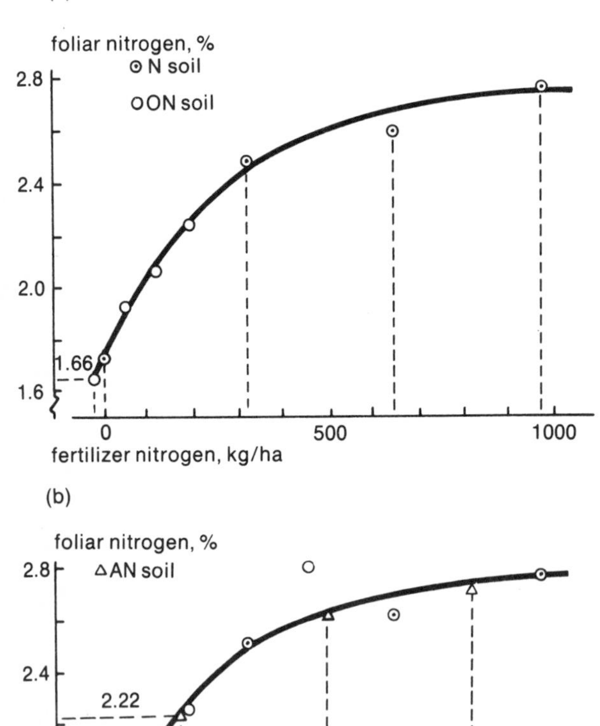

The adjusted values are plotted in Figure 8.7b, and the foliar nitrogen levels of red oak from *AN* soil which received the same nitrogen fertilization as the *N* soil are also included. These points fall on the same response curve and the relation of the control *AN* soil foliar nitrogen (2.22 per cent) to the nitrogen-supplying ability of the soil indicates that soil *AN* has a nitrogen supplying ability of 187 kg/ha. The three soils can be quantitatively compared in terms of their nitrogen-supplying ability for red oak.

General fertility considerations

Considerations of soil fertility are almost inevitably related to conscious attempts by man to change the fertility with respect to one or more nutrients. Some aspects of the effects of changes in soil fertility are:

Plant use of growing season. It is generally found that increase in growth from improvement in fertility results from an increase not only in the rate of growth but also of the period during which growth can take place. This is illustrated in Figure 8.8. If the extension of the growing period exposes the plant to hazards such as low-temperature injury, increase in fertility may be detrimental, but in many situations this does not happen.

Wood quality. Change in soil fertility can alter the proportions of spring and summer wood and the relative thickness of cell walls and thus alter specific gravity and other properties. As a result of nitrogen additions, Klem (1972) noted that the percentage of summerwood in spruce and pine was not changed, although there was an overall growth response. The specific gravity of the spruce was reduced by 5-10 per cent but that of the pine was not. Gladstone and Gray (1973) noted that wood from potassium-fertilized red pine and nitrogen-fertilized Douglas-fir was more uniform than unfertilized wood and fibre and paper qualities were improved. In hardwoods, Saucier and Ike (1969) found that no differentials in specific gravity or fibre length occurred in response to nitrogen, phosphorus, or potassium treatments, although growth increases did occur in response to nitrogen.

Forage and browse. Increase in fertility usually results in a growth response by vegetation generally, including the crop species. Often there may be differential responses by certain species but the effect in lesser vegetation has been found to be shortlived - one to two years for forage species in slash pine stands (Hughes et al., 1971), although the first-year response was a fivefold increase. Fertilization of the crop species may enhance browsing, as was found for potassium-fertilized red pine (Heiberg and White, 1951) and nitrogen-treated Douglas-fir (Oh et al., 1970).

Figure 8.8 Effect of fertilization on seedling dry weight of white spruce during their second growing season. From K.A. Armson in *The Forestry Chronicle* (1966)

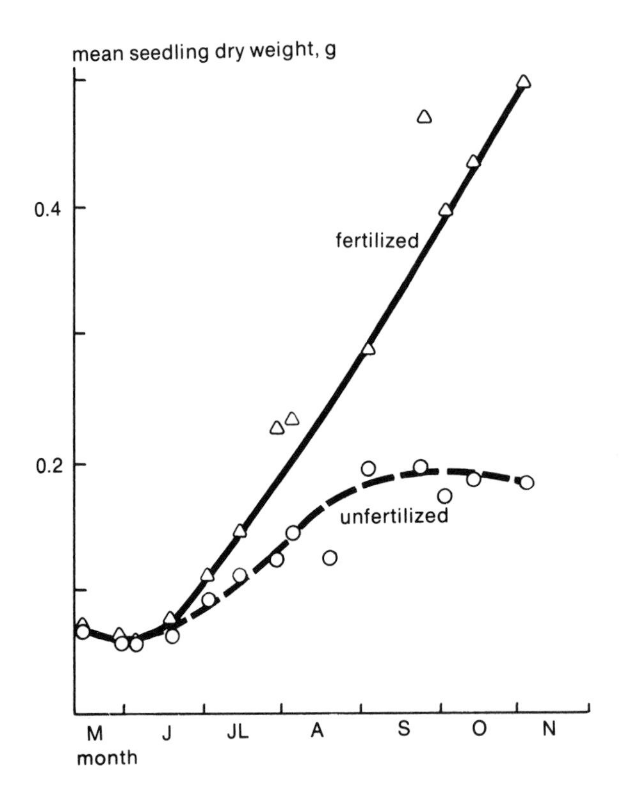

Insects and disease. There are no general patterns that would seem to apply. The effect of change in fertility would appear to be largely indirect and acts by altering the plant's physiological state and the chemical nature of both internal constituents and exudates. The precise mechanisms, however, are largely unknown.

Goyer and Benjamin (1972) found that in soils supporting jack pine plantations the incidence of pine root weevils (*Hylobius rhizophagus* Millers) was less, the greater the total nitrogen and available phosphorus in the upper 15 centimetres of soil; they found experimentally that applications of nitrogen and phosphorus fertilizers drastically reduced larval feeding and increased larval mortality. Xydias and Leaf (1964) found in a ten-year-old plantation of eastern white pine on potassium deficient soil that nitrogen fertilizer applications reduced growth and white pine weevil (*Pissodes strobi* Peck) damage but potassium treatments increased both height growth and weevil damage. Evidence of the effect of fertilizers on sucking insects is somewhat conflicting, but Carrow and Betts (1973) found that ammonium nitrate foliar-applied to grand fir reduced aphid (*Adelges piceae* Ratz) numbers compared with controls, whereas urea applications resulted in an increased aphid population. In Europe considerable success has been achieved in reducing sawfly populations by fertilizer applications (Baule and Fricker, 1970). Smirnoff and Bernier (1973) found increased mortality of Swaine's sawfly (*Neodiprion swainei* Midd.) on jack pine fertilized with urea and, when a virus was introduced, it was more effective in increasing sawfly mortality on the fertilized than unfertilized trees.

Hesterburg and Jurgensen (1972) reviewed the incidence of disease and fertilization and concluded that effects would vary with organisms and growing conditions. Canker disease such as fusiforme rust (*Cronartium fusiforme*) or blister rust of jack pine (*C. comptoneae* Arth.) appear to be stimulated by fertilizer applications which increase cambial growth. Low levels of potassium in plant tissues have been found to enhance attacks by such parasitic fungi as snow mould (*Phacidium infestans*) and some leaf rusts. Development of root rot (*Poria weirii*) in northwestern North American forests appears to be inhibited by growth of red alder and this has been attributed to the inability of the parasite to use nitrate nitrogen (Li et al., 1967; Nelson, 1970). Lavallée (1972) in eastern Canada reported that *Polyporus tomentosus* was observed more frequently in nitrogen-treated plots than in control or other fertilized plots.

Water. Consideration of the interrelations between soil fertility and water yield and quality is clearly related to specific locations and conditions. As might be expected, the more fertile a soil the greater the content of dissolved elements occurring in the ground water. Much depends, however, on the pattern of water movement in the hydrologic cycle. Usually the intensity of root development is greater in more fertile soils and soil horizons and this usually results in greater efficiency of water absorption from the soil.

Fertilizer materials and their use

There are many types of fertilizer materials available for use in forestry, but for practical reasons of handling, application, and cost, inorganic substances have been used most commonly. There may in the future be local areas in which organic wastes are applied, but this will be primarily for disposal purposes rather than as a fertilizer addition.

TABLE 8.5
Some commonly used fertilizers and related properties

Fertilizer	Nutrient elements (%)						Remarks
	N	P	K	Ca	Mg	S	
Ammonium sulphate $(NH_4)_2SO_4$	20	–	–	–	–	23	Very water soluble crystals; has a residual acidic effect on soil
Ammonium nitrate NH_4NO_3	33	–	–	–	–	–	Very soluble, available in prilled form; nitrate ion readily leached
Potassium nitrate KNO_3	13	–	37	–	–	–	Very water soluble crystals; nitrate ion readily leached
Urea $CO(NH_2)_2$	45	–	–	–	–	–	Very water soluble, fine crystalline but usually formed into spherical pellets of desired size; subject to volatilization losses under certain soil conditions
Anhydrous ammonia NH_3	82	–	–	–	–	–	Gas, stored under pressure in liquid form, injected into soil with special equipment; loss is least in finer-textured soils
Phosphoric acid H_3PO_4	–	24	–	–	–	–	Liquid, which can be applied either to irrigation water or injected with special equipment into soil. Phosphorus is usually fixed by calcium, iron, and aluminium in relatively insoluble forms
Orthophosphate $Ca(H_2PO_4)_2 . H_2O$ $CaSO_4 . 2H_2O$	–	7 – 9	–	18 – 21	–	10–12	phosphorus approximately 85% water soluble; size of granule can affect availability; usually only slight movement of
Triple superphosphate $Ca(H_2PO_4)_2\ H_2O$	–	19 – 23	–	12 – 14	–	–	fertilizer phosphorus in most mineral soils
Monammonium phosphate (MAP)	11	21	–	–	–	–	Water soluble; granular materials; gives acid reaction in soil

Diammonium phosphate (DAP)	18	20	–	–	–	–	As for MAP except that it gives an alkaline reaction in soil
Ammonium phosphate – sulphate	16	9	–	–	–	14	Water soluble; granular; gives acid reaction in soil
Ammonium polyphosphate	15	26	–	–	–	–	Water soluble; has the ability to sequester metal ions and hence add micronutrients
Potassium chloride KCl	–	–	51	–	–	–	Water soluble; granular; widely used but at high rates of application the Cl^- ion may be toxic to some plants unless leached from soil
Potassium sulphate K_2SO_4	–	–	44	–	–	17	Water soluble; granular
Gypsum $CaSO_4 . 2H_2O$	–	–	–	22	–	19	Sparingly water soluble, crystalline form
Calcium carbonate $CaCO_3$	–	–	–	35 – 40	–	–	The most critical factor in the effectiveness of these materials is their particle size.
Calcium magnesium carbonate Ca $Mg(CO_3)2$	–	–	–	variable	–	–	The finer the particle, the more effective the application
Magnesium sulphate $MgSO_4 . 7H_2O$	–	–	–	–	10	13	Water soluble; crystalline form

Inorganic fertilizers may be *single*, containing only one nutrient element, as does ammonium nitrate (NH_4NO_3), or *mixed* where there is a mechanical or chemical combination of two or more essential elements. Nitrogen, phosphorus, and potassium are the most abundantly used nutrients in the fertilizer industry and it is customary for the quantities of these elements to be expressed in a 'formula' placed on each fertilizer material. Three numbers are used representing the percentage contents by weight: N,P_2O_5 and K_2O. In order to convert P_2O_5 and K_2O to elemental quantities, consult Table 7.1. A fertilizer designated as 16-48-12 contains 16 per cent nitrogen, 48 per cent P_2O_5, and 12 per cent K_2O, or more properly 16 per cent N, 21 per cent P, and 10 per cent K. Fertilizer compounds will often contain nutrient elements other than nitrogen, phosphorus, and potassium but these are not usually listed. For example, ammonium sulphate $(NH_4)_2SO_4$-20-0-0 contains 24 per cent sulphur, and ordinary superphosphate-0-20-0 contains both calcium and sulphur. Adding a fourth number to the fertilizer formula, representing the per cent sulphur, is increasing in use.

Although the choice of fertilizer reflects many factors, particular properties of the material usually are of major importance. These are 'grade' or analysis, nature of the compound, means of application and transformation in the soil. A listing of some of the commonly used fertilizers is given in Table 8.5.

Grade or analysis. The cost per unit of nutrient element is usually less with increase in analysis. The costs of transportation, storage, handling, and distribution are equal whether for a metric ton of low analysis or high analysis material. However, other properties of fertilizer materials such as form, solubility, and fate in the soil environment may be important and outweigh a decision based simply on grade alone.

Nature of fertilizer. Fertilizers may be obtained as solids, liquids, or gases. In most forestry applications solid materials are used, but liquid or even gaseous fertilizers may be used in special instances as in forest nurseries or other intensive productions of young trees.

The form of the solid fertilizer will vary with the chemical nature to some degree, but uniformity of size is necessary if efficient even distribution is desired. Ammonium sulphate is available as small crystals whereas many other fertilizers can be manufactured to produce specific sizes of granules or particles. The size of the particle may be a critical factor in distribution; for example, where aerial applications are made, small particles are more difficult to apply uniformly than larger uniform spherical materials. The size of particle, as with certain phosphatic fertilizers, may also be critical in terms of its effectiveness in the rooting zone. Generally, greater plant response is achieved when water-insoluble phosphates are applied in fine granular or powdered form and mixed with the soil; conversely, effectiveness is greater with coarse particles than fine particles when water-soluble phosphates are used (Tisdale and Nelson, 1966).

The chemical nature of fertilizer materials can exert a profound influence on their action in the soil. Thus the sulphur coating of readily soluble urea pellets has been used to delay the rate of nitrogen release. Polyphosphates, particularly ammonium polyphosphate, have been developed since the 1950s; their higher analysis and physical properties have rendered them useful in bulk handling and mixing.

Means of application. The techniques and equipment used to apply fertilizer materials in forestry are usually those derived from agriculture. In open land or on young plantations, application is usually made from the ground, using mechanical equipment to spread or blow the fertilizer (Baule and Fricker, 1970). Where established stands have developed to the stage that ground access is not feasible and particularly where large areas are involved, aerial application is the most effective means of fertilizer distribution. Uniformity of application is a matter for concern, especially in aerial application and may be critical in estimates of growth response (Armson, 1972). In forest tree nurseries injection of water-soluble materials into irrigation lines may frequently be used (Armson and Sadreika, 1974). Placement of fertilizer, common for many agricultural crops, has only limited use in forestry and then primarily in the application of phosphate fertilizers to nursery soils.

Transformation in the soil. The processes that affect the fertilizer material when it is added to the soil are often most critical in determining its effectiveness. Thus, volatilization of nitrogen from urea added to forest soils can be expected to be least with the use of large particle sizes on acid soils and greatest with fine particles on alkaline soils. When water-soluble materials are added, some ions may be retained in available form by the soil exchange complex, or leached. Other materials, particularly phosphates, may be held in slowly available form as iron, aluminium, or calcium phosphates, although this will not occur when phosphatic materials are applied to organic soils. When an application of a fertilizer material is being considered, the properties of the soil to which it is to be added should be assessed as being equally as important as the nature of the forest crop whose growth response is the justification for the treatment.

9
Soil classification

It is the nature of man to name, to categorize, and to attempt a systematic approach in any field of study; in effect to develop a 'scientific' classification. Classifications evolve in relation to increase in knowledge and new and different groupings developed from time to time in turn establish other relationships. What may have begun as a simple separation of objects to facilitate description or use can develop so that the arrangement itself provides insights about the objects and stimulates the development of new concepts about them.

All schemes of classification are man-made and therefore arbitrary and in a relatively new field of study there will be many different forms of classification develop on regional, geographic, and national bases. As greater exchanges of information about soil occur and the number of soils which are observed increases, the various schemes of classification become fewer and a small number become predominant. This is the present state of soil classification.

It has been stated that problems in soil classification relate to definition, nomenclature, and classification proper (Manil, 1959). Although the soil individual - the pedon - is not discrete in the sense that a plant or animal is, comparisons are often drawn between the Linnean system of classification and one for soils. Probably a clearer analogy would be between the classification of forests and that of soils. The character of a forest stand may be defined in terms of its quantifiable properties - species, dimensions, density, etc., and in a comparable manner a soil may be defined in terms of its quantifiable properties - texture, colour, chemical, etc. Some soil classifications have placed emphasis on categorizing the soil-forming processes which is comparable to classifying forests on their mode of formation. The selection of measurable properties, whether it be for a forest or a soil, therefore becomes of great importance and in this way the problem of definition of the soil is intertwined with the mechanics of classification.

Problems of nomenclature - the application of names to soils - do have a close parallel with nomenclature of the plant and animal kingdom. Colloquial and vernacular names are impediments if more than one language is involved and indeed can perhaps be even more confounding within a language. The ash (*Fraxinus* spp.) of North America bears no close relation to the ash (*Eucalyptus* spp.) of Australia. The Linnean system and its nomenclature has done much to remove this type of problem. Another difficulty related to nomenclature arises when names which have had a certain history of usage (not necessarily uniform) are employed in a restricted sense in a new classification. Human habits are often difficult to break and word usage is one of them. People will tend to give the old word in its new setting some of its old meaning and thus diminish the value of the new classification. This kind of difficulty can be overcome by creating new words and then, of course, soil scientists must learn the new vocabulary. Kubiena (1953) stated that a nomenclature should be clear, convenient to handle, and easily applicable in all languages, a desirable goal - more readily stated than achieved in soil science.

The problems of classification proper derive from the decisions as to what properties are to be used and this has inevitably involved the objectives for which the soils are to be classified. Concern for the practical use of a system is so much a part of many soil classifications that it is as if von Linné had made a taxonomy for the plant kingdom on the basis of plants that were edible or non-edible, or could be used for building and so on. An example of the problem is given by Buol et al. (1973); when speaking of the new system, the 7th Approximation, they state, 'In our present soil classification, we try to approach a natural classification system as an ideal, though we tend to give weight to properties of higher agricultural relevance.' There is nothing amiss in a classification whose primary purpose is categorize soils in relation to one or more particular uses - agriculture, road-building, or the like - but it cannot then lay claim to being a 'natural' or 'genetic' system. A common objective of many systems is to provide a rational organization of objects so that a person wishing to identify objects unknown to him can, by becoming familiar with the scheme of classification, more readily identify the objects. When a classification is used this way it is a key. Most universal schemes of classification, such as those for plants and animals, are usually poor keys. They are poor because, being universal, they embrace too large a population of objects and many of the properties which may be critical in terms of the classification are frequently of little use in field recognition. Thus for regional or national purposes a key or field guide to soils will be more useful for general recognition purposes in the same manner that a field guide to flowers, based on flower colour or to trees, based on twig arrangement is more functional than a botanical flora. The flora usually precedes the keys in development.

Lastly, it should be appreciated that classifications are effective only in so far as they serve the function for which they were established. There is nothing sacred about either the system or its nomenclature but, if a system is created to perform a function, its weaknesses and strengths can only be determined by usage.

History

The earliest attempts to classify soils took place three to four thousand years ago in China (Simonson, 1968), and there is ample evidence in Europe that from Greek and Roman times certain soils were considered more suitable for a particular crop than another. The soil was viewed only in terms of being a medium for growth of plants, a conception that is still valid. The soil can also be viewed as a mantle of weathered rocks and minerals and this geological perspective of soils was common when a Russian geologist, Dokuchaev, provided a third and new conception of soil. He considered soil to be a natural body which could be described, classified, and mapped in an analogous manner to geological deposits. He recognized that horizons of a soil profile were the diagnostic features that separated soil from being considered as only modified geological material. The A,B,C designation was introduced. Initially it referred to somewhat different horizons, but the beginnings of soil classification on a rational, scientific base began with Dokuchaev in the late 1800s. A student and co-worker, Sibirtsev, developed the concepts of Dokuchaev, particularly the ideas of zonal,* intrazonal, and azonal soils and the factors of parent material (geological), organisms, topography, climate, and time as being primary in determining the character of a soil. Coincidentally, Müller during the same period was describing the upper humus layers of forest soils and classifying them in relation to forest conditions, yet it was not until well into the twentieth century that there was a full coalescing of these two perspectives.

The Russian workers laid more stress upon the soil-forming processes than they did on the soil itself (Basinski, 1959) and for this reason many systems of classification which subsequently developed were more attempts to classify soil-forming factors than soils themselves, although ostensibly it was still the soils that were being labelled. Efforts in the first few decades of the 1900s to delineate podzolization† and laterization as distinctive processes resulting in different climatic regimes, illustrate this approach.

* Zonal soils were those considered to reflect primarily the broad influence of climate. Intrazonal soils reflected the dominating influence of a local factor on soil development such as topography, geology; and azonal soils were those lacking a well-developed profile morphology.

† Podzolization is a term used to describe the processes resulting in the formation of podzols and podzolic (spodic) soils. Similarly, laterization is used to describe the processes creating lateritic (oxisols) soils. Both terms are very general and embrace many similar chemical reactions.

Concurrently with the Russian work on soils in the 1880s and 1890s there occurred in the United States an interest in soils which was developed by an American geologist, Hilgard. Quite independently he came to view soil as a natural body in much the same manner as Dokuchaev. In the first decade or two of the 1900s more interest centred around soil surveys. Credit for the first consistent application of a scientific rationale to soil description and classification in the United States must go to C.F. Marbut. He emphasized the soil profile and laid the groundwork for the tremendous development in soil surveys which followed from the 1920s to the 1940s. The main use of the soils' information was in agriculture and concern with forest soils was minimal, although paradoxically the cleared lands of the northeast, southeast, lake states, and many parts of the central and western states in which farming was taking place originally supported forest prior to settlement. Soil was the plow layer and what might occur below it to a depth of three feet (approximately 1 metre) - no more, and often less.

During this same period, the conception of soil development with time was that it proceeded from the initial weathering of rock through intermediate stages to a final equilibrium state - the mature soil. This was similar to the theory of climax vegetation. One had primarily to search for these mature soils, describe them, and thus lay the basis for a classification. Increasingly, the naive simplicity of this view of soil development from a juvenile condition to steady-state maturity has been challenged. While one property of a soil such as the forest floor may reach a steady state, others such as the translocation of sesquioxides may not. Infrequent disturbances appear to be the rule rather than the exception in many areas of the world and these alter the nature and direction of soil processes.

While some classifications, such as those in Russia and the United States, employed the soil profile and soil-forming processes as the bases for categorization, elsewhere other systems placed emphasis on other aspects of the soil, especially chemical properties. All, however, were essentially conceived and to a large degree employed on a national or regional basis. It was already evident in the 1940s and 1950s that the confusion in terminology and the inability of most classification systems to cope with soils being described in various parts of the world was a serious impediment to the development of soil science. One of the first major examples of a reorganized classification was regional - Kubiena's *The Soils of Europe* (1953) in which he employed soil names already used but ordered their use to emulate the binary form of the Linnean system. The first division of soils is based on the A B C horizon development. Kubiena also provided keys to the identification of soils based on observable field features.

The most profound change in soil classification came in 1960 with the publication of the 7th Approximation by the United States Department of Agriculture. It and its successor - 'Soil Taxonomy' - will be described in more detail later, but the salient features were that it attempted to be a global classification and employed a new terminology in order to minimize the misconceptions which arose when old terms were used with new or restricted meanings. The new words, often created from classic Greek or Latin roots, were, therefore, considered to be readily assimilated by a large number of European languages without involving preconceptions. Although the initial placing of a soil is based on the diagnostic horizons, the subsequent breakdown largely uses specific chemical and/or physical properties. The emphasis is on those which can be objectively determined and quantified. Generally, classifications even for limited areas have placed increasing emphasis on using measurable properties of the soil itself to place it within the classification; an example of this is the Canadian System of Soil Classification (1974).

Another approach engendered in part by the inevitable difficulties in placing soils within a rigid system of classification, and made possible by the use of high speed computers is numerical taxonomy (Sneath and Sokal, 1962). One of the first attempts to use a numerical approach was made by Hole and Hironaka (1960) who took a system of ordination used to categorize vegetation, and applied it to soils. Both laboratory data and information from field profile descriptions were employed. The data were ranked on a scale from 0 to 100 and indices of similarity were calculated. The advantages and background logic for the use of numerical taxonomy have been well stated by Bidwell and Hole (1964) who stress its emphasis on objective quantification. There have been three types of soil properties used in numerical classification which Rayner (1966) treated as dichotomies, alternatives, and scales. Dichotomous properties are those which may be present or absent. Alternatives are multistate properties which may be unranked as, for example, kind of structure - prism, granular, etc. - or ranked as when the size of structural aggregate is measured. Some properties are continuously variable; they exist as a scale of values, for example, soil pH and amounts of exchange cations. The base data are normally transformed and numerical measures of similarity derived which then enable the soils to be compared. The results are frequently expressed graphically in the form of *dendrograms* (Figure 9.1). In this example Grigal and Arnemann (1969) segregated forty profiles, each of which was named according to the 7th Approximation, on the basis of physical data relating to horizon texture and thickness. Other dendrograms could be constructed using more soil properties. The dendrogram will vary depending upon the set of data used. It is, in a way, a sophisticated form of the simple dichotomous key used in plant identification. Numerical taxonomy should not be viewed as a separate and competing manner of classifying soils but rather as a useful technique which can be helpful in establishing relationships for soils within and between other systems of classification.

Figure 9.1 Dendrogram based on horizon texture and thickness for 40 forested Minnesota soils. Euclidian distance is a measure of similarity for the particular property. Adapted from D.F. Grigal and H.F. Arnemann in *Soil Science Society of America Proceedings* (1969), by permission Soil Science Society of America

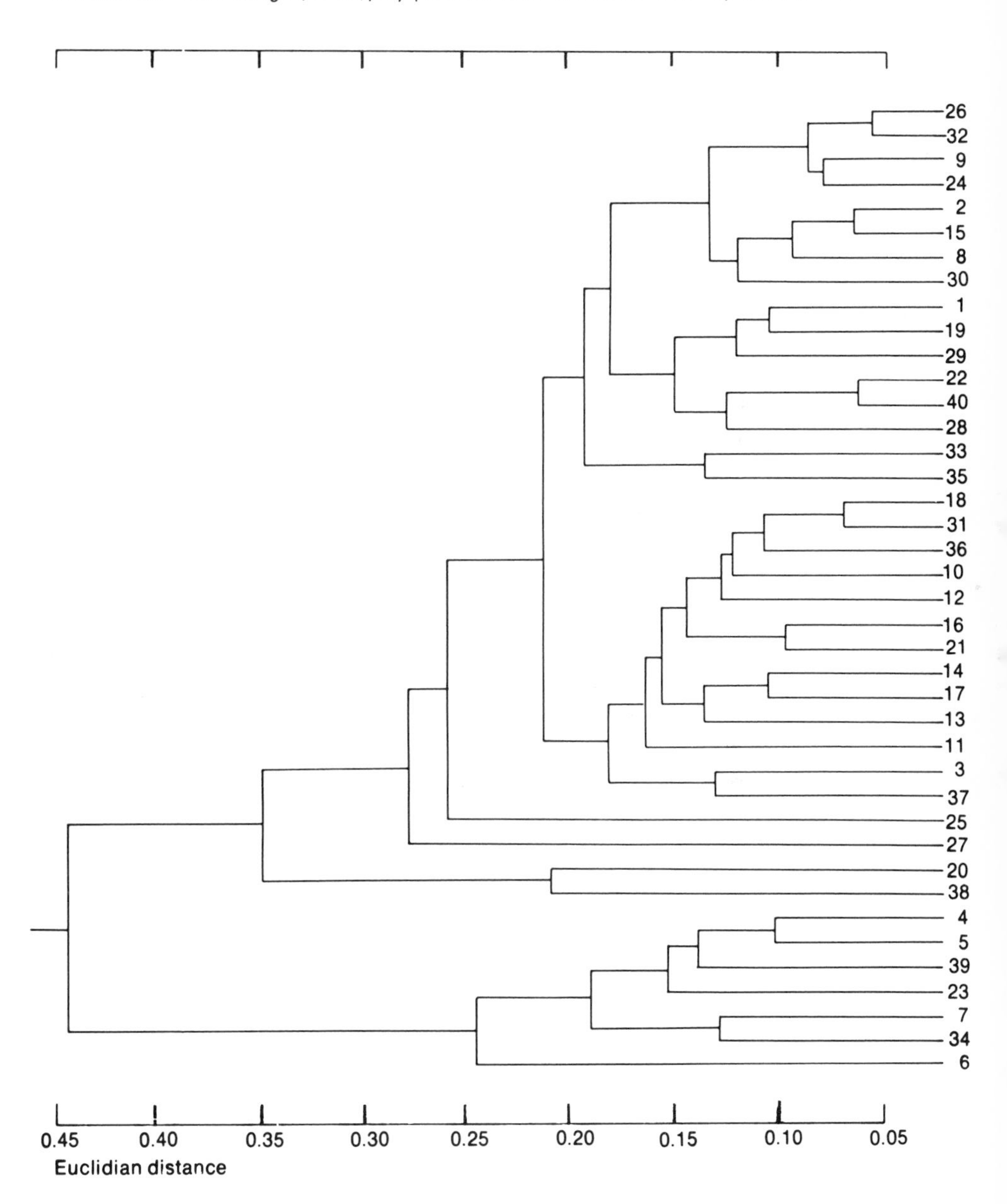

Units of soil and classification

In Chapter 1 an arbitrary definition of the soil individual - the pedon - was given. It is useful to consider the nature of the object to be classified:

1 / It is *anisotropic*; that is its properties vary in terms of direction as evidenced by the horizons in the solum.

2 / It is not a discrete body; only in exceptional instances does it occur with finite boundaries.

3 / Its physical, chemical, and biological properties are seldom static. Many vary both over short term periods (hours and days) and long term periods (months and years). Commonly over periods of several years (decades and centuries) soils are subjected to major disturbances both natural and man-caused.

4 / Soil is a component of the landscape but it should not be confused with it. Although the natural vegetation of an area is usually related to the landscape on which it occurs, a classification of vegetation does not involve the properties of the landscape. The same principle is true in a consideration of soils.

In addition to these attributes of soils, a major distinction in comparison with our knowledge of plants and animals and geological formations, is that soil science is still acquiring knowledge of 'new' soils in the same sense that the field of biology was characterized in the nineteenth century by the discovery of new plants and animals. It is difficult (and misleading) to create a system of classification that is 'all-embracing,' when only two-thirds of the population to be classified have been described. This is why regional or local classifications of soils for a limited but known population have been very functional whereas world-wide systems have been objects for debate.

The reasons for classifying soils are fivefold (Buol et al., 1973) - in order to: (1) organize knowledge; (2) bring out and understand relationships; (3) remember properties of the objects classified; (4) learn new relationships and principles in the population; (5) establish groups or classes of the objects in a manner useful for practical or applied purposes.

To arrive at a single classification to serve these purposes equally well is a formidable task and unlikely of achievement. It is as if von Linné had as a major consideration of the plant kingdom the possible uses to which each species might be put. It is most convenient if a classification can provide information that may be practical in the same way that a forest inventory can provide information useful in planning the utilization of species for a wide range of purposes, but the purposes should not determine the basis of a general, scientific study.

Given the basic nature of soils and the general purposes for classification, it can be realized that what is being categorized are 'classes' of properties - texture, structure, pH, and the like which we then relate to classes of soils based on morphology. Historically the recognition of morphology came first and it was after this that measurement of specific properties was made. In fact, we still indulge in this process of describing a profile and then determining its properties. As Bunting (1965) has pointed out, the basic concept underlying classification should be an upward and integrative one. Much of the earlier work especially was descending, something akin to building a temple and then looking around for a form of religious worship to put in it.

The measurements of individual soil properties come into intelligent use when they are viewed as statistics; soil pedons can then be considered as arrays of populations of properties about each of which quantitative statements can be made concerning their size. Quite clearly most soils are stratified - they have horizons - and this stratification provides a means for segregating the data of properties either within or between individual soil pedons. The pedon is then seen for what it actually is - a sampling unit - not a soil individual.

A classification should be of use to anyone studying soils in the field. If the identification of many soils is dependent on sophisticated laboratory measurements, it will find little general applicability. The array of properties which can be assessed in the field have been found to be the most generally useful ones in the methodical description of the overall morphology of the soil, and most classifications of any field value begin with the soil morphology as a first division. A numerical classification may begin this way as does Northcote's key (1965) to Australian soils.

Since the time of Dokuchaev most attempts at soil classification have emphasized factors of soil formation and, as a consequence, have considered properties which reflected modes of formation or are themselves the causes of other properties. Where the attempt is primarily to define the soil in terms of soil-forming processes (Jenny, 1941), the result is hardly a classification. The philosophic arguments for the selection of soil properties to be used in classification have been stated clearly by Robinson (1936) who also emphasized the provisional nature of soil classifications.

The objective evaluation of soil properties should provide the evidence from which we may deduce or infer modes and processes of formation, rather than try to place soils in schemes of classification which have been based on some preconceived notion of formative processes. The hypotheses concerning formation should have substantive bases and these are the quantitative measurement of soils.

The 7th approximation and soil taxonomy - a world system

THE BASIS

The 7th Approximation (1960) and its successor Soil Taxonomy (1975), represent a classification that focuses attention on the soils themselves in a quantitative rather than a qualitative manner. In attempting to provide a system for all

TABLE 9.1
Illustration of sequence of names at various levels in the 7th Approximation and Soil Taxonomy

Classification level	Name
Order	Spodosol
Suborder	Aquod
Great group	Humaquod
Subgroup	Alfic humaquod

soils in the world, it has been arranged so that it can accommodate new soils which may be discovered; associated with this is a certain flexibility of the scheme to allow for modification without confusion in terminology. In the selection of properties to be measured, those which affect or result from soil genesis were used. It might be mentioned that on this basis it would be difficult to reject any soil property. Where a choice is to be made between two properties of equal genetic significance, the one of greater importance to plant growth is to be used. Within the system, subdivisions in a given category are not based on a common property or set of properties but rather on the properties most useful for that particular division in the classification. For example, in one group of soils at a particular level of subdivision, the differential feature may be the presence or absence of a horizon of clay accumulation, but in other soils this feature may be the presence or absence of permafrost. The changing nature of soils has been recognized by attempting to ensure that the system is flexible enough to provide for such soils and bases the differentiation at most levels of the classification on horizons below the surface ones. This is the part of the soil likely to be altered least by factors such as fire, cultivation, and erosion which alter surface layers readily. The most obvious feature of this classification is the nomenclature. A new set of words was coined, mostly from ancient Greek and Latin roots, which could be used in most modern languages using the Latin alphabet. A rationale was adopted in the forming of terms so that the term indicates the level of categorization. The prime division is into ten orders; the name of each ends in 'sol.' The second level (suborder) name consists of two syllables, the first of which is suggestive of the property of the class and the second of the order in which it occurs. The next level of division is the Great Group, in which names are made by placing a prefix in the suborder name. Subgroups are named by placing a modifying adjective before the suborder name. An illustration of this sequence of terminology is given in Table 9.1. The adjective naming the subgroup may reflect a property characteristic of a great group in the same suborder, in a different suborder but the same order, or of the great group in another order. Adjectives may also reflect features not related to characteristics related to great group, suborder, and order.

DIAGNOSTIC HORIZONS

Many of the separations of soils into categories are based upon features of the solum and diagnostic horizons may be one of two forms depending on their vertical position.

Surface horizons (epipedons)

These include the upper part of the soil darkened by organic matter or the upper eluvial horizons or both. It may include part or all of a B horizon if darkening by organic matter extends from the surface into the B. Six types of surface horizons or epipedons are described: (1) *mollic*, (2) *anthropic*, (3) *umbric*, (4) *histic*, (5) *ochric*, (6) *plaggen*. The mollic and umbric horizons both contain organic matter but, whereas the mollic has high base (cation) saturation, the umbric is low in base saturation. Ochric horizons have little organic matter; histic horizons normally have high contents of organic matter and are seasonally saturated. The anthropic and plaggen horizons are man-modified by presence or cultivation.

Subsurface horizons

These horizons are generally synonymous with B horizons, but they may include part of the A horizon. There are 17 types: *argillic, agric, natric, sombric, spodic, placic, cambic, oxic, duripan, fragipan, albic, calcic, gypsic, petrocalcic, petrogypsic, salic,* and *sulfuric.* The features used to distinguish these horizons relate primarily to accumulation of clay and chemical compounds and physical form. For example, silicate clay accumulation is the dominant feature in argillic and natric horizons, but the natric has a columnar and prismatic structure and more than 15 per cent exchangeable sodium. The agric horizon has accumulation of both clay and organic matter and is formed under cultivation. The spodic horizon has accumulation of free sesquioxides (principally of iron and aluminium) and organic matter in various combinations and amounts. An oxic horizon has oxides of iron and aluminium, is at least 30 cm thick and has a high content of 1:1 lattice clays. A cambic horizon is one where there may be accumulation of iron, aluminium, sesquioxides, and some clays, but not enough to place it in other categories such as argillic or spodic horizons. A duripan is a silica-cemented horizon, while a fragipan is seemingly cemented when dry but only moderately to weakly brittle when moist. A placic horizon is usually a thin pan cemented by iron, iron and manganese or by an iron-organic matter complex. If horizons are cemented into hard massive forms by carbonates, they are termed petrocalcic; if

TABLE 9.2
Soil orders in 7th Approximation and approximate equivalents from an earlier classification

Soil Taxonomy and 7th Approximation	Approximate equivalent
1. Alfisols	Gray-Brown Podzolic, Gray Wooded, non-calcic Brown soils, Degraded Chernozem, and associated Planosols and some Half Bog soils
2. Aridisols	Desert, Reddish Desert, Sierozem, Solonchak, some Brown and Reddish Brown soils and associated Solonetz
3. Entisols	Azonal soils of little to no horizon differentiation and low humic gley soils
4. Histosols	Bog soils
5. Inceptisols	Ando, Sol Brun Acide, some Brown Forest low humic gley and humic gley soils
6. Mollisols	Chestnut, Chernozem, Brunizem (Prairie), Rendzinas, some Brown, Brown Forest, and associated Solonetz and Humic Gley soils
7. Oxisols	Laterites and latosols
8. Spodosols	Podzols, Brown Podzolic soils and Ground-Water Podzols
9. Ultisols	Red-Yellow Podzolic, Reddish-Brown Lateritic soils of the U.S., and associated Planosols and Half-Bog soils
10. Vertisols	Grumusols

the cementing materials are sulphates they are petrogypsic. Albic horizons reflect the light colour of primary sand and silt particles which remain after clay and free iron oxides have been removed or are segregated.

The diagnostic features for Histosols (organic soils) are depth and the nature of the organic materials. Three kinds of organic materials are distinguished: *Fibric* are the least decomposed and consequently have very low bulk densities. *Hemic* materials are intermediate in decomposition between the fibric and the most highly decomposed - the *Sapric* organic materials.

CATEGORIES

The ten orders which form the first division in the system are listed in Table 9.2. The basis for placing a soil in one of the orders relates primarily to the soil profile and usually the presence or absence of specified diagnostic horizons or features that are marks in the soil of differences in the degree and kind of the dominant sets of soil-forming processes that have gone on.

Alfisols show evidences of processes that result in the translocation of silicate clays without excessive base depletion and the absence of a mollic epipedon. Essentially, all soils which are Mollisols have a surface horizon which is either a mollic epipedon or meets its requirement to a depth of 18 cm, after mixing; but not all soils with mollic horizons are necessarily Mollisols. One order, Vertisols, is marked by processes that mix the soil regularly and thus prevent the development of diagnostic horizons which otherwise might develop.

The suborders, great groups and subgroups are differentiated by various criteria involving diagnostic and other horizons as well as specific physical or chemical properties. The pattern of nomenclature has already been described. It should be noted that keys to identification of orders and suborders, in addition to great groups and subgroups, are an integral part of this system.

The Canadian soil classification system

Prior to 1955 several systems of soil classification were used in Canada and were largely a reflection of the interest in soils developed on a provincial basis. Between 1955 and 1960 a proposed classification was worked on and adopted subject to modification for the next eight years. It was codified into a system in 1970, modified somewhat as a result of use and revised (Canada, 1974). This type of evolution even within one country is illustrative of the pattern of development of many soil classification systems.

Unlike the 7th Approximation, the Canadian System makes no attempt to be all-embracing and it may be assumed that the nature of the populations of soil which were to be included was generally known. There are three major divisions within the system - *order, great group*, and *subgroup.* The nomenclature used includes many of the terms which have historical precedents in naming soils, particularly in Europe. Although emphasis is placed on the horizon sequence and soil morphology, particularly at the order level, specific properties may be used to differentiate soils. A fundamental distinction between the Canadian system and the 7th Approximation is the manner in which soils (other than those which are organic) are dealt with where saturation with water, either continuously or periodically influences the profile. In the Canadian system such soils are placed in a separate order (gleysolic), whereas in the 7th Approximation they are segregated within each order at the suborder level. Within the Canadian classification, separations may be made on the basis of internal comparisons within a solum. For example, a Bt is defined by its clay content in relation to that in the upper

Ae and the definition of a spodic B is related to a comparison of oxalate-extractable iron and aluminium levels in the B and IC.

CATEGORIES

Orders. There are eight recognized orders; all soils within the first six are well to imperfectly drained.

1 / Chernozemic. These have Ah, Ahe or Ap horizons at least 9 or 15 cm thick respectively. The Munsell colour value must be one unit darker (moist or dry) than in the C and the values should be darker than 3.5 when moist and 5.5 when dry. The organic matter must not exceed 30 per cent.

2 / Solonetzic. The B horizons are characterized by a columnar or prismatic macrostructure which can be broken into a blocky meso structure. The ratio of exchangeable Ca^{+2} to Na^{+} is 10 or less. The C horizon is saline.

3 / Luvisolic. An eluvial Ae and an illuvial Bt characterize this order. The Bt must contain a minimum of 13 per cent clay and at least 3 per cent more than in the Ae if the clay content in the Ae is less than 15 per cent. If the Ae clay content is between 15 and 40 per cent, the Bt must be 8 per cent higher. In any situation the Bt should be at least 5 cm thick.

4 / Podzolic. Soils in this order have podzolic B horizons which are defined as being similar to spodic horizons of the 7th Approximation (i.e., an accumulation of iron and aluminium sesquioxides and/or organic matter).

5 / Brunisolic. These are soils which have (L-H) and Ah horizons and may have eluvial Aej or Ae horizons. All have a brownish Bm and none a Bt or spodic horizon.

6 / Regosolic. The horizon development is not sufficient that the soil can be placed in any of the other orders.

7 / Gleysolic. These are saturated soils and under reducing conditions seasonally or continuously during the year. The soil matrix colours within 50 cm of the surface are of low chroma and within this zone there must be a layer at least 10 cm thick which has certain specified soil colour chromas. The thickness of the surface organic layer must be less than 40 cm if the bulk density is greater than 0.1 g/cm^3 or less than 60 cm if the bulk density is less than 0.1 g/cm^3.

8 / Organic soils contain 30 per cent or more organic matter to (a) a depth of at least 60 cm if the surface layer is mainly fibric Sphagnum moss, or (b) a depth of at least 40 cm for other types of organic material, or (c) to a lithic contact at depths of greater than 10 cm but shallower than (a) or (b). Mineral material less than 10 cm thick may overlie the lithic contact but organic materials must be more than twice the thickness of the mineral layer; and (d) may have no mineral layer as thick as 40 cm at the surface or if less than 40 cm have at least 40 cm of organic soil beneath it; and (e) may have mineral layers thinner than 40 cm begin-

TABLE 9.3
Summary of Orders and Great Groups in the Canadian System of Soil Classification (1974)

Order	Great Group	Simplified basis for distinguishing Great Groups
Chernozemic	Brown	Ah $\geqslant$15 cm; Ah colour value when dry darker than 5.5
	Dark Brown	Ah $\geqslant$15 cm; Ah colour value when dry darker than 4.5
	Black	Ah colour value when dry darker than 3.5
	Dark Gray	L-H over Ah or Ahe
Solonetzic	Solonetz	AB transition abrupt ($<$2.5 cm); Bnt surface very hard
	Solod	Distinct AB transition ($>$2.5 cm) or Bnt breaks readily
Luvisolic	Gray Brown	Ah $>$5 cm
	Gray	L–H
Podzolic	Humic	Bh $\geqslant$10 cm; ratio, org. matter : free Fe $>$20 in B
	Ferro-humic	Upper B has $>$10% org. matter; ratio, org. matter : oxalate-extractable Fe $<$ 20 in B
	Humo-ferric	Upper B has $<$10% org. matter; ratio, org. matter : oxalate-extractable Fe $<$20 in B
Brunisolic	Melanic	Ah 5 $>$cm; mull; Bm base saturation (NaCl) 100%
	Sombric	Ah 5 $>$cm; moder; Bm base saturation (NaCl) 65–100%
	Eutric	L–H; pH(CaCl) $\geqslant$5.5; Bm base saturation (NaCl) 100%
	Dystric	L–H; pH(CaCl) $\leqslant$5.5; Bm base saturation (NaCl) 65–100%
Gleysolic	Humic Gleysol	Ah $>$8 cm
	Gleysol	Ah 0–8 cm
	Eluviated Gleysol	Aeg and Btg horizons; Ah may be present
Organic	Fibrisol	Dominantly fibric middle tier
	Mesisol	Dominantly mesic middle tier
	Humisol	Dominantly humic middle tier
	Folisol	L–H $\geqslant$10 cm; either lithic contact at $<$130 cm or L–H is on fragmental materials

ning within 40 cm of the surface within the organic soil. A mineral layer or layers cumulatively thinner than 40 cm may occur within the upper 80 cm.

Great groups and subgroups
Within each order there are several (usually not more than four) great groups. Their distinguishing features are in the main morphological and readily observed in the field, but in certain orders (the podzolic and brunisolic) chemical properties may be used. The orders and their great groups together with simplified distinguishing features are given in Table 9.3.

The subgroups within each great group are distinguished by very specific horizon designations, with some quantitative measures of certain chemical properties. For the most part the profiles can be identified and named on the basis of field descriptions.

Family and series

In both the 7th Approximation and the Canadian System the levels of classification below that designated as subgroups are in descending sequence *family* and *series*. The family has never been adequately defined, but has been generally considered to be a level in a system at which groups are designated largely on those properties that influence plant growth. Considering the large variation in plants and their requirements, a grouping on this basis would not appear feasible and this no doubt accounts for the relative lack of development of this level of classification. More recently stress has been placed on the family as a grouping of soil series with broadly similar attributes.

The series is a category that has been relatively well-defined and occurs in both the older and new systems of soil classification. A series consists of soils that are essentially alike in all major profile characteristics except the texture of the surface (usually the A horizon). Within a series the soil morphology - the nature and arrangement of the horizons - and physical, chemical, and biological properties are considered to be uniform and the solum is usually developed in similar geological materials. Although the definition of series as a classification unit is clear, the degree of uniformity of the soil properties which will place a soil in one or other series is the most critical feature to be established. Soil series are given geographic place names together with a textural description of the surface soil layer. For example, a Hinckley sandy loam, a Cecil clay loam are series names. The series name apart from texture is purely a label and it provides no information about the soil. Each soil series can be described using terminology at a higher level in the classification; thus a Hinckley can be described as a brown podzolic, the Cecil as a red-yellow podzolic or typic hapludult. In many soil surveys it is the soil series that are the basic mapping units.

10
Soil surveys

A soil survey involves the systematic examination, description, classification, and mapping of soils in an area. Although soil classification is an integral part of any soil survey, they differ in that a survey deals with the areal distribution of soils and classification does not. This basic difference is reflected in the fact that the scale of mapping must be decided before a survey can proceed. Once that is done, the information may be coalesced into groups of smaller scale, but the maximum scale is that at which the work is originally undertaken. Soil geography is, in fact, the study of soil distribution on a global scale.

Soils exist as part of the earth's landscape and the mapping of soil usually involves the recognition of landforms and other landscape features. Soil surveys may often include information about certain landscape features in addition to that of soils. Conversely surveys of land may be the vehicle for the inclusion of soil survey information.

A survey is an inventory of soils information and its use depends basically on two features. The first is the nature of the data in the inventory. For example, the distribution of soils by texture classes requires that the areas of sands, clays, loams be shown. A soil survey listing soils by great group designations will not be of much assistance for this purpose. The second feature is a knowledge of relationships between sets of soil information and between soil information and other sets of data such as plant growth, bearing strengths, microorganism requirements, and so on. If the amounts and nature of clay and organic contents are known from the data provided by a soil survey, some general statements can be made about the buffer capacity and cation exchange capacity. Inferences may be drawn about certain aspects of soil fertility in relation to a particular plant species and cultural treatments such as fertilization. These are interpretations of soil survey information and the larger the number that can be made and the greater the probability of their accuracy, the more useful the soil survey. The use to which a soil survey may be put depends not only upon the state of knowledge and relationships in the field of soils but also upon the knowledge from other fields - forestry, biology, agriculture, engineering, and so on - and their relationships to soils. Many of the disappointments experienced by persons from other disciplines when they try to interpret soil survey information for their own purposes results from a lack of understanding of their own discipline in relation to soils as much as from an inadequacy of the soil survey itself.

One important aspect of soil surveys is the manner in which hypotheses concerning soil formation and development may be more intelligently made once the spatial distribution of soils in the landscape is known. It is the link which the survey provides in relating soils to the landscape which makes this possible.

In summary, a soil survey of specified scale must provide objectively determined quantitative and qualitative information about soils and their distributions in a given area. It may also provide certain interpretative information in relation to various uses, but it should not do this at the expense of basic soil survey information.

Soils surveys and the landscape

The most commonly used range of scales at which soils are mapped in the field are: 1:10,000 to 1:60,000. These scales are convenient for the mapping of soils and landforms and they are also the range of scales at which soils and other inventory data such as vegetation types can be used in the general level of intensity of management practices. In areas where practices may be more intensive, larger scales of 1:5,000 or 1:2,000 may be desired; for general planning, small scales of 1:250,000 are usually satisfactory.

Assuming that the objective of a survey is to describe and map soils of similar morphology and identify them as series, the first step is to delineate the landforms in the area since this will be the most useful first stratification. A landform is defined by the geological nature of the materials which constitute it and by associated topography. Aerial photographs are the most useful tool for this purpose, together with any relevant geological information that is available. Various types of film may be used and each will have advantages or disadvantages for soil survey but, for general terrain features, black and white photography is usually adequate (Valentine et al., 1971). An example of an aerial photograph in which the landforms have been delineated is shown in Figure 10.1. The reconnaissance and establishment of the landform stratification must take place in the field and it is useful to examine and describe soils fully at this time. As the survey proceeds, not only will the differences in soil morphology between soils on the various landforms become evident, but also the variation within a landform. This variation may result from many different factors but, where the soils form a

Figure 10.1 Aerial photographs (stereopair) showing various landforms (photos courtesy, N. Keser)

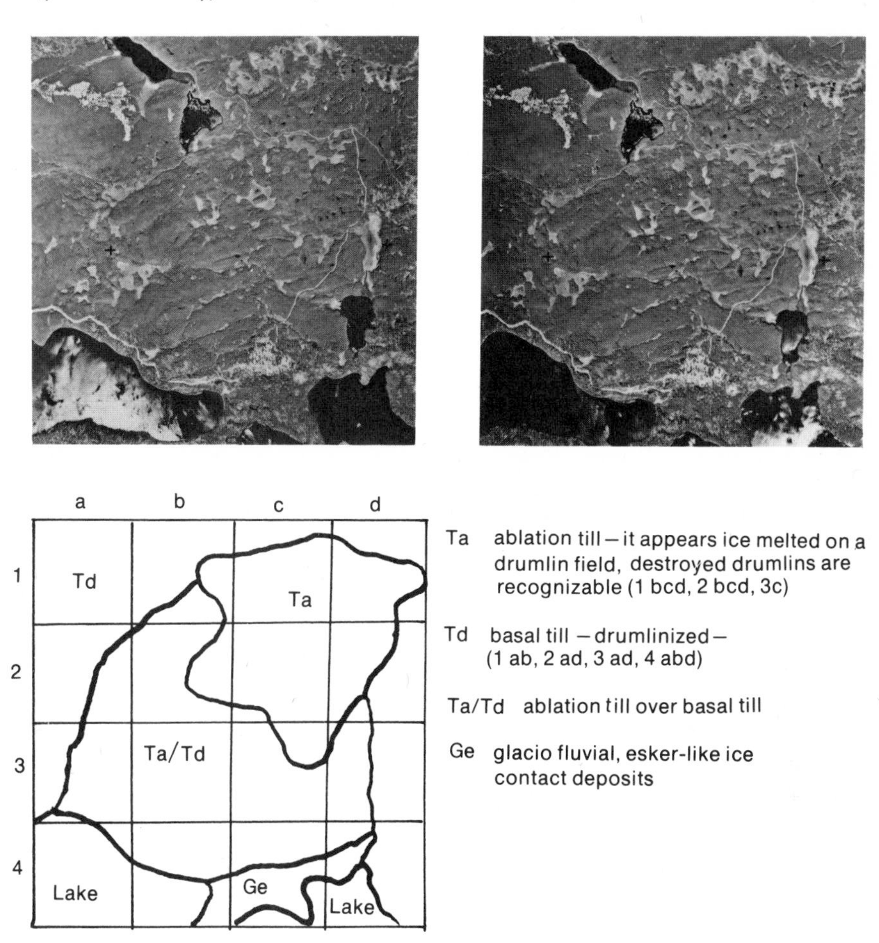

sequence due to variations in relief and drainage and they are about the same age, derived from similar parent materials and occur under similar climatic conditions, they are said to form a *catena.* Two or more series may constitute a catena and the concept is useful both in the survey work and for certain interpretive uses of the survey when completed. The definition of a particular catena does not imply that all the individual series which are its members will be present at each location. Erosion, changes in slope or depth to water-table will all vary and one or other components may not be present.

In the conventional soil surveys undertaken in North America soil series have been subdivided into *soil types*; the soil type name is the series name plus the textural class for the surface soil (the upper 15–18 cm – the zone usually cultivated in agriculture). Undoubtedly, textural differences within the solum are important, but whether the arbitrary setting of a surface 15–18 cm in depth for their delineation is justifiable for a soil survey of land whose uses may be largely non-agricultural is debatable.

Once the landforms and the catenas have been tentatively identified and described, the survey usually proceeds rapidly. In forested areas the major features which serve as a guide in designating the areas of soil series on aerial photos are drainage, relief, and vegetation. The soils with some exceptions are not mapped; only the surveyor's interpretation of visible features on the photographs are mapped and given soils' names. The basis for the interpretation is the surveyor's knowledge of the soils of the area and their relationships with landform, vegetation, and cultural or land use practices that may be present or have occurred in the past. The strengths and weaknesses of a soil survey are often a reflection of the degree to which these relationships are comprehended.

Frequently, it is while these relationships are being developed during the course of a survey that insights and evidence of the soil-forming processes occur. Sometimes a soil survey is undertaken from the outset to provide only limited information about soils in relation to one or a few uses as, for example, texture and depth to bedrock for road building purposes or specific soil properties which have been correlated to the growth of a tree species. Such a survey not only will be of limited use but it is unlikely that relationships or insights will be possible to the degree that they are when a survey based on soil morphology and associated properties is undertaken. Within a conventional survey, a subdivision of the mapping unit, soil series, or type may be undertaken in relation to some feature that may be important in management, or some use, actual or anticipated. The degree of stoniness, slope, or amount may be indicated as a soil *phase.*

When a soil survey is completed, it should consist of an appropriate description of the area, its geology, and any other relevant information historical or otherwise, descriptions of the soils mapped and their properties, together with an account of the relationships which were established and used in the making of the survey. The precise nature of much of this type of information is given in the United States Soil Survey manual (Soil Survey Staff, 1951) and in Clarke (1957). In many soil survey reports it is customary to include information and particularly interpretations of the soil information in relation to various uses. These may encourage a wider usage of soil surveys by non-soil scientists and are very desirable.

Soils may be described as soil *associations.* These are groupings of soils, with the definition of the grouping variable. In Canada a soil association is synonymous with the term catena (Anon, 1970). In the United States, association has two meanings (Soil Science of America, 1973): (1) a group of defined and named taxonomic (Classification) units occurring together in an individual and characteristic pattern over a geographic region; and (2) a mapping unit used on general soil maps in which two or more defined taxonomic units are combined because the scale of the map or the purpose for which it is being made does not require delineation of the individual soils.

Examples of soil surveys

The nature and extent of soil surveys vary considerably. In undeveloped regions of large size and sparse population it may be necessary for economic reasons to obtain as much information as possible about many landscape features of which soil is only one. Such surveys are often extensive and the information may be presented at a small scale. In contrast to this is the survey made to provide information at a large scale in which variations in soil properties within a soil series may be recognized and mapped; such surveys are usually limited in extent because of high cost, but they may be undertaken to determine the extent of variation in soil properties within a limited area as well as for some specific use.

The following examples are taken, with one exception, from published reports and the range is by no means all-inclusive.

COMPREHENSIVE SURVEYS WHICH INCLUDE SOIL SURVEYS

In several countries, particularly those with large land areas much of which may be undeveloped, surveys usually at relatively small scales are made. These include landform vegetation and soils. The smallest unit is termed a *land component* and it is defined as an area where climate, geological material, topography, soil, and vegetation are uniform within limits significant for a particular form of land use. Individual land components are often too small to be mapped at the scale of mapping used, 1: 125,000 or smaller, and it is possible to delineate patterns of components as a *land-unit.* For large areas and at smaller scales, where many land-units are involved, a less intensive subdivision is made into *land-systems.* At the smallest scale, similar land-systems may be grouped into *land-zones.*

A diagrammatic illustration of a land system for an area in the southwestern portion of the state of Victoria, Australia is shown in Figure 10.2 (Gibbons and Downs, 1964). The map sheet for this area shows land systems at a scale of 1:250,000.

Figure 10.2 Diagrammatic representation of the Nelson Land System in south-western Victoria, Australia. From F.R. Gibbons and R.G. Downes, *A study of the land in southwestern Victoria* (1964), by permission Soil Conservation Authority, Victoria

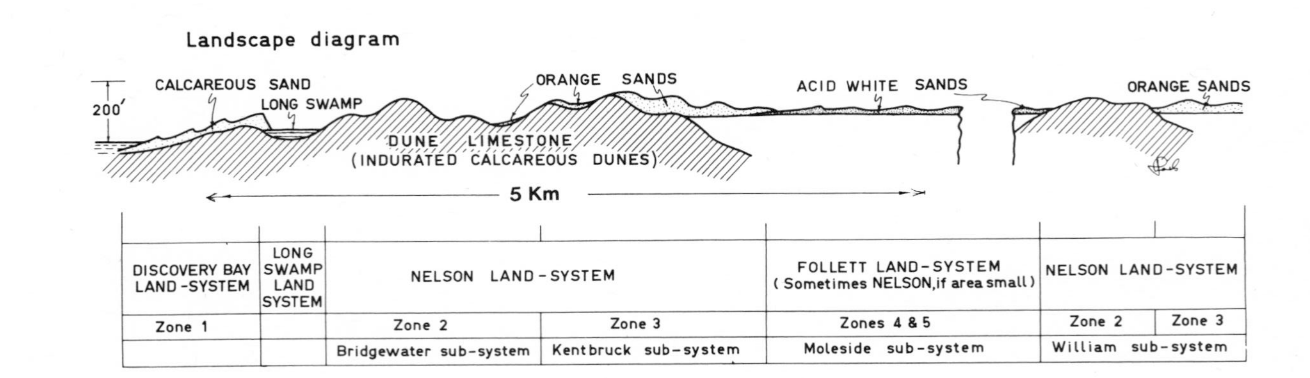

DISCOVERY BAY LAND-SYSTEM	LONG SWAMP LAND SYSTEM	NELSON LAND-SYSTEM		FOLLETT LAND-SYSTEM (Sometimes NELSON, if area small)	NELSON LAND-SYSTEM	
Zone 1		Zone 2	Zone 3	Zones 4 & 5	Zone 2	Zone 3
		Bridgewater sub-system	Kentbruck sub-system	Moleside sub-system	William sub-system	

Land-system diagram

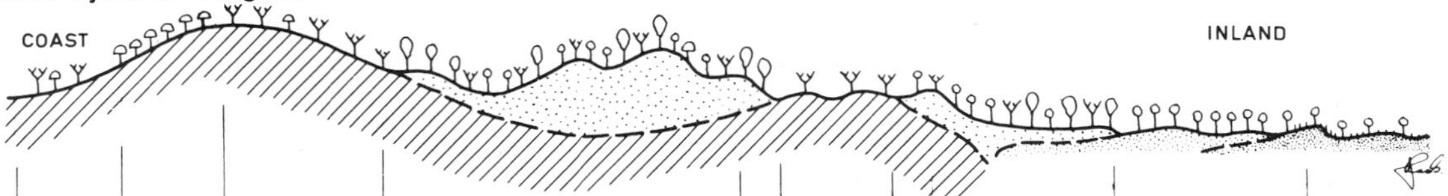

COAST — INLAND

		Zone 2			Zone 3						Zone 4	Zone 5
		Bridgewater sub-system			Kentbruck sub-system			William sub-system			Moleside sub-system	
CLIMATE		70-90 cm average annual rainfall, most areas receiving above 75 cm ; marked winter incidence ; warm summers, cold winters, exposed in parts										
PARENT MATERIAL		Indurated calcareous dunes (" dune limestone ")			Orange sands			Dune limestone	Orange sands ←——→ intermediate ←——→ acid white sands			
TOPOGRAPHY	Land-form	Steeply rolling						Gently rolling			Undulating	
	Position	Lower slopes	Exposed position	Middle slopes	Swale	Complex of summits and slopes		Complex of summits and slopes				
SOILS	Sub-group	Rendzina		Terra rossa	Sandy iron leptopodsol		Sandy brown earth	Terra rossa	Sandy brown earth	Sandy iron leptopodsol	Intermediate	
	Type, Series or Family	*Bridgewater sandy loam*		*Nelson sandy loam*	*Kentbruck loamy sand*			*Woakwine sandy loam*		*Kentbruck loamy sand or Caroline sand*	" Grey-brown variant of *Caroline sand* "	
	Features	Very dark brown friable sandy loam surface soil with good crumb structure, merging into brown loamy sand at depth ; limestone floaters present ; abruptly overlying broken limestone at variable depth		Dark reddish-brown surface soil, friable, porous, becoming less dark with depth ; abruptly overlying hard limestone with an irregular boundary	Dark greyish-brown friable and porous loamy sand surface soil, becoming yellowish brown at 23 cm with coarse brown occlusions ; then merging gradually into light brown sand ; overlying at 120 cm strong brown massive loamy sand which extends to 9 feet ; there overlying mottled grey and brown sandy clay			As Nelson sandy loam		As to left	Dark grey sand A_1 merging into light brownish-grey sand A_2 ; B_1 horizon at 90 cm of brown sand lightening in colour with depth, overlying mottled grey and brown sandy clay at 2.7 m	
VEGETATION	Formation	Tall woodland	Dry scrub	Tall woodland	Tall woodland or dry sclerophyll forest			Tall woodland		Tall woodland or dry sclerophyll forest		
	Alliance	*E. viminalis —E. ovata*	*Melaleuca pubescens*	*E. viminalis —E. ovata*	*E. baxteri*			*E. viminalis —E. ovata*		*E. baxteri*	*E. baxteri*	
	Association or Chief Species Present	*E. viminalis A. melanoxylon Casuarina stricta*	*Melaleuca pubescens*	*E. viminalis A. melanoxylon C. stricta*	*E. baxteri E. viminalis C. stricta, B. marginata, Poa australis*		*E. vitrea*	*E. viminalis A. melanoxylon E. pauciflora*	*E. vitrea*	*E. baxteri E. viminalis B. marginata Poa australis*	*E. baxteri Pteridium aquilinum L. juniperinum*	
LAND-USE	Potential	Cross-bred wool-growing with or without fat lambs			Mixed farming, pine-growing, S.Q.R. II to IV			As to left	As to left		Possibly pines, not farming	
	Present	Some rough grazing ; Merino and cross-bred wool-growing			Rough grazing, a little mixed farming, some pine-growing			As to left	As to left		Native hardwood	
EROSION	Hazard	Low wind erosion			Some low to moderate wind erosion			As to left			Low to moderate wind erosion	
	Actual	Low wind erosion			Some low to moderate wind erosion			As to left			Low to moderate wind erosion	
PROBLEMS		Deficiencies of cobalt, copper and zinc ; exposed positions ; shallow soils			Deficiencies of copper and zinc			As to left			Low fertility ; query pines	Pines doubtful

Follett and Kanawinka Land-systems

In Canada a Bio-physical Land Classification (Lacate, 1967) has been used to differentiate and classify ecologically significant segments of the land surface in order to satisfy the need for general inventory purposes. The unit of land mapped at a scale of 1:250,000 is termed a bio-physical unit and it may be subdivided into components (not usually mapped) consisting of combinations of soils, relatively uniform in morphology and geological materials, vegetation, and moisture regime. The bio-physical units may in turn be grouped into bio-physical systems, each of which consists of patterns of complex units. The system may be combined at smaller scales to form regions. An example of a bio-physical classification for a large area is that of Jurdant et al. (1972) for a portion of Quebec.

Combined soil vegetation surveys have been made in California in which soils are classified on a soil series basis but, in addition, vegetation is categorized by species and abundance, and site quality for dominant tree species is also determined. Zinke and Colwell (1965) have reviewed some of the general relationships among forest soils for this area and noted that for each parent rock type there was a predictable sequence of soil profiles of increasing stage of development. The recognition of stage of development could be based on field features of colour, stoniness, and clay content.

GENERAL SOIL SURVEY - USDA

The soil survey for Scotland County, North Carolina (Soil Survey Staff, 1967) covers an area of 82,102 ha of which more than half (55,919 ha) is in woodlands and uses other than cropland or pasture. The report presents a general map of the area (scale 1:190,080) showing the extent of five soil associations; an association is considered to be a landscape with a distinctive pattern of soils. Each association can be represented diagrammatically to provide a visual image of the land (Figure 10.3). The soil series and slope classes are shown at a scale of 1:15,840 on monotone aerial photomosaics (Figure 10.4). The use of photomosaics as a mapping base has a number of advantages over the base map sheet. Land use patterns as well as the normal landmarks (roads, railways, etc.) are clearly evident. The type of vegetation and general species composition may also be interpreted. The major portion of the report consists of a general description for each soil series and a more complete description of a representative profile for the series and a section on the use and management of soils for crops and pasture, woodlands, wildlife, and engineering uses. This section on use and management is primarily interpretive but does provide some limited data on physical properties, especially those that are relevant to engineering uses. A final short section deals with the general nature of the county, its geology, physiography, history, land use development, and climate. The report does not provide any routine data from physical, chemical, and biological analyses for each soil series.

Figure 10.3 Soils of the Marlboro-Norfolk-McColl Association in Scotland County, North Carolina. From Soil Survey of Scotland County, North Carolina, by permission United States Department of Agriculture, Soil Conservation Service

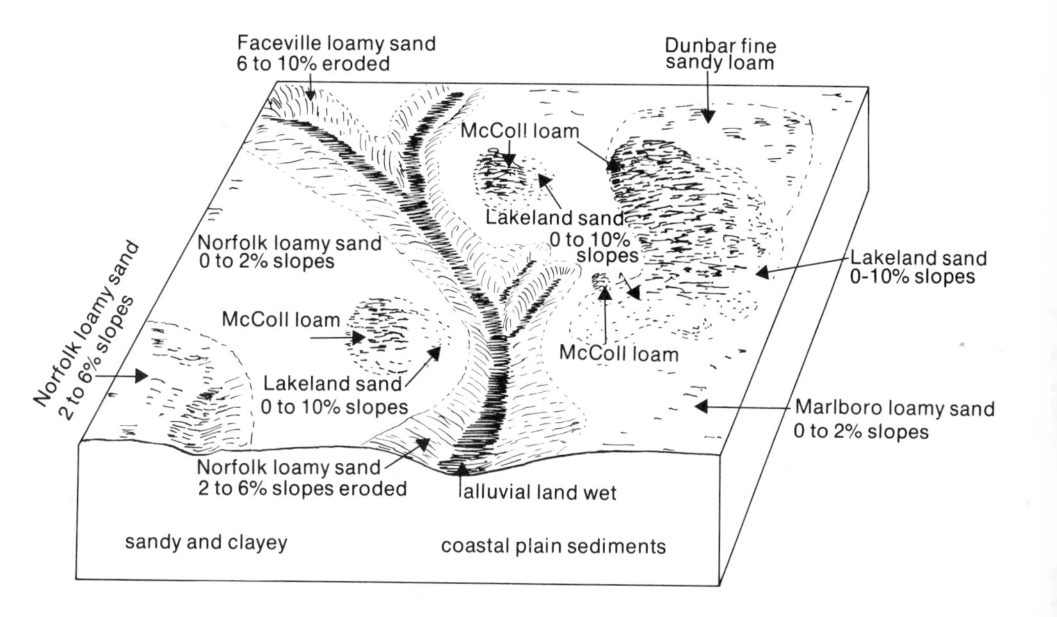

GENERAL SOIL SURVEY – CANADA

The soil survey of southeast Vancouver Island and Gulf Islands, British Columbia (Day et al., 1959) is for an area of approximately 287,690 ha, much of which has been cleared of forest for agriculture and is in the coastal plain on the west side of the insular mountain zone. The report includes a description of the area-location, history, natural resources and industries, population, transportation, social services and agricultural extension services, physiography, relief and drainage, climate, and native vegetation. The geology and general relationship between geological materials and the occurrence of soil series is portrayed diagrammatically. Following a brief description of the great soil groups represented in the area, soil profile descriptions are given for each series; a short section follows which deals with the agricultural use of the soil. Results of certain physical and chemical analyses for representative profiles are given. The soil series are mapped at a scale of 1:63,360 (Figure 10.5) and descriptive information about the soils is given on the map sheet legend.

Figure 10.5 Portion of a map sheet for a soil survey for southeastern Vancouver Island. From J.H. Day, L. Farstad, and D.G. Laird, *Soil Survey of southeast Vancouver Island and Gulf Islands, British Columbia* (1959).

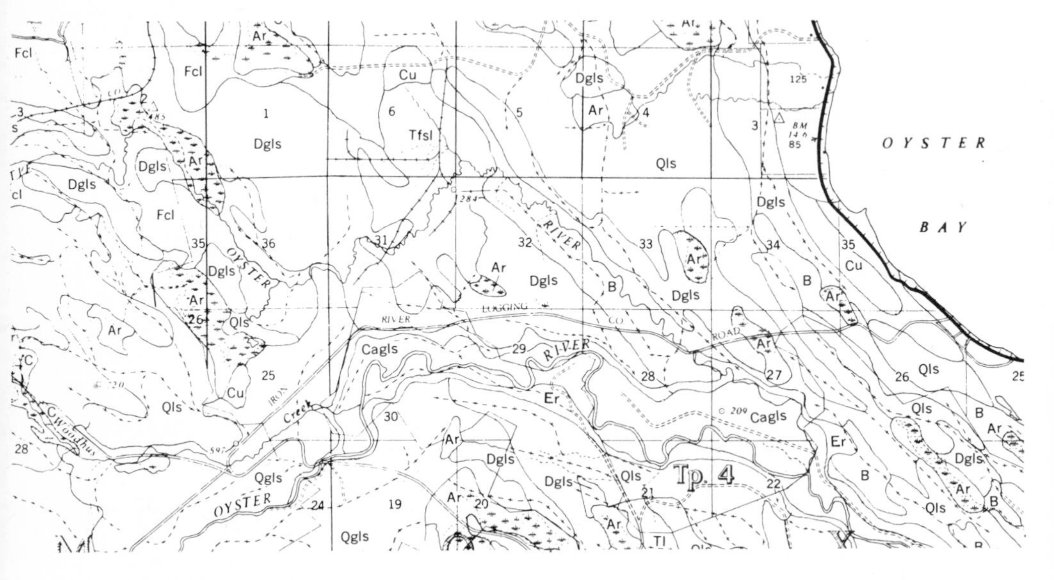

Soil Series
Ar: Arrowsmith, peat
B: Bowser, loamy sand
Cagis: Cassidy, gravelly loamy sand
Cu: Custer, loamy sand
Dgls: Dashwood, gravelly loamy sand
F: Fairbridge, silt loam to silty clay loam
Qls: Qualicum loamy sand
Tfsl: Tolmie fine sandy loam
Er: eroded land

FOREST SOIL SURVEY - JAPAN

Forest soil survey reports and maps published by the Government Forest Experiment Station of Japan provide illustrations of forest soils mapped at a scale of 1:20,000 and growth data for tree species. The soils are categorized on the basis of profile morphology. At a broad level they are distinguished by essentially great group categories - podzols, brown forest, black soils. Within each of these major groupings they are subdivided by soil horizon development which may reflect differences in moisture regimes or other properties. Descriptions and scale drawings of the profiles are important features of the report together with physical and chemical analytical data for the profiles. Figure 10.6 shows drawings of a

Figure 10.4 A portion of survey sheet for Scotland County, North Carolina, showing soil series and slope classes. Scale 1 : 15,840. From Soil Survey of Scotland County, North Carolina, by permission United States Department of Agriculture, Soil Conservation Service

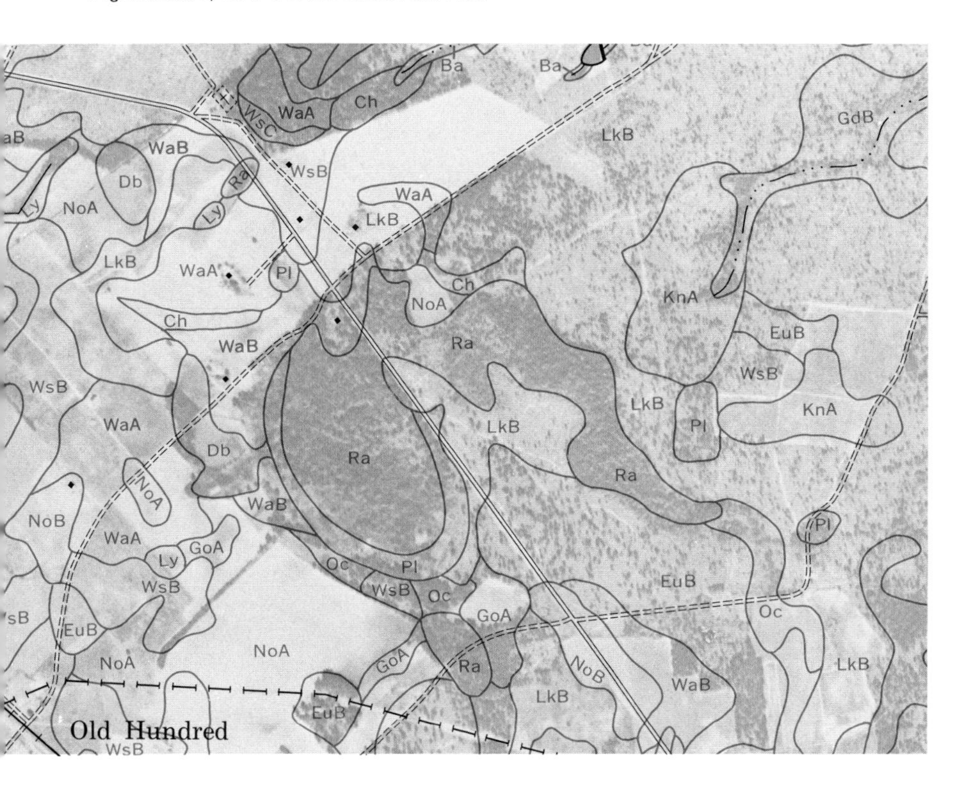

Figure 10.6 Diagrammatic illustrations of a brown forest and podzol profile from Yogawa National Forest, Japan. Adapted from S. Hayashi and A. Shiomoide in *Forest Soils of Japan* (1954)

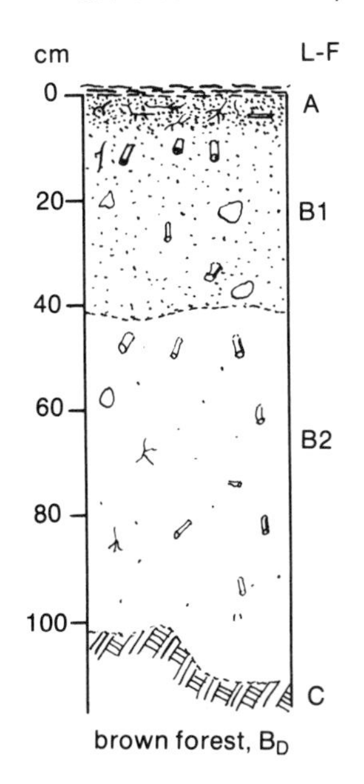

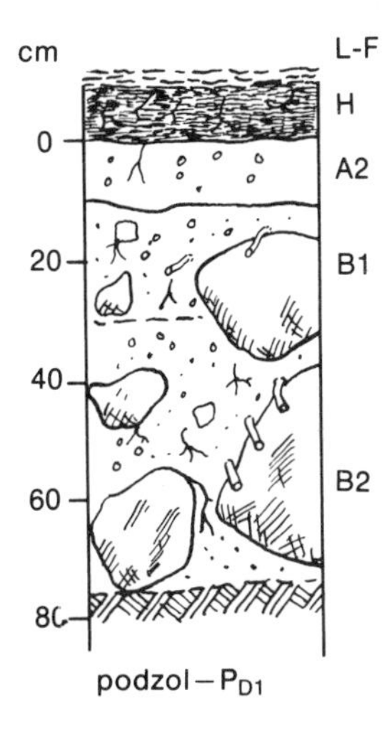

brown forest and a podzol profile. Intensity of rooting, nature of horizon boundaries and abundance of coarse fragments are immediately apparent.

An integral part of these reports is the progression of growth of some of the major tree species for the major soils described. Usually, height, diameter, and volume growth curves with age are given on both a total and periodic basis. Examples of curves for sugi on two brown forest soils (B_C and B_D) in the Tokyo National Forest are given in Figure 10.7.

BOREAL FOREST SOIL SURVEY

In areas where staff and facilities to undertake a conventional soil survey are limited, there may be a need for a simple inventory of soils information which may be used in forest management. Such an inventory can be provided so that it meets two objectives: (1) provide objective information about the soils which may be utilized in management; (2) assemble soils data to provide a basis for more detailed soil surveys which may be undertaken in the future.

An example of such a survey is one that was undertaken in the boreal forest of northeastern Ontario. An area of approximately 121,400 ha was surveyed to

Figure 10.7 Height and volume growth curves over age for sugi (*Cryptomeria japonica*) growing on two brown forest soils (B_C --- and B_D —) in Tokyo National Forest. Adapted from T. Kizaki and T. Watanabe in *Forest Soils of Japan* (1954)

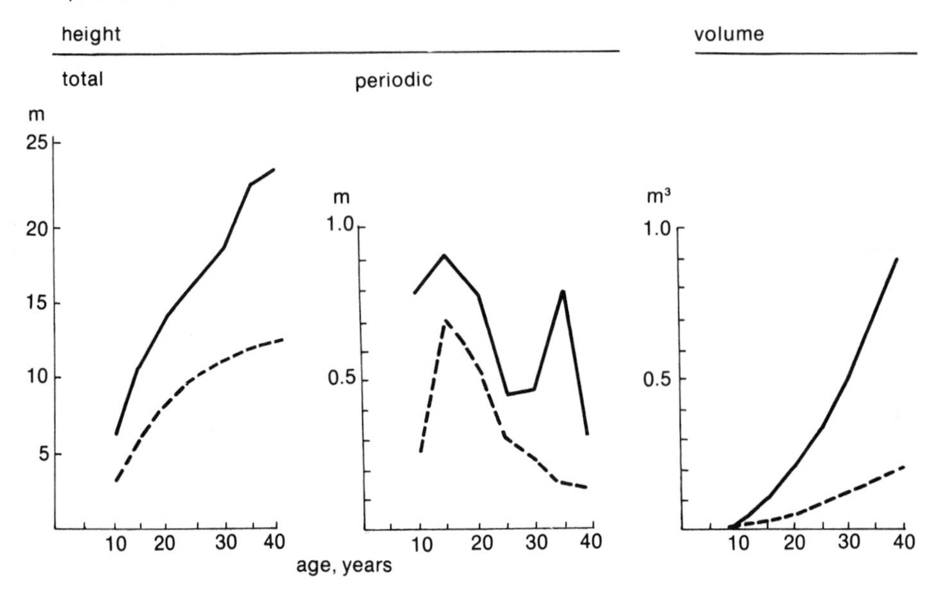

TABLE 10.1
Categories used in boreal forest soil survey and mapping symbols

Soil materials		Soil drainage	Slope
Gravel	G	*Well-drained* – w;	
Coarse sand	cSa	no evidence of gleying	0 - 5% —
Medium sand	mSa	or mottles in solum	
Fine sand	fSa	or upper C horizon	
Sandy loam	SaL	*Moderately well-drained* – m;	
Loam	L	some mottles in lower part	5 - 30% ⌒
Silt	Si	of B or upper C horizons	
Silt loam	SiL	*Imperfectly well-drained* – i;	
Clay	C	mottles common in B, some mottles	
Clay loam	CL	in Ae and C horizons	>30% Λ
Bedrock	R	*Poorly-drained* – p;	
Organic	O	solum strongly gleyed	
		Very poorly-drained – vp;	
		water table within 30 cm of surface	
		most of the year	

Figure 10.8 Diagram of representative soil materials in Davidson Township, Ontario. For key to symbols see Table 10.1.

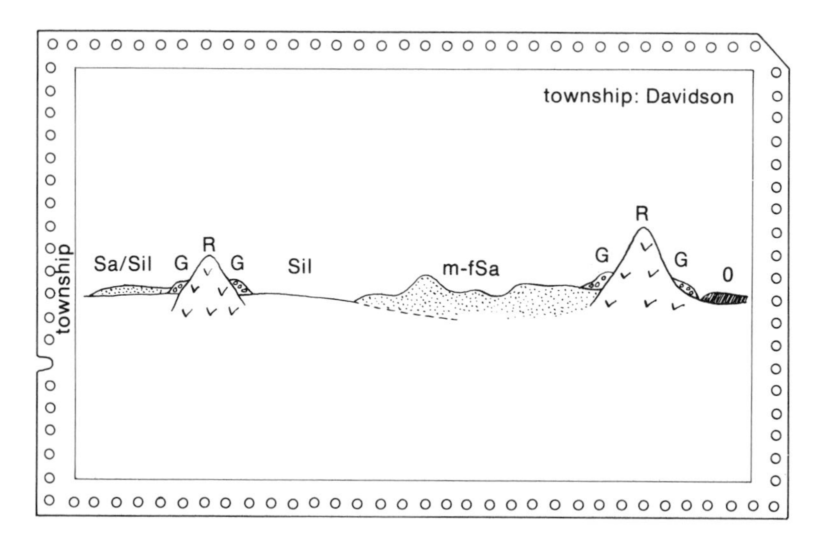

provide data on soil materials, soil drainage, and slope. The details of these categories are given in Table 10.1. The various materials were mapped at a scale of 1:15,840 on aerial photographs and detail was then transferred to base map sheets. Areas less than 20 ha were not delineated and were estimated to the nearest 10 per cent and these values were incorporated into the symbol for each unit mapped. For example a map symbol might be

$$\frac{\text{SiL}_8\ -\ ;\ \text{R}_2\Lambda}{\text{m}_6\text{i}_4\ -\ ;\ \text{w}_{10}}$$

This would indicate that 80 per cent of the area is silt loam with slopes of 0–5 per cent and that 20 per cent of the area is rock outcrop with slopes greater than 30 per cent. In the area of silt loam, 60 per cent is moderately well-drained and 40 per cent is imperfectly drained. The area of rock outcrop is all well drained. In addition to the maps, a diagram of the soil materials was portrayed on a needle-sort card for each township. This enables a manager to obtain a visual image of the land at a glance (Figure 10.8). Further, the information for each map type was placed on a needle-sort card so that selection of areas by location, soil materials, and drainage class for such a large area could be done in a few minutes by the manager on location. These cards also permit the addition of information about management practices in relation to soil mapping units.

Figure 10.9 Profile in dune sand under Scots pine and broadleaved trees. Note buried organic layers. From J. Schelling *Stuifzandgronden Inland-dune sand soils* (1955), by permission

During the field work, full soil profile descriptions were made for each of the classes of soil materials and drainage and these are not only reference value for current use but provide a base for any future soil survey in which soil series may be delineated.

THE NETHERLANDS – A LARGE-SCALE SURVEY

An example of a large-scale soil survey is that of Schelling (1955) which provides information for soils developed in blow sand in the Netherlands. Differences in the soils were reflected by dramatic differences in growth of tree species used to reforest the area. The survey was concerned with quantitatively evaluating the soil properties and their distribution responsible for the tree growth differences.

The origin of the dune development began with the removal of forest cover by man and the restriction on vegetation by grazing of livestock. As the sands moved, the slightly moister depressions, usually vegetated, were filled with drift sand from the surrounding drier, bare areas. During periods of less active movement, vegetation became established and formed horizons of organic accumulation which may subsequently be buried or exposed and wind eroded (Figure 10.9). The higher moisture-holding capacity and fertility (primarily in nitrogen and phosphorus levels) of soils with increasing humus content was established by analysis. Other factors were the presence or absence of a buried profile (podzol) and the presence of a gleyed zone in the soil. The three major categories of soils which were then mapped at a scale of 1:2500 were (1) blow sand over a C-layer; (2) blow sand over a buried profile, and (3) blow sand with gley on a buried soil. Within each one of these three categories, the areas were subdivided by depth classes of blow sand and relative humus content. A portion of a completed map is shown in Figure 10.10.

The use and interpretation of the survey is for growth of conifers, particularly Scots pine, and an illustration of the height growth of this species in relation to type of subsoil and colour value (degree of darkness which had been previously correlated with organic matter content) is shown in Figure 10.11. This illustration is simplified from the original map which includes areas of shallow and deep tillage and also the areas with profile development to 30–50 cm.

Interpretation and use of soil surveys

There are an infinite number of interpretations and uses to which soil survey information may be put. It has already been mentioned that the type and amount of data and the scale at which the information is available are important factors when use is to be made of a soil survey.

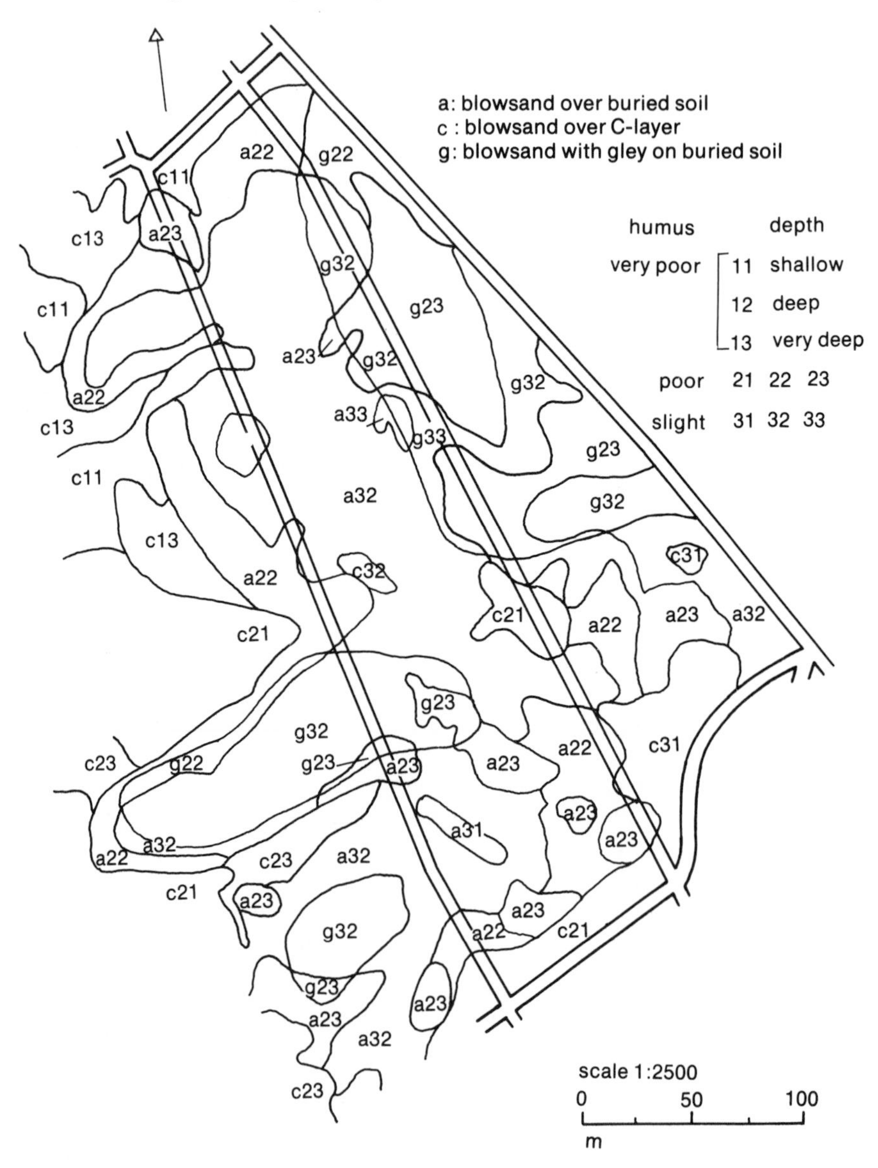

Figure 10.10 Portion of detailed soil map for 'Nunspect,' Netherlands. Adapted from J. Schelling, *Stuifzandgronden, Inland-dune sand soils* (1955).

Figure 10.11 Relations between tree height of 45-year-old Scots pine and depth of blowsand: (a) blowsand over buried profile, (c) blowsand over a C-layer. Munsell Color value numbers 3-6. Adapted from J. Schelling, *Stuifzandgronden Inland-dune sand soils* (1955)

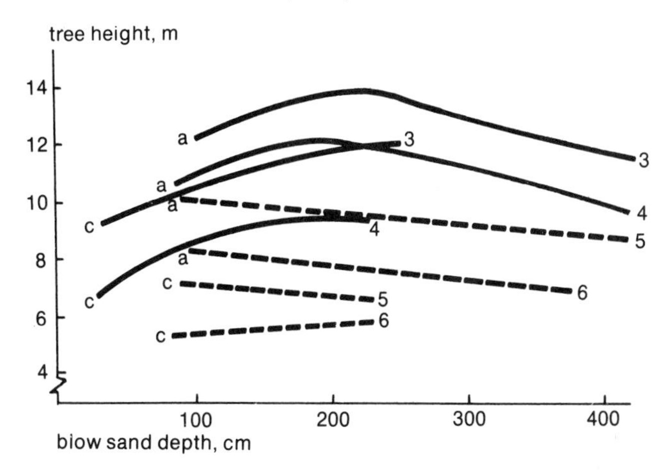

One of the common interpretations is that for *capability*. Capability classes have been established primarily on the basis of limitations that a soil may have for agricultural use. The placing of a soil in any capability class will depend on the current state of knowledge and agricultural practices and will change with time. Each capability class may contain soils of quite different kinds, and management practices and treatments may be quite dissimilar. In the United States eight capability classes are designated and in Canada seven are used. For both systems class I lands are those with soils that have no significant limitations for agricultural use. They are productive soils, level or with only gentle slopes, relatively free from erosion hazards. Limitations and erosion hazards increase with capability class number. The first three classes of the USDA system are lands suitable for cultivation; class IV is suitable only for limited production and classes V to VIII are unsuitable for cultivation, but may be suitable for forestry, grazing, or wildlife purposes.

The interpretations of soils information may vary considerably from region to region and the types of correlations which are established and function well in one area may not be useful in another. Coile (1948) developed mathematical relationships for the estimation of site index values (mean height of dominant and co-dominant trees at a specified age, usually 50 or 100 years) in relation to

Figure 10.12 Relation of site index to soil drainage class. Dotted lines indicate 95% confidence limits; dashed lines 90% interval for prediction from a single observation. Drainage classes: p – poor, i – imperfect, m – moderate, w – well, and e – excessive. Figure reproduced from Experiment Station Bulletin 1020, 'Influence of Soil and Site on Red Pine Plantations in New York,' Part II, soil type and physical properties, by J.A. DeMent and E.L. Stone, published by the Agricultural Experiment Station of the New York College of Agriculture and Life Sciences, a Statutory College of the State University, Cornell University, Ithaca, New York

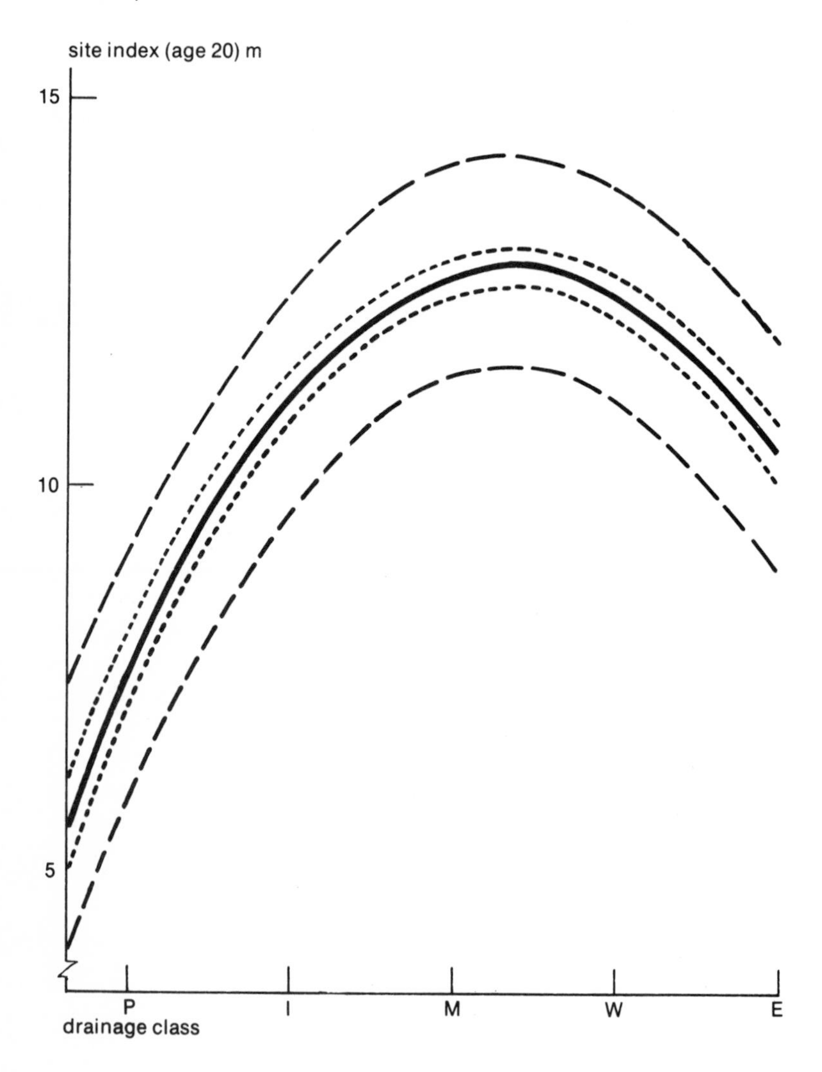

Figure 10.13 Soil map showing soil series for a 405-ha tract, Aroostock County, Maine. From P.H. Montgomery and F.C. Edminster in *Soil Surveys and Land Use Planning* (1966), by permission Soil Science Society of America

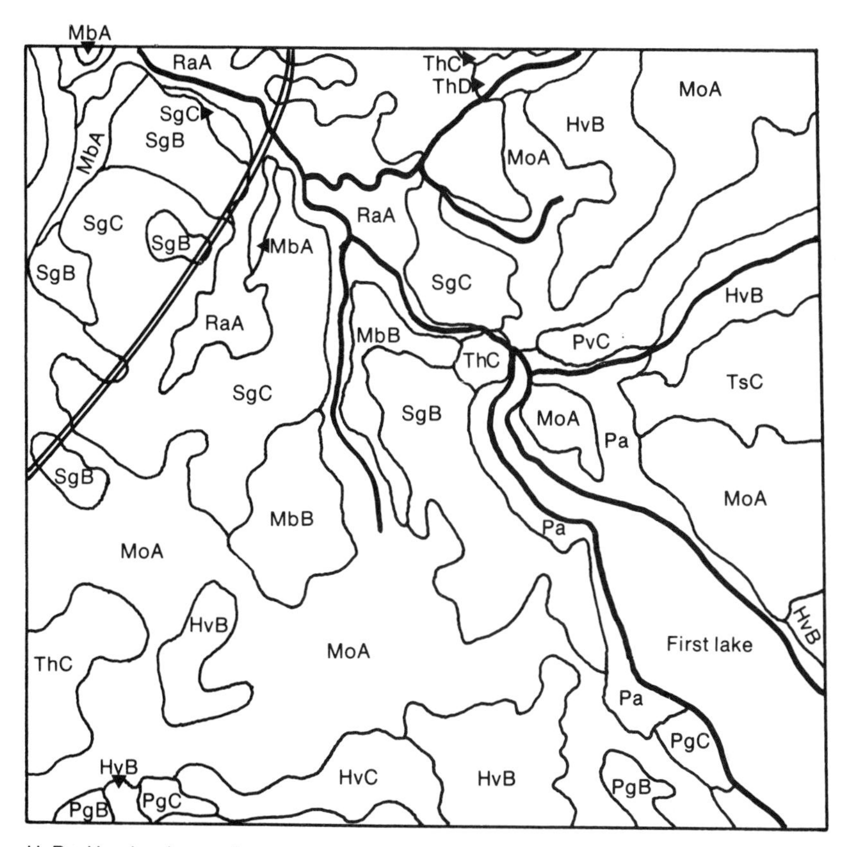

HvB – Howland very stony loam, 0-8% slopes
HvC – Howland very stony loam, 8-15% slopes
MbA – Madawaska fine sandy loam, 0-2% slopes
MbB – Madawaska fine sandy loam, 2-8% slopes
MoA – Monarda and Burnham silt loams, 0-2% slopes
Pa – peat and muck
PgB – Plaisted gravelly loam, 2-8% slopes
PgC – Plaistead gravelly loam, 8-1.5% slopes
PvC – Plaistead and Howland very stony loam, 8-15% slopes
RaA – Red Hook and Atherton silt loams
SgB – Stetson gravelly loam, 2-8% slopes
SgC – Stetson gravelly loam, 8-15% slopes
ThC – Thorndike shaly silt loam, 8-15% slopes
ThD – Thorndike shaly silt loam, 15-25% slopes
TsC – Thorndike and Howland soils, 8-15% slopes

Figure 10.14 Map same area as shown in Figure 10.13, illustrating soil limitations for intensive picnic use. From P.H. Montgomery and F.C. Edminster in *Soil Surveys and Land Use Planning* (1966), by permission Soil Science Society of America

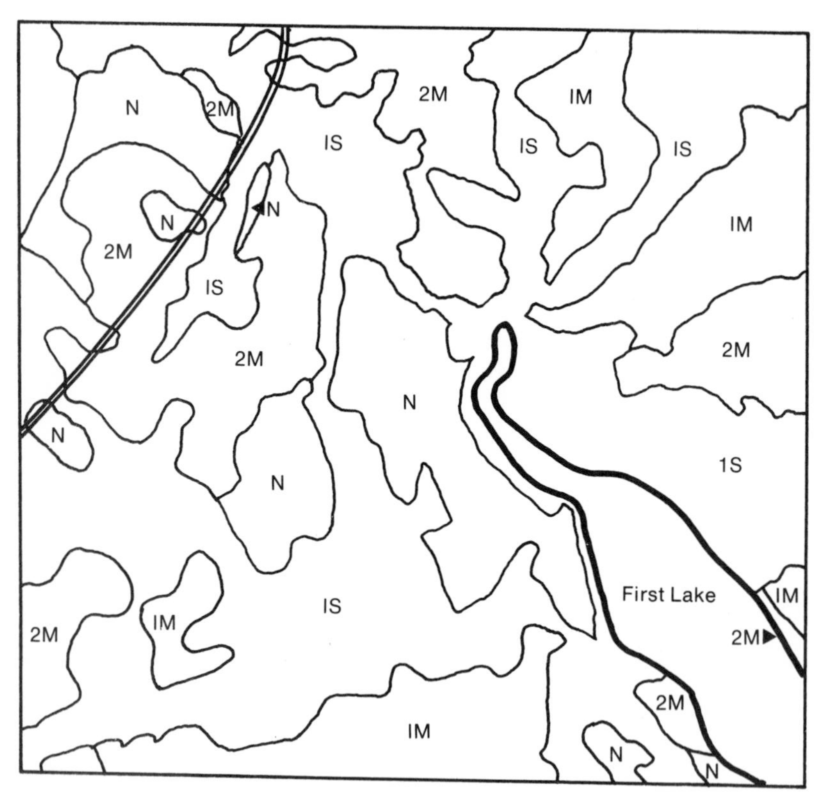

Soil limitations for picnic areas (intensive use)
None to slight
N well-drained, level and gently sloping soils
Moderate
1M Very stony, sloping (2-15%) soils
2M Sloping (8-15%) soils
Severe
1S wet soils
2S steep (15-25%) soils

thickness of the A horizon and a soil water value of the B horizon for loblolly and shortleaf pines on residual soils in the Piedmont plateau region of the southeastern United States. The soil water value was found to be characteristic of a soil series within rather narrow limits. This mathematical approach has been found to be useful elsewhere. Steinbrenner (1965) in a study of Douglas-fir in western Washington found that total depth, effective depth, and depth of A horizon were significant and positive in their relationship with site index, whereas gravel content and macroscopic pore space of the B horizon were significantly negatively related to site index. When the soils data were stratified into high and low elevation classes, the relationships were accentuated.

The establishment of relationships between growth and soil properties for a tree species in one region is well-illustrated by studies of Richards et al. (1962) and DeMent and Stone (1968) for red pine plantations in New York. It was established that site index for this species was strongly correlated with soil drainage class as shown in Figure 10.12. It was possible from further investigations to relate physical characteristics of the soils to site index and describe the component soil series in tabular form.

Another example of the use of soil survey information is in relation to recreational development. Figure 10.13 shows a portion of a soil survey in which soil series and slopes have been mapped for an area in the northeastern United States. Figure 10.14 is a map of the same area interpreted for possible development of picnic areas, and for this purpose the degree of limitation largely based on slope and soil moisture conditions was the basis for the interpretation. Interpretations for camping areas, trails and paths, and building development were also made for the same area (Montgomery and Edminster, 1966).

These examples illustrate two approaches that are used in making interpretations of soils surveys. One proceeds by a process of inclusion; soils are ranked by their degree of suitability for some purpose - for example when an area is ranked for growth of a tree species. Another proceeds by excluding areas progressively on the basis of limitations as was done in the interpretation for picnic area use (Figure 10.14).

11
Roots and soil

Apart from the matrix of soil materials, the most conspicuous components of forest soils are roots. Living and dead roots provide pathways for the movement of water, nutrients, and air in the soil and their presence modifies the soil to such an extent that they can be considered to have a major influence on soil profile development. Root development is modified and controlled by soil properties and the degree of control may vary with the tree species.

On a weight basis, roots constitute approximately 20 to 25 per cent of the total tree biomass but, because of their inconspicuous location in the soil and the laborious and usually destructive nature of most root studies, detailed knowledge of their form and functions in soil is for most species not well documented.

In a manner similar to the formation of associations and communities of plants above ground, the roots of the plants form an interwoven fabric within the soil. Although much of the information about roots comes from observations of individual trees, it is the sum total of all roots of a forest stand which is of most significance.

The form and amount of roots show considerable variation. Some root systems are longlived and extend several metres horizontally and vertically within the soil, whereas others may occupy only limited zones. A majority of root studies have focused on the variation in form and amount of roots as well as the interrelationships between soil properties and root development. The rate of development of root systems in association with above-ground development of the forest is of interest where the health of plants is a concern. Roots are the means by which water and nutrients are absorbed and transported within a tree and the ability of any soil to supply the materials, although variable, is a finite one, whereas the demand by vegetation of ever-increasing size is not. Frequently, therefore, the capacity of a soil to sustain a particular form of vegetation may be more clearly perceived if the progress of root development within the soil is known.

While the most obvious addition to the soil from roots is in organic content, the modification of the zone of soil immediately about the root surface - the *rhizosphere* - is less obvious. Changes in microbiological population and consequently their activities appear to be one of the most important differences between this zone and the remainder of the soil. Roots are subject to insects and diseases and may be weakened by attacks of these organisms to the extent that individual trees or sometimes an entire stand may be killed. Any soil-root conditions which increase the opportunity for such attacks are of obvious importance.

The roots and the soil, taken together, determine to a large degree not only the possible nature and development of the vegetation but also the development and character of the soil itself.

Form and abundance of roots

METHODS OF STUDY

The simplest technique is exposure of a root system by removing the soil from about the roots by hand. The chief advantage of this process is that as a particular root is exposed, its precise place in relation to the soil can be noted and any features associated with it can be recorded or sampled. Disadvantages of hand excavation are magnified as increasingly large root systems are exposed. The root excavation may have to start some distance away from the tree under study and the trench or hole will have to be deep enough that soil can be worked loose from about the main body of the roots and allowed to fall to the bottom of the pit. This means that considerable energy may be spent in removing the material and for large and deep root systems this can be a major task. One way of minimizing this disadvantage is to dig a large pit by machine at the outer extremity of the root system to be studied and then proceed to move the soil excavated from the roots into the pit.

It is seldom possible to expose large root systems without destroying some roots. Where quantitative information is required, one of the most useful procedures is to expose the roots in rectangular sections of soil, for example 2 m wide, 50 cm thick, and the depth of the roots. These roots may be drawn to scale or more easily photographed by a camera against a grid or scale. The roots in the volume of soil may then be removed by being cut off flush against the soil from which they protrude; they may be sorted and/or bagged for subsequent measurement. This procedure ensures that soils and related roots may be sampled for measurement on a soil volume basis and a progressive pictorial record of the root system can be obtained.

A second method is root-washing by hydraulic means. A water pump is set up and a high pressure jet of water is used to wash soil from about the root system. Large quantities of water may be used and both the supply of water and disposal of the slurry of soil from about the roots can present problems. A major disadvantage is that fine roots are usually torn off by the force of the water and any possibility of noting abnormalities of root development in relation to local soil conditions is minimal. For small root systems or portions of larger root systems, exposure of roots with fine jets of water allows a more detailed examination and eliminates loss of even the finest roots.

A third procedure is the removal of intact blocks or cores of soil from the sampling site. The roots can be extracted at a central location by mechanical and washing procedures. Cores or blocks may be used in conjunction with whole root system excavations in order to provide data on a stand as well as an individual tree basis. A block sampling technique used successfully with agricultural crops is the pin or nail-board in which the sample block, usually a thin rectangle in form, is penetrated by uniformly spaced nails mounted in a board. The block and board are removed intact and the soil can then be washed from the roots which are held by the nails in the same relative position as they were in the soil.

Virtually all the information on the quantity and distribution of the roots in forest soils has been obtained from studies involving variations of the three destructive sampling procedures described. Non-destructive techniques involving the use of radioactive compounds or special observations of growing roots can provide specific information about the extent, movement of elements, and response of root growth to experimentally determined soil conditions, as well as the rates of various physiological activities.

In tracer studies ^{32}P, ^{131}I, ^{86}R, ^{14}C, ^{45}Ca, and ^{137}Cs have been used primarily to determine the extent of root systems or the movement from roots of one plant to those of another as in natural grafting of roots. The isotope may be placed in the soil at measured distances from a stem and its movement monitored, or it may be injected into the stem and its subsequent translocation into roots and adjacent stems measured.

In order to study and measure responses of root systems and indeed whole plants under experimental conditions, installations called *rhizotrons* have been constructed. Plants, particularly perennials, are grown in soil containers with specially fitted sloping glass walls through which portions of the root system may be observed and measured. Sensors to measure soil moisture, temperature, and other properties are installed and the soil can be treated experimentally. The sampling of the soil-root system to determine the presence and nature of root exudates and microorganism populations is greatly facilitated by the rhizotron. An interesting type of rhizotron is one which is constructed on a temporary basis for use in forest conditions. A shed-like building is constructed over a trench, excavated in forest soil, and roots from trees in the surrounding forest stand are then grown under desired conditions.

Figure 11.1 Diagrams illustrating three main tree root forms - tap, heart and flat.

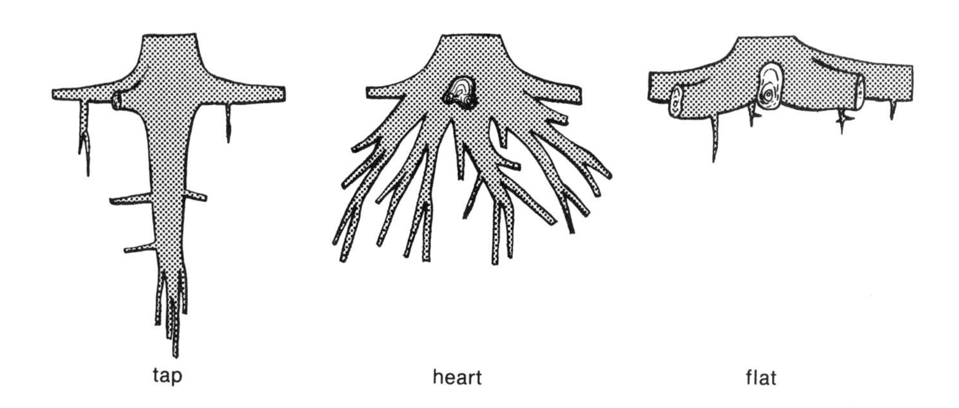

Schuurman and Goedewaagen (1965) and Yorke (1968) both provide comprehensive reviews of techniques for studying root systems.

FORMS OF ROOTS

Three classic forms of roots have been recognized - tap, heart, and flat (Büsgen and Münch, 1929). These are illustrated in Figure 11.1. As mentioned by Sutton (1969), although these three descriptive terms are convenient, considerable variations in form can exist, even within a species and it is more profitable to examine the components of the root form than to discuss it in terms of categories. The form which a root system takes is determined partly by internal controls, primarily genetic, and partly by environmental conditions. Immediately following germination, initial root development of many species of plants is similar; a primary or taproot develops from which lateral roots arise. Within a short time after germination the pattern of root development commonly associated with each species or its inherent form is usually apparent. In some species such as longleaf pine, which has a highly developed tap root form, there may be little variation, while in other species such as red maple the root form may be highly variable. For most forest tree species the degree of variation which might be governed genetically and its heritability are not known. Some species are consistently shallow rooting, while others exploit the soil to considerable depth. An example of the extreme depth to which roots may grow is the 53 m cited by Phillips (1963) for roots tentatively identified as belonging to *Prosopsis juliflora* (Swartz) D.C. which were exposed during a mining excavation in southern Arizona.

Figure 11.2 Brush or candelabra form of root of red pine. Numbered divisions of scale (cm). Photo courtesy D.C.F. Fayle

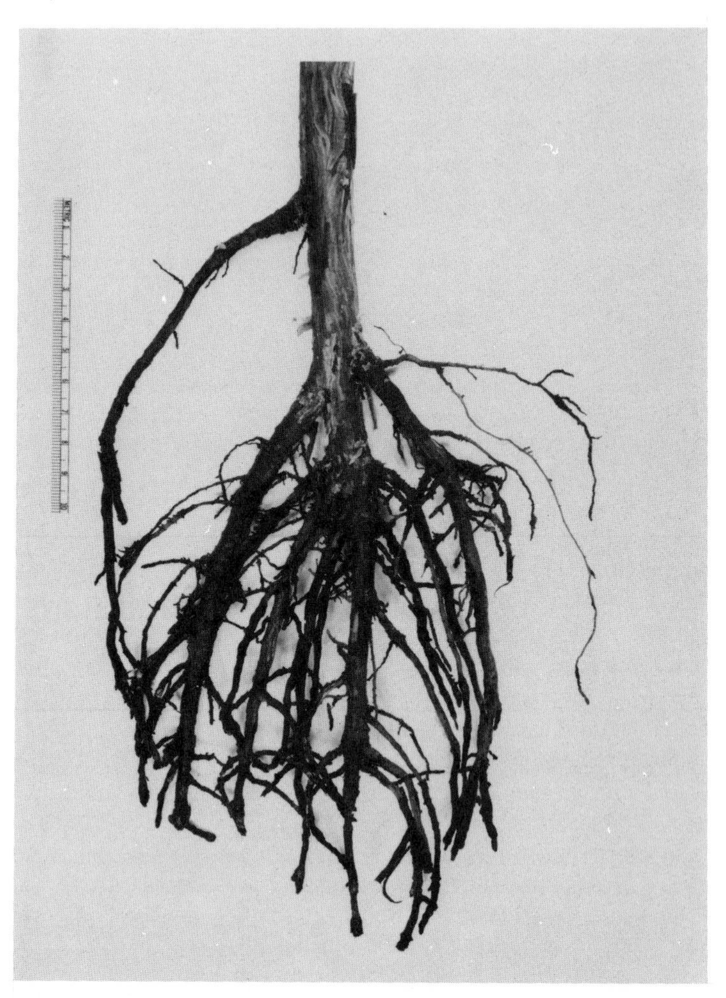

In addition to the tap root, vertical roots or *sinkers* develop from horizontal roots. In some species, roots develop downwards between the horizontal and vertical and these are often termed *oblique* roots. When downward-penetrating roots are impeded by a mechanical barrier of firm soil, by a water-table or similar adverse soil condition, a series of secondary roots may develop which then

Figure 11.3 Drawing of root system of a black spruce growing in peat and derived from a layering of another older tree. From Stanek (1961), by permission

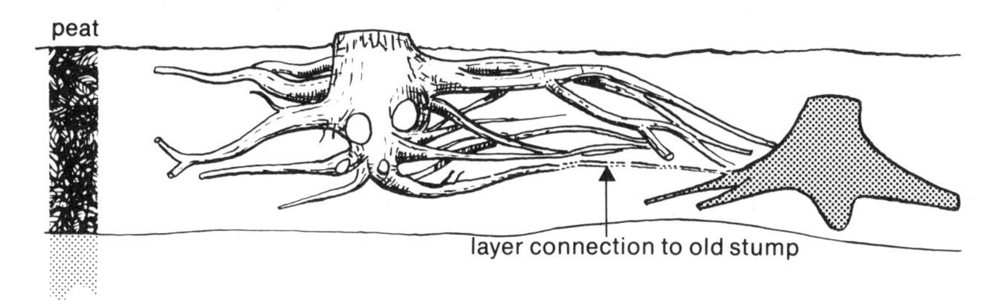

grow obliquely down to the adverse zone where they are halted and another set of roots grows again. After several sequences the roots at the adverse zone form a brush or candelabra form (Figure 11.2) which is particularly evident in conifers.

Some general characteristics of root forms in relation to certain soil depths and general horizon properties can be considered. For species or vegetation with inherent shallow or flat root systems, deep permeable soils with little or no restriction to rooting within the solum offer no advantage. Conversely shallow soils or those in which fluctuating water-tables may restrict deeper roots can be successfully exploited by flatrooted species. For example, black spruce may be commonly found over a wide range of soils - thin soils over bedrock, organic soils with a fluctuating water-table within 30 cm of the surface and on deep well-drained soils of various textures. In a wide range of conditions the roots of this species were found mainly (80 per cent) in the upper 60 cm and in many stands were not found below 45 cm (Schultz, 1969). This same species is also adapted to the increase in thickness of organic peat in which it may be growing by the production of adventitious roots from stem and branches. The rooted branches or *layering* can then develop into new individuals. Figure 11.3 shows the forms of a black spruce root system in a peat soil.

In soils where root penetration is restricted by stones and shallowness over a fissured bedrock, those species in which the root form consists of laterals, some oblique roots and sinkers have an advantage over other root forms. This is true for eastern white pine, in which the root system is capable of exploiting both lateral and vertical fissures in a fractured bedrock (Figure 11.4). Tree growth by this or other species with similar rooting ability on what is apparently a shallow soil with virtually no solum thickness can be much greater than would normally be expected. Although rooting volume is restricted, the fissures are often filled with mineral fragments and decomposing organic debris from both litter sources

Figure 11.4 White pine roots growing in a fissure in granite gneiss bedrock. Photo by K.A. Armson

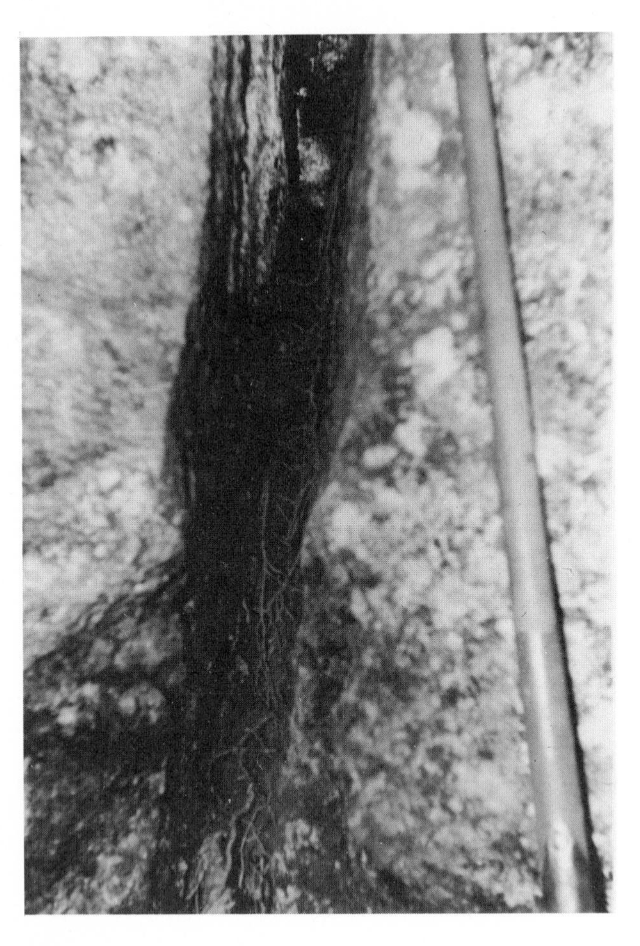

Figure 11.5 Comparison of root systems of similar-sized sugar maple (left) and yellow birch (right) growing on similar soils. Upper drawings are of sapling stage. Middle drawings are for older trees on stony sandy loam and lower are for older trees on a fine sandy loam over a soil zone with increased silt plus clay content. From D.C.F. Fayle (1965), by permission Canadian Forestry Service

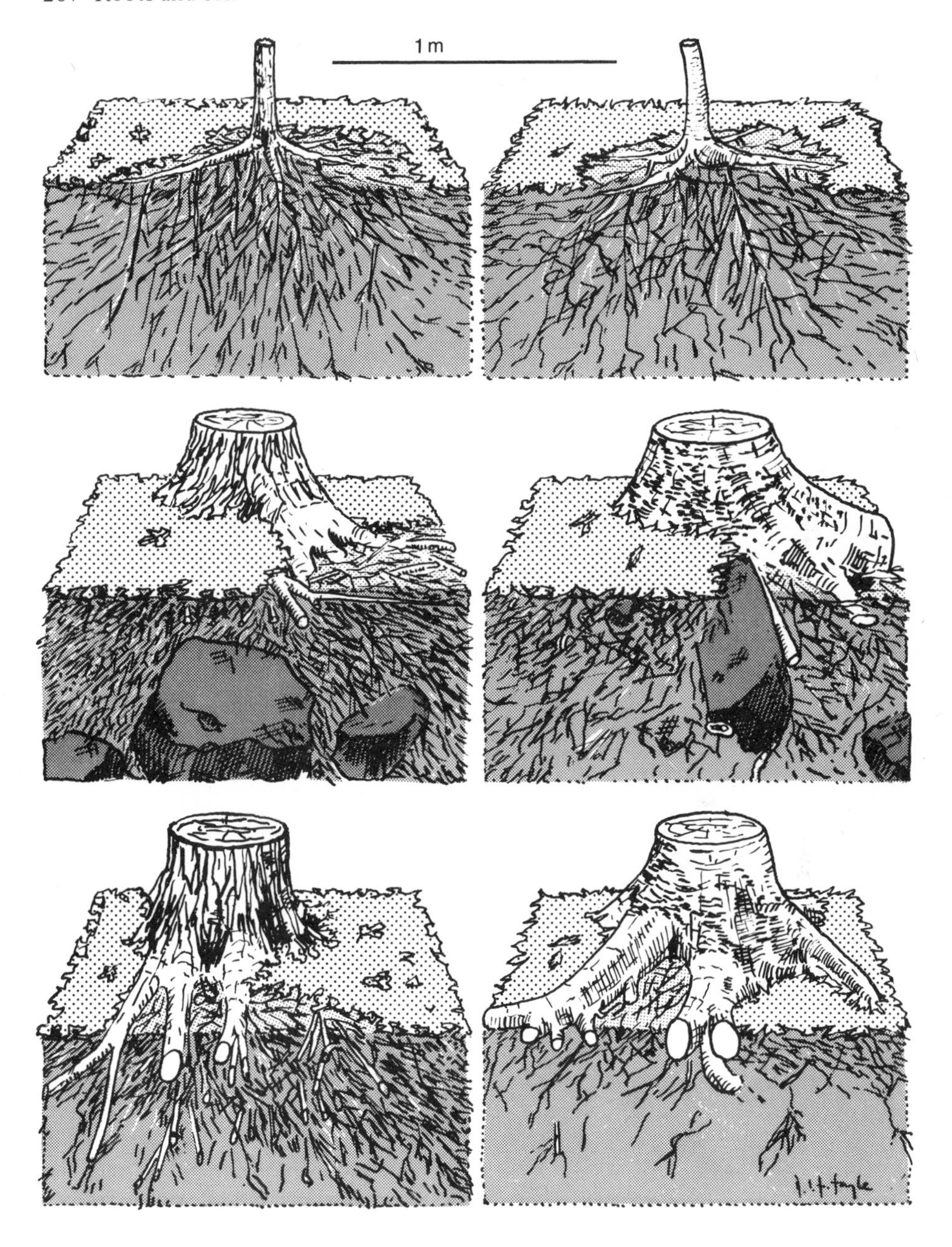
1 m

Figure 11.6 Volcanic soils, recent on top and two buried soils beneath with associated roots from ponderosa pine. Photo courtesy C.T. Youngberg

in the upper crevices and more abundantly the remains of previous generations of tree roots. Precipitation at the surface moves readily into the fissures where the high organic level ensures a large water retention per unit rooting volume and, apart from the water which moves downward out of the root zone, that which is retained is primarily available for absorption by roots and enters the transpiration stream. Species such as eastern white pine and Douglas-fir which form a heart root system, sometimes with a tap root, are adapted not only to stoney soils and fissured bedrock but also to deep well-drained soil materials.

Sugar maple is a species in which the root form is characterized by intense development in the central root crown area. Although this form may be modified to some degree by the soil, it contrasts (Figure 11.5) with the root system of yellow birch which has a more open system and a main network of fairly prominent laterals which maintain their identity throughout the life of the tree (Fayle, 1965). Trees with highly developed central root systems are much less able to exploit heterogeneous soils and do best in more uniform materials.

Figure 11.7 Multilayered root form of white spruce developed in response to lacustrine deposition. Adapted from J.W.B. Wagg (1967)

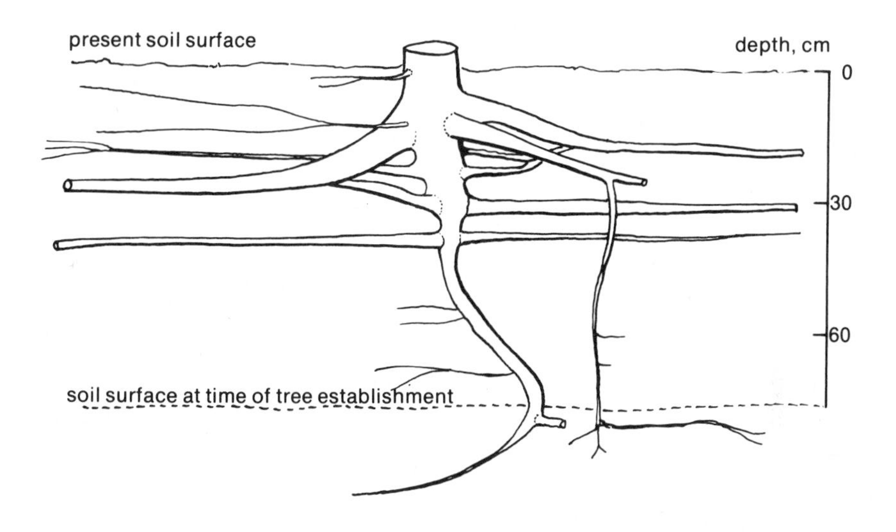

Some soils may exhibit highly developed stratification and, where this is associated with major differences in rooting properties, species which have a strong vertical form and a capacity to develop horizontal systems at depth, where more favourable horizons occur, have an advantage. An example is the rooting of ponderosa pine in buried surface horizons covered by pumice deposits from volcanic activity (Figure 11.6). White spruce can develop multilayered horizontal root systems in response to surface deposition by flooding (Jeffrey, 1959; Wagg, 1967) - Figure 11.7. This ability also exists in black spruce and eastern larch.

Longleaf pine has a well-developed tap root form (Heyward, 1933) and this form is shown by a number of species of both coniferous and deciduous trees. Bur oak and black walnut show a very pronounced tap root development which is normally persistent, although sinkers may also develop from laterals (Biswell, 1935). DeByle and Place (1959) compared the depth and distribution of roots of a 25-year-old jack pine and 30-to-35 year-old bur oak and northern pin oak and found that the oak tap roots penetrated more deeply than the roots of the pine. A number of species which have persistent tap roots are found on soils which often have water-tables at depth; slash pine and bur oak are two such examples. Roots do not usually continue growth into the zone of permanent saturation although McQuilkin (1935) reported that pitch pine roots developed below the water-table and were even mycorrhizal; it would appear from his description that the water was not stagnant but moving and thus aerated to some degree. Certainly soils in which species with persistent tap roots make their best development are deep, permeable, and well-drained.

Little is known of the variation in root form for many species. With increasing emphasis on the selection and breeding of individuals for growth and other desirable attributes, it would be advantageous to find out more in a quantitative manner of the variation that exists. Schultz (1972) compared the root systems of vegetatively propagated (airlayers) and seedling slash pine grown for ten years under different experimental treatments. Whereas all the seedlings had distinct tap roots, the airlayers did not but they had a few (one to three) well developed sinkers, arching from the root collar to the maximum depth of rooting. The seedling and airlayer stock did not differ significantly in the total root surface area.

ROOTING VOLUME OF THE SOIL

Occupation of soil by roots is neither uniform nor static and this applies to individual plants as well as communities of plants. As plants develop, both seasonally and over their lifespan, there is continual change in the root system. Some roots extend into 'new' soil, while mortality causes a reduction elsewhere in the root system. Also the growth and development of roots is controlled not only by soil environment factors such as temperature, moisture, and fertility, but also by the condition and physiological status of the above-ground portion of the plant on which the root system is dependent for photosynthate and growth substances.

If, in a simplified manner, the increase in size of a forest stand as it develops is considered to create an increased demand for water and nutrients, then an appropriate increase in supply can occur in one of two ways. First, there can be an absolute change in the supply as, for example, by management practices of irrigation or fertilization, or it may occur as a result of a period of above or below normal precipitation under natural conditions. Whatever the origin, the increase or decrease in supply of essential materials from the soil to the plants is altered; this represents a change in soil *quality*. Second, the root systems of the plants may extend farther into the soil or intensify the exploitation of volumes of soil in which roots already exist. This represents a relative change in soil *quantity* with respect to the supply of raw materials. To some degree a reduction in soil quantity may be offset by an increase in quality but there is a limit to this because living organisms such as forest trees must continually increase in size – it is their inherent biological nature. The space or volume available for root development in soils is finite and thus sooner or later there is a limit in the capacity of a soil to provide an adequate supply of materials to the higher plant. Day (1955) has defined this in a pathological sense – 'In order to remain healthy a tree must by its nature develop continually in size and to do this it must continually increase its demand on the soil.' The increase in supply to meet such a demand is met primarily by an increase in root extension, that is soil quantity, within a given soil, but differences in supply abilities between soils are frequently associated with levels of soil quality. The demand by vegetation for water and nutrients is not constant but varies within a growing season as well as from year to year. Thus the rate of supply in relation to demand rate by the vegetation is often more critical than the balance of total demand and total supply - 'The continued satisfaction of the demand on soil supply by the developing tree depends therefore on the condition of minimum supply being greater than that of maximum demand' (Day, 1955).

The assessment of a soil in terms of its ability to support plants, particularly in terms of moisture supply and fertility is, therefore, not a simple one of measuring soil properties as static entities, but should take into account both the nature of the demand and supply rates.

Although depth of rooting or depth of solum are frequently used to describe rooting space, more useful approximations of rooting space are:

Root system sorption zone - the volume of soil occupied by roots which is determined by using a segment of a sphere with mean radius of rootspread and depth of rooting. This provides a quantitative approximation of the total soil volume of root exploitation but not of root intensity.

Root surface sorption zone - the volume of soil within one cm of any root surface determined by calculating the volume of soil associated with root lengths by diameter classes. It is an indirect measure of rooting intensity.

The use of a 1-cm distance to define the root surface sorption zone is somewhat arbitrary. In terms of water supply, this distance will vary in effectiveness with both moisture level in the soil and the pore system as reflected by texture and structure. For nutrients, the distance will be significantly less for phosphorus which in most mineral soils moves only slightly, compared to potassium which can move to a root surface over considerably greater distances. Nye (1968) emphasizes the need to view the processes of soil supply to the root surfaces in terms of rates and fluxes and also the importance of defining the effective root zones for each component of the soil supply. Illustrations of amounts of phosphorus (Table 8.3) and calcium and potassium (Table 11.1) demonstrate the great differences in nutrient supply that are associated with each of these zones.

AMOUNTS AND DEVELOPMENT OF ROOTS

The amounts of roots within a soil are very much a function of the stage of development of the vegetation and the choice of parameter - root length, root surface area, root volume, or root weight - can be of consequence. Table 11.2 gives

TABLE 11.1
Amounts of calcium and potassium in young (5-9 year-old) pitch pine trees and soil root system and surface sorption zones (from Voigt et al., in Soil Science Society of America Proceedings, 1964, by permission Soil Science Society of America)

Nutrient – mg	Ca	K
Total in tree	415	247
Annual tree uptake	53	28
Exchangeable in Root Surface Sorption Zone	198	35
Exchangeable in Root System Sorption Zone	1.64×10^4	3.53×10^3
Total in Root Surface Sorption Zone	3.23×10^4	6.51×10^3
Total in Root System Sorption Zone	2.80×10^6	5.60×10^5

TABLE 11.2
Total amounts of roots of Scots pine measured as total root length per surface soil area (m/m^2) and associated root volume per cent of soil volume (data of Kalela, 1950)

Stand age (years)	Total root length / m/m^2	Root volume (%) / Soil volume
10	104	–
20	192	–
30	262	0.14
40	323	–
50	356	0.22
60	370	–
70	367	0.29
80	347	–
90	320	0.37
100	293	–
110	266	0.45

the total root lengths and the root:soil volume ratios for Scots pine stands of different ages (Kalela, 1950). It will be noted that the maximum intensity of roots measured as lengths occurs at 60 years, but the root:soil volume ratio increases continually. In a study of loblolly pine roots, Coile (1937) found that the num-

Figure 11.8 Numbers of fine roots (diameter < 2.5 mm) in relation to loblolly pine stands of various ages and by soil horizon. From T.S. Coile in *Journal of Forestry* (1937), by permission Society of American Foresters

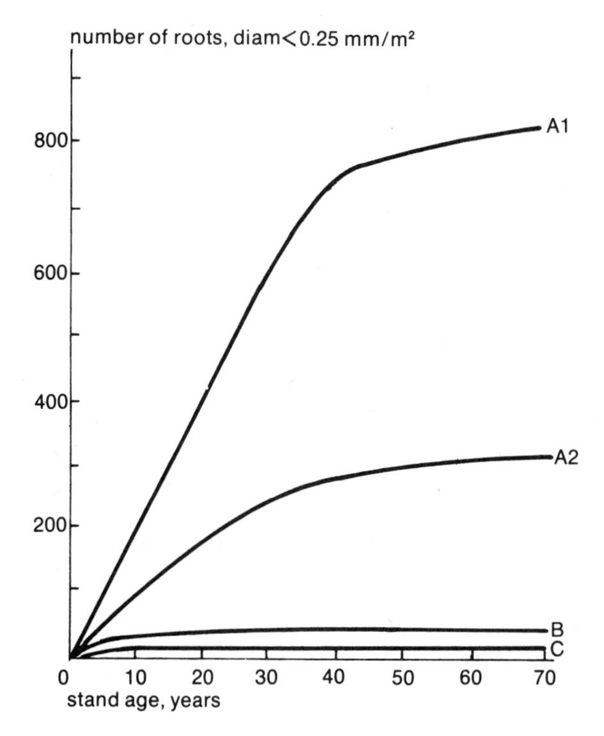

ber of fine roots (diameter < 0.25 cm) increased rapidly especially in A1 and A2 horizons over the first 30 to 40 years but showed relatively little increase with further increase in stand age (Figure 11.8).

The degree of development of a root system is related to the dominance of the tree. McMinn (1963) measured root lengths of Douglas-fir of different crown classes in stands of different ages in British Columbia and found that root lengths of trees which were dominants were markedly greater than intermediate or suppressed trees (Table 11.3). Since the lengths were for roots > 1 cm in diameter, the pattern of increase with stand age will not be similar to that shown for loblolly pine in Figure 11.8. Reynolds (1970) studied the root distribution of plantation Douglas-fir in Great Britain and found the total root length in a 36-year-old stand to be 7.69 km/m^2 in a soil depth of 107 cm and the total root dry weight was 1.37 kg/m^2 for the same depth.

TABLE 11.3

Length of root systems of Douglas-fir of various ages and crown classes (from McMinn, reproduced by permission of the National Research Council of Canada from the Canadian Journal of Botany, 1963)

Age and Crown Class	Number trees sampled	Average Ht – m	Average diam – cm (b.h.)	Av. age – years	Av. length roots > 1 cm diam – cm				
					Primary	Secondary	Tertiary	Quaternary	Total and range
10-yr-old stand									
Dominant	6	2.5	2.5	8	99	38	–	–	137 (65–187)
Intermediate	4	1.8	1.0	7	31	–	–	–	31 (20–44)
Suppressed	7	0.9	–	7	7	–	–	–	7 (2–17)
25-yr-old stand									
Dominant	2	20.0	15.7	23	1,129	887	156	–	2,172 (1,747–2,561)
Intermediate	2	10.0	8.1	21	484	257	10	–	751 (530–971)
Suppressed*	1	5.0	5.1	16	55	51	–	–	106
40-yr-old stand									
Dominant	2	26.0	20.4	40	–	–	–	–	4,337 (4,000–4,692)
Intermediate	1	23.0	16.0	40	–	–	–	–	1,820
Suppressed*	1	21.0	11.7	30	–	–	–	–	484
55-yr-old stand									
Dominant	1	39.0	45.7	54	2,015	6,400	3,945	1,555	13,915
Intermediate	1	26.0	26.0	54	1,270	2,200	1,245	350	5,070

* Tree dead for several years prior to excavation.

TABLE 11.4
Oven-dry weights of fine roots (<3 mm diameter) per cm^3 with depth in a 39-year-old white spruce plantation in Maine (from Safford and Bell, reproduced by permission of the National Research Council of Canada from the Canadian Journal of Forest Research, 1972)

Soil horizon and depth (cm)		Fine-root concentration (mg/cm^3)		Coefficient of variation (%)
		Mean	Range	
F–H	0 – 5	5.9	3.6 – 13.0	46
Ap	6 – 15	3.1	1.6 – 5.6	39
B	16 – 30	0.4	0.1 – 1.2	72
B	31 – 45	0.2	<0.1 – 0.5	90

Safford and Bell (1972) noted when comparing their own work on white spruce with studies of other conifers that the oven dry weights of fine roots (roots <3 mm in diameter) were very similar, ranging from 2–3 kg per tree for Monterey pine, Scots pine, and Norway spruce. The distribution of fine roots in a 39-year-old white spruce plantation is given in Table 11.4. The great variation to be expected when sampling roots using a cylinder is evident; similar large variation was also noted by Reynolds (1970) for Douglas-fir. Kohmann (1972a), using a 4.8 cm sampling tube to examine the roots of Scots pine, determined that between 4 and 8 samples were required to obtain standard errors of the means of less than 20 per cent for roots <1 mm in diameter when length and dry weight are the parameters. For larger diameter roots, the numbers of samples for a similar required standard error increases greatly.

The total biomass of roots can vary considerably in different forest conditions although it will often approximate 20 to 25 per cent of the total biomass. Will (1966) found for an 18-year-old Monterey pine stand that the total root biomass was 11 per cent of the standing crop. Morrison (1974) found that the root biomass for 30-to-35-year-old natural jack pine was only 13 to 14 per cent and it may be that for fast-growing early successional species the proportion of the total biomass constituted by the roots is less than for later successional species.

The general pattern of development for conifer roots is rapid extension of both tap roots and horizontal roots. Horton (1958) found for lodgepole pine that maximum tap root growth was reached long before maximum stem height growth was attained and that length of lateral roots increased linearly with height up to heights of 7.6 m; total overall rooting extent was a maximum when tree age was 35 years. In natural jack pine Shea (1973) found that often 75 per cent of the lateral root extension was achieved in the first eight years of growth and similar rapid elongation was found for the tap root development. Subsequent vertical root development by sinkers originated in later years and extension was often less rapid. Eis (1974) found a somewhat similar pattern of extension in Douglas-fir roots. He noted that relatively little overall extension occurred after a tree was 15 to 20 years of age and that a 35-year-old tree had a root system which extended over a soil surface area of 122 m^2 and a soil volume of 107 m^3. Western hemlock and western red cedar root systems had similar forms to that of Douglas-fir; 68 per cent of the total number of sinker roots initiated at a given tree age, up to 30 years for all species, occurred at ages between 11 and 22 years and only 8.4 per cent were initiated at ages beyond 23 years. This pattern of development contrasts with the extension of lateral roots of sugar maple which show a slow rate of about 15 cm a year or less and a fast rate of 30 to 60 cm per year over a period of 30 to 60 years (Fayle, 1965; Stout, 1956).

Root growth and soil properties

The general pattern of root distribution is for the largest number of roots and the majority of fine roots to be located in the upper portion of the soil profile. This is exemplified by the results of a classic study by Lutz et al. (1937) on the distribution of eastern white pine roots in gray brown podzolic and podzolic soils in the northeastern United States (Table 11.5). It will be seen that, although the maximum number of roots occurs in the A horizons, the greatest density of roots is in the H layer, but because of its thinness the total number in this organic layer is only about one tenth the number of roots in the much thicker mineral A horizons.

The intensity of rooting in upper soil horizons often reflects greater soil fertility, temperature, aeration, and the presence of soil moisture from precipitation during the growing season. The proliferation of root growth in finer-textured (non-clay) strata occurring in coarser-textured soils is usually evidence of a response to increase in soil water supply, without reduced aeration. Often species' differences in tolerance to reduced aeration are important; for example, Zak (1961) demonstrated that loblolly pine had a greater tolerance to poor soil aeration than shortleaf pine, and red pine in eastern North America is known to be more sensitive to reduction in aeration than other conifers. Frequently, it is the change in moisture-aeration conditions related to seasonal patterns of precipitation or microtopographic differences which may be of significance in affecting root development. Day (1959) observed that planted eucalypts in Cyprus on deep, moist soils grew rapidly and the increase in crown size resulted in increased transpiration and a greater drying out of wet soil at depth with a concurrent

TABLE 11.5
Distribution of eastern white pine roots by soil horizon for each 3 m length of profile face and per unit profile area (approx. 0.1 m^2) (from Lutz et al., 1937, by permission Yale School of Forestry)

Horizon	Ave. number of roots per 3 m length	Ave. number of roots per 0.1 m^2
F	88 ± 20[1]	137 ± 37
H	156 ± 22	407 ± 37
A	1,448 ± 160	322 ± 23
B	1,282 ± 187	107 ± 12
C	373 ± 77	11 ± 2

1 Standard error of the mean.

deeper penetration by roots. In a climatic condition where cool, wet periods alternate with hot dry periods, a time will be reached in the development of the eucalypts when root dieback will occur during an extended hot dry season. Following this there is a reduction in foliage, since water supply is reduced and also a decrease in transpiration. If this is followed by a period of high precipitation, there is likely to be an excess supply of soil water – the smaller crowns are transpiring less and part of the root system has died back. Reduced aeration in the rooting zone can result in even greater root mortality. In a subsequent period of moisture stress, the tree is even more susceptible and its decline proceeds more rapidly.

Lorio et al. (1972) studied loblolly pine root systems on mound and intermound flat or slightly concave sites in Louisiana. For trees of similar aboveground dimensions those growing on the intermound areas had significantly less dry weight and surface area of fine roots ($\leqslant$5 mm) in the upper 15 cm of the soil. It had previously (Lorio and Hodges, 1971) been established from studies of tree growth and soil moisture regimes that growth was slowed during rapid moisture depletion even when soil moisture content was high and that fast growing trees on intermound sites were disproportionately stressed compared to those on mounds. It was concluded that the development of deficient root systems and intermittent severe moisture stress on the intermound sites probably contributed to premature tree decline and susceptibility to bark beetle (*Dendroctonus frontalis* Zimm.) which had been observed in such stands.

It is not uncommon to observe long horizontal roots with few branches and little taper in the surface soil horizons. Lyford and Wilson (1964) termed these *rope-like roots.* Day (1962) noted in Sitka spruce this same form of root and considered that, while the long root was able to persist and grow, the ability for secondary roots to do so was more variable and there was considerable mortality in new side roots. Such rope-like roots can occur where they are subjected to considerable short-term variation in moisture-aeration conditions, particularly where internal drainage may be impeded. Leaphart and Wicker (1966) found that, when soil moisture was deficient, root systems of Douglas-fir, western white pine, grand fir, western larch, and red cedar permeated a greater volume of soil than when adequate soil moisture was provided, but that the density of roots in the upper 15 cm of the soil with adequate moisture was greater than where moisture was deficient. It was concluded that among these species western white pine could not compensate for low available soil moisture supplies by increase in fine root : foliage ratio to the same degree that the other species could and that this may contribute to its susceptibility to pole blight.

The relationship between root growth and soil fertility is best illustrated by responses to experimental changes in nutrient supply. In Europe, Scots pine growing in swamp soil responded to fertilization with nitrogen, phosphorus, and potassium by an increase in root length in the upper 12 cm (Paavilainen, 1961). Kohmann (1972b) noted that nitrogen application generally resulted in an increase in the fine root component but that the effect was variable, depending upon level of nitrogen and position in the soil; an increase in fine roots was more consistent in the upper 10 cm of mineral soil than in the surface raw humus layers. Hoyle (1971) noted that although a nitrogen deficiency inhibited general root growth of yellow birch, deficiency of calcium at lower soil depths inhibited primary root development. An increase in fertility can affect root development not only in the treated zone of soil but also in untreated soil as is illustrated by Table 11.6. The banded phosphorus treatment results in greater absorption within the treated soil zone and a greater 'foraging' ability by the increased root system in the remainder of the soil.

Although not based on observations of forest species, Cornforth (1968) proposed some general conclusions relating soil volume and nutrient accessibility. He considered that an increase in the total soil volume used by roots was associated with increase in yield but decreased intensity of rooting. Increase in nitrogen supply to a crop was found greater with increase in soil rooting volume than increase in phosphorus. Phosphorus uptake increased with rooting intensity and relatively greater responses were obtained to phosphorus fertilization in larger soil volumes, whereas for nitrogen the relative response was greater in smaller soil volumes. He concluded that, when a soil rooting-volume is restricted, it is more important to apply a mobile nutrient such as nitrogen than one that is readily immobilized such as phosphorus, although to a degree the value of each depends on the supply of the other in compensating for inadequate rooting volume.

TABLE 11.6
Root surface areas for white spruce seedlings (70 days-old) grown in a sandy loam soil in which phosphorus had been mixed with soil at the 2-4 cm depth at the rate of 112 kg/ha as triple superphosphate. Soil extractable[1] phosphorus at end of experiment is given

Banded			Control	
Mean seedling total dry weight	45 mg		21	
Mean seedling ht	4.4 cm		2.6 cm	
Mean seedling Root surface area (mm^2) by 2 cm depths		Extractable soil P mg/g		Extractable soil P mg/g
0 - 2	138.2	3.4	76.7	3.3
2 - 4	190.6	27.7	111.4	3.6
4 - 6	149.8	3.5	65.7	3.3
6 - 8	66.8	3.7	54.1	3.4
8 - 10	15.7	3.9	3.5	3.2
Mean seedling total root surface area	561.2		311.4	

1 Bray and Kurtz. No. 2 extracting solution.

The general increase in root growth with temperature to some maximum, followed by a decline, has been established for a number of tree species (Ladefoged, 1939) Root growth was slow for all species (ash, beech, Norway spruce, and silver fir) up to 10–14°C, but increased more rapidly to a maximum which varied somewhat for the species, being lowest at 24°C for the beech and highest at 32°C for the silver fir. Maxima for the other species were intermediate. Lyford and Wilson (1966) found the optimum temperature for red maple root growth to be 12 to 15°C, although the rate of elongation increased up to 25°C; the roots were more subject to decay at the higher temperatures. For temperature, as for other soil properties, any root response to changes in the levels of these factors cannot be completely independent of the other portions of the tree above ground. Richardson (1956), for example, found that changes in night temperature of the shoot system of silver maple seedlings resulted in changes in root elongation rate. At shoot-night temperatures above 10°C the root elongation was inversely related to temperature.

Root growth is generally restricted at some level of compactness by mechanical impedance. One of the major difficulties in field examination of soils or in experiments is to separate the effects of a simple mechanical barrier from related changes in soil moisture status and aeration. It is not always appreciated that porous, coarse-textured sands and fine gravels constitute a very rigid soil system and can create as much mechanical impedance to a root as a compacted or cemented soil horizon such as a fragipan. Obviously, the two examples can present extremes in aeration and soil moisture supply – the coarse-textured soil with low water-holding capacity but porous and well-aerated and the fragipan with impeded drainage and seasonally reduced aeration. Reduced growth of planted Douglas-fir seedlings has been attributed to compaction following tractor logging. Youngberg (1959) found that growth of seedlings was significantly greater on a cutover where soil bulk densities to 30 cm were between 0.87 and 0.98 g/cm^3 than growth in roadways or berms where soil bulk densities to the same depth ranged from 0.88 to 1.73 g/cm^3. Red pine seedlings showed reduced primary root development in a sandy loam at a bulk density of 1.45 g/cm^3 compared with growth in the same soil material with a bulk density of 1.02 g/cm^3 (Armson and Williams, 1960). In many soils, although compact or impermeable layers may restrict general root development, there are often places where a vertical root manages to grow through and, if more favourable soil zones exist under the impermeable layer, the root will proliferate in this region. Once the original root channel has been created, succeeding generations of roots may make use of it.

Ecological and silvicultural aspects of roots

The general nature of the root biomass has already been discussed and the annual contribution to the soil of dry matter is quite considerable. Nye (1961) estimated that under a moist tropical forest the net annual contribution of dead roots was 2,578 kg/ha. The current annual dry matter production in an 18-year-old Monterey pine stand in New Zealand was estimated at 3,026 kg/ha, but this did not include annual fine root mortality or roots below 90 cm depth (Will, 1966). The *in situ* addition of this organic material provides a large substrate for soil organisms. The nutrient content of roots, although not large, can locally within a soil form a significant source of organically-bound elements and the contribution that roots make to the nutrient cycle in the ecosystem can be significant.

Roots are the major water-absorbing organs of plants and their abundance and distribution within the soil are important in relation to soil water movement. Although responsible primarily for removal of water from the soil, roots to some degree function as conduits for water movement into soil (Voigt, 1960). This occurs particularly with tap and heart roots when stemflow from high intensity precipitation passes down the external surface of the roots from the root collar. Such water is rapidly channelled to depths in the soil reflecting the distribution of these vertical roots. Stemflow can result in significant changes in physical and chemical soil properties at the base of the tree (Gersper and Holowaychuk, 1970a,b).

The rhizospheric effect of tree roots would appear to be different in magnitude, if not qualitatively, from effects of non-woody lesser species. Fisher and Stone (1969) suggested that the rhizosphere of conifers has a capability to mineralize or extract nitrogen and phosphorus from old-field soils which was not associated with the grass vegetation. They speculated that the effect was not common during the first decade of stand establishment. Richards (1973) considers that dinitrogen-fixation takes place in the rhizosphere of slash and hoop pine and from fertilizer trials deduced that interplanting hoop pine with loblolly pine resulted in increased nitrogen accumulation (Richards, 1962). As has already been mentioned, metabolites produced by tree roots can influence not only the micropopulation of the rhizosphere but also, if phytotoxic, can inhibit development of other species.

The extension and expansion of a root within the soil changes the microfabric to some degree. Under deciduous hardwood stands Blevins et al. (1970) found that within a 0–0.4 mm distance at the root-soil interface there was a relative decrease in grains and pores less than 20 μ diameter and an increase in elongated grains and oriented clay. They considered that the orientation of the grains was due to root pressure and that the alignment was not similar to argillans,* commonly found in clay accumulation horizons.

The network of roots beneath a forest stand may be an important factor in minimizing mass soil movement or *solifluction.* This is especially likely on steep slopes in mountainous areas where at times the soil may become so wet that it may tend to flow. The ability of tree roots to retain soil under such conditions is less following clear felling due to exposure of the surface and removal of the interception effects of the tree canopy. The clear felling will also result in higher soil moisture status with the removal of transpiration losses. O'Loughlin (1974) has determined that under these conditions Douglas-fir and western red cedar roots rapidly lose 50 per cent or more of their tensile strength within three to five years of the death of the parent tree and hence their stabilizing effect on slopes is lessened.

Windthrow of stands and of individual trees within a stand will result both in mixing of the upper mineral soil layers and the development of a mound and hollow microtopography. The amount of soil disturbance depends very much on the form of root system of the tree species. Root systems that are of the heart form usually result in maximum disturbance and those of flat root form, the

* Argillans are clay films or skins.

Figure 11.9 Distorted roots of jack pine, seven years after planting as a 2-year old seedling. Root distortion resulted from bending at time of planting. Photo by K.A. Armson

minimum; often only the surface organic layers may be modified. Species with a well-developed deep tap root tend not to be subject to windthrow. Many trees are more subject to stem breakage, for example, the aspen poplars, than to windthrow.

Effects of various silvicultural treatments may affect root development. Yeatman (1955) reported in detail on the effects of cultivation and other types of site preparation in the establishment of Japanese larch, Sitka spruce, Scots, lodgepole, and Corsican pines in Great Britain. Any treatment that increased rooting volume by improving aeration, as subsoiling, assisted growth. Strip cultivation results in orientation of roots in the direction of cultivation (Yeatman, 1955; Tucker et al., 1968). Planting, especially of pines often results in distortion of roots which persists for several years. Often these roots grow in one plane in relation to handplanting in slits (Little and Somes, 1964) or are twisted when compared with root systems of direct-seeded trees (Figure 11.9). In a study of the effects of site treatment and planting on slash pine, Schultz (1973) found that 35 per cent of the tap roots were deformed by planting, but that root deformation could not be correlated with tree growth as much as 12 years after planting.

The general effect of increase in fertility in stimulating root growth in many instances has already been noted. In a study in which both depth to water-table and soil fertilization were regulated in slash pine, it was found (White et al., 1971) that over a five-year period slash pine fine root (0.5 cm diameter) biomass was increased 145 per cent by fertilization with diammonium phosphate at 392 kg/ha. Control of the water-table at depth of 46 and 92 cm had no effect on fine root biomass production, but did result in an increase in total root mass.

The presence of root fusions or grafts between individuals of the same species (intraspecific) focuses attention on the interrelationships that exist in many forest stands and emphasizes the vital links that can exist between individuals. Graham and Bormann (1966) in a comprehensive review reported evidence of root grafting existing for more than 150 species of both conifers and deciduous trees. In some species such as white spruce intraspecific grafts may occur in seedlings only a few years of age whereas in another species, e.g. jack pine, grafts may be uncommon. For some species, incidence of root grafting increases when thinnings are made over a period of several years (Armson and van den Driessche, 1959).

Root grafting enlarges the effective root system for residual trees after a thinning, but photosynthate is often used in maintenance of live stumps. Certainly the transmission of a number of root diseases can be facilitated by root grafting and herbicides applied to stems or stumps can be transmitted to other individuals (Cook and Welch, 1957).

12 Fire and soil

Vast areas of the world are burnt annually by both man-set and natural fires. A large proportion of these areas is in forest cover extending from the equator to the limits of tree growth in both southern and northern hemispheres.

Fuel is necessary if burning is to take place and the living vegetation and accumulated organic material of the soil, primarily the forest floor, are the two fuel sources. In many natural forests plant species occur which are ecologically adapted to the occurrence of fire and some even depend upon it as a factor in their normal cycle of growth and reproduction. The effect of fire may be direct as when heat from fire is required to open serotinous cones and allow seed dispersal or it may be indirect by consuming a part or all of the forest floor and providing a suitable seedbed which would not normally occur. Similarly the effects of fire on soils are both direct and indirect. The duration of a fire is very short, usually a matter of minutes or a few hours on any given location and hence the indirect effects which may persist for years following the fire are more obvious.

Historically, man has extended the areas which have been burnt, using fire largely as a tool in clearing and in maintaining certain desired vegetation; indeed, fire and the ecology of man are interwoven (Komarek, 1967). The use of fire by man has often resulted in an increase in the frequency of occurrence compared with that of natural fires for an area, and often what may be considered harmful results from fire are due not so much to the fire but to its increased occurrence on the same area. Although, generally, man's actions are associated with greater extent and frequency of fires, there are areas of forest land, particularly in the present century, on which fire occurrence is less than would occur naturally. These are areas in highly developed countries, forests that have been set aside for specific uses such as wood production and recreation or wildlands which are 'protected' on the principle that fire is destructive. It is increasingly apparent that for most forest areas where fire has been a natural ecological factor, reduction or prevention of occurrence may be equally as disruptive to the environment as an increased intensity of fire may be, although the manner of the ecological dislocation would be different. Often the ecological effects of fire which may be desirable can be achieved by other types of management practices or by a use of fire as a prescribed tool in which the objectives are known and the fire effects foreseeable.

The effects of fire on soil are neither good nor bad. They are extremely variable but can be considered in several ways. The most direct effect is the change in energy form of the organic material and its dissipation as heat, but associated changes in physical and chemical soil properties, especially close to the soil surface, may be both short- and long-term. Changes in biological properties, particularly of soil micropopulation, are common since the organic substrate which forms a food base is destroyed or modified. Many of the indirect effects are associated with the process of revegetation. Where the surface is covered rapidly the erosional processes are minimal and temperature fluctuations less than when, for whatever reason, revegetation of the soil surface is slow or prevented entirely.

Much of the variability of results from fire is related not only to differences in the amount of fuel but also to the intensity of the fire as it is affected by weather conditions at the time of occurrence. For wildfires these are not foreseeable to any meaningful degree and, therefore, the effects of fire can only be forecast in general terms. Where fire is used as a prescription tool in forest management, not only can the amount and condition of the fuel base be controlled but also a choice made in the weather conditions during which burning takes place. Hence a knowledge of the effects of fire on forest soils is important.

Effects of fire on forest soil properties

Fire affects physical, chemical, and biological properties of the soil. Changes in these properties inevitably modify the soil both for plant growth and soil fauna activity. Ahlgren and Ahlgren (1960) made a comprehensive review of the ecological effects of forest fires, with particular emphasis on north temperate forests, and Batchelder (1967) has given a concise account of the occurrence of fire in tropical areas of the world. Both the extent of fires and the variability in effects are clear from these studies.

PHYSICAL PROPERTIES

Temperature

During a fire the temperature above, at, and below the surface of the mineral soil depends on the amount of fuel, the burning conditions, and the form of soil organic layers at the soil surface. The larger the amount of fuel, the higher the temperatures and this is particularly evident where slash has accumulated after cutting. Where there is a heavy slash accumulation, temperatures may increase to some depth in the soil. Under windrowed slash in Australia (Humphreys and Lambert, 1965) surface temperatures ranged from 350 to 900°c, but at 5 to 10 cm beneath the surface the temperatures were approximately 100°c under these severe conditions of burning. In North American west coast forests where large slash accumulations exist after felling, surface temperatures may be in excess of 540°c even in light slash and temperatures may rise to 65°c at 12 cm depth (Neal et al., 1965). However, Tarrant (1956) has observed that severe burning may only occur on a small proportion (five per cent) of the area. The increase in mineral soil temperature will be least where residual mor humus layers remain unburnt and under these conditions little or no change in soil temperature may be observed. Where roots burn down to a depth in the soil, zones of soil adjacent to them may be heated for considerable periods of time and, under these conditions, 'firing' of soil minerals has been observed.

After a fire the minimum and maximum soil temperatures are usually greater than previously. This results from the removal of both vegetation and part or all of the surface organic layers which act to insulate the soil. The greater insolation received by the soil and its darkened surface will increase the heat absorbed and this will cause temperatures to increase. The increase will be larger with increase in soil slope facing the sun, and surface temperatures can often become lethal to newly germinated seedlings. The season at which the fire occurs and the weather conditions for a succeeding period during which revegetation takes place may be critical in determining whether plant regrowth is rapid or not. Aspect and slope which are major contributing factors can be identified and measured. Under certain conditions, changes in soil temperature regimes may result in more permanent changes. In areas of permafrost when fire destroys the surface insulating organic layers, the permafrost table will be lowered, resulting in subsidence of the ground surface. On flat areas this may result in ponding and on slopes the melted soil materials may flow downwards (Zoltai and Pettapiece, 1973).

Organic matter

Fire results in loss of soil organic matter, but the amount and location of the loss is dependent on fire intensity. A light, early spring (April) surface fire in mixed wood forest of white pine and sugar maple in the Great Lakes-St Lawrence Region of Ontario resulted in no significant reduction in surface organic matter (oven dry weight per 100 cm^2 of L and F layers of equal depth, 26.5g unburned; 29.0g burned). More usually some or all of the surface organic material is destroyed and sometimes a portion of the incorporated organic matter in the upper

TABLE 12.1
Means and standard errors of means (cm) for thickness of soil organic layers (L–H) in boreal forest soils, Ontario, Canada, for unburned (fire more than 50 years prior to measurement), one fire and two fires at varying intervals (from Armson et al., 1973)

		Two burns separated by (years)			
Unburned	One burn	30 - 50	20 - 30	10 - 20	1 - 10
13.4 ± 1.6	5.6 ± 0.6	5.4 ± 1.1	6.5 ± 0.7	5.0 ± 2.0	3.4 ± 2.0

mineral soil may also be lost. In eastern Canadian boreal forests, Armson et al. (1973) found that natural fires reduced the depth of surface organic layers (L-H) by approximately 50 per cent (Table 12.1).

In the Adirondack region Diebold (1942) made a detailed study of the effects of fire and found considerable reduction in the thickness of the F–H layers but, except under pure hardwoods where it was removed completely, there was an average of 5 cm thickness remaining after burning. Prescribed burning results in losses of organic matter, similar to those for wildfires of the same intensity. Van Wagner (1963) in a prescribed burn in red and white pine found that 48 per cent of the forest floor (L-H) was consumed, but that the amount of bared mineral soil was variable, ranging from less than 2 to 56 per cent of the surface area, depending on fire intensity. In ponderosa pine in California, Sweeney and Biswell (1961) found that in no instance was all the surface organic matter destroyed, but 76 per cent of the litter and 23 per cent of the duff* layer were burned. Twenty years of litter burning beneath a red and white pine stand in the northeastern states had a small stimulating effect on growth (Lunt, 1951). One of the more thoroughly studied areas of prescribed burning is that of the Santee Forest in South Carolina. Here, after 20 years of prescribed burning, Wells (1971) found that, although there was some reduction in the forest floor when the organic content of the upper 10 cm of mineral soil was included, the principal effect of burning was not a reduction in organic matter but an overall redistribution (Figure 12.1). It should be noted that the soils were poorly or very poorly drained.

Soil moisture and porosity

Following a fire, soil moisture regimes are usually changed drastically because the transpiration loss by the vegetation is removed. The time of year that the fire

* An imprecise term used to refer to partially decomposed surface organic layers

Figure 12.1 Organic matter in forest floor, 0–5 cm and 5–10 cm depths of mineral soil for control (CK), periodic winter (PW), periodic summer (PS), annual winter (AW), and annual summer (AS) burns. Periodic burns at 5-year intervals over a 20-year period. From C.G. Wells in *Prescribed Burning Symposium Proceedings* (1971), by permission United States Department of Agriculture, Forest Service

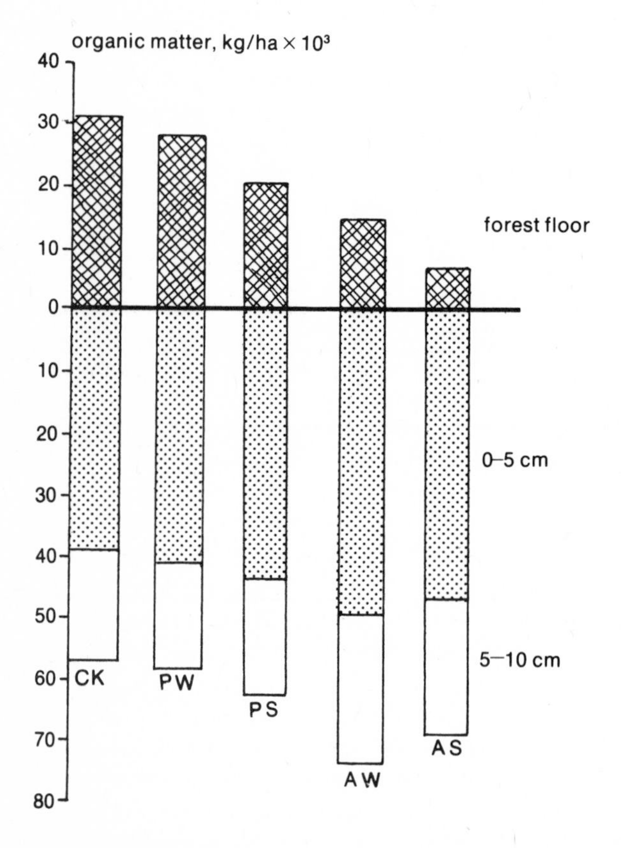

occurs will be important in determining the degree and rate of revegetation, as also will be the presence or absence of residual surface soil organic layers. Where a surface organic layer covers the mineral soil and revegetation occurs within days, or a few weeks, the soil moisture status is usually higher during this period because of continued high infiltration and percolation rates but minimal transpiration losses. Where a water-table exists within the rooting zone, it will generally rise, reflecting the reduced transpirational loss.

When little or no organic cover remains over the mineral soil, conditions may be quite different. During burning certain organic constituents are vaporized, some portion of which moves downward in the soil to form a well-defined water-repellent layer within a few centimetres of the surface (Krammes and De Bano, 1965). This hydrophobic layer reduces the rate at which water may move through it and the thickness of the zone increases as the silt plus clay contents of the soil decrease (De Bano et al., 1970). Following severe burns it has been noted that the macropore space at the surface is reduced and bulk density increased (Beaton, 1959; Moehring et al., 1966; Tarrant, 1956). Compaction of surface soil layers can occur as a result of exposure to the beating action of raindrops (Fuller et al., 1955). The net result of both reduction in large pore space and compaction is that infiltration rates are decreased and with this there is increased susceptibility to erosion, depending upon slope and intensity and duration of precipitation.

In some soils severe burning may affect soil structure. Dyrness and Youngberg (1957) found that there was a 21 per cent decrease in degree of aggregation of soil structural aggregates where heavy accumulations of slash were burned. A similar decrease in soil structure occurred in soils of the Coconino National Forest, Arizona, after severe burning (Fuller et al., 1955).

The occurrence of erosion by water movement is the single most important process by which the soil is permanently altered. As has been mentioned, the susceptibility to erosion will vary with many factors - intensity of burn, time of burning, soil texture, slope, intensity and duration of precipitation following the fire, presence or absence of other activities such as logging. Thus even in one location erosion may vary from year to year. Pase and Lindenmuth (1971) found for annual burning that virtually no erosion took place when 70 per cent of the surface litter cover remained and over a four-year period virtually all the erosional sediment originated during two years when litter cover after fire was 44 and 51 per cent. On slopes, the presence of roads will often provide the opportunity for gully erosion to occur and following fire this erosion is usually accentuated. Time of year can be very important especially in areas where winter snowfall is considerable. If late summer or autumn fires occur in such areas, little to no revegetation may take place and the depth of frost penetration will vary depending on the degree of removal of vegetation and reduced thickness of surface organic insulating material. Since the depth of snow accumulation will be greater than if forest cover were present and in the spring surface runoff will be a maximum, the probability of erosion damage will be great. An area burned in the spring or early summer will often be revegetated to a great enough extent that ground cover will provide for greater surface stability against erosion.

Figure 12.2 Relationship between pH values for surface soil layers and years since last burn - boreal forests - central Canada

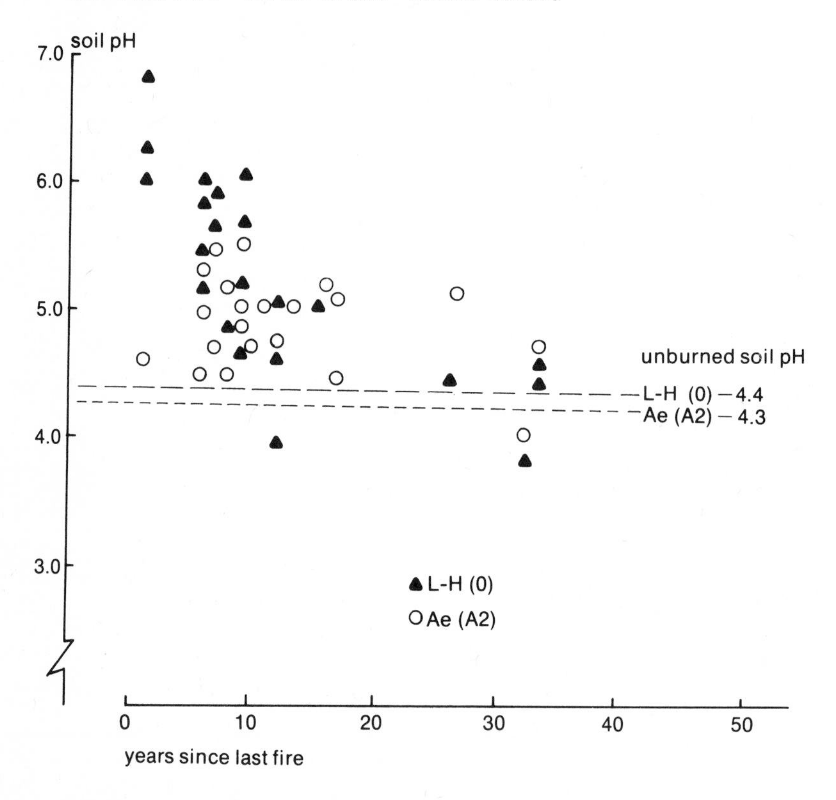

CHEMICAL PROPERTIES

Soil reaction – pH

Generally fire results in an increase in the pH values of the surface soil layers. The extent and duration of the increase will depend on the intensity of the fire, the amount of organic matter consumed, and the residual buffer capacity of the soil. The increase in soil pH reflects the presence of relatively soluble bases in the residual ash from the fire. The range of pH increase can be small or large. Heyward and Barnette (1934) found it ranged from 0.15 to 0.48 of a pH unit greater in burned than in unburned soils; Tarrant (1956) found that severe burns increased the pH values from 4.4 to 7.2. Where fires are frequent or prescribed, it appears that increases of the order of one half to a full pH unit are to be expected, but after intense burning an increase of one to two pH units may commonly occur. The duration of increased pH is variable but appears to be longest in forest soils with mor humus types. Figure 12.2 shows the pH values for boreal forest soils in central Canada related to number of years since natural wildfire. It would appear that there is a period of at least ten years before the pH values of the surface organic horizons (L–H) return to about the same level as those in the unburned soil. In New Jersey, where pine barren soils were subjected to burns at intervals of one to fifteen years, Burns (1952) found no difference greater than one half a pH unit between values of the A horizon in relation to number of burns or years since the last burn. Increases of the order of one half a unit were found by Wells (1971) for a very acid, poorly drained soil in the southeastern United States, where pH values of the F and H horizons were increased from 3.5 to 4.0 after 20 years of burning. Most of the increase occurred in the first ten years.

The degree of change and its duration can be expected to vary with the cation exchange capacity associated with the organic and clay components of the soil. Soils with little clay or organic content will show marked changes in pH immediately following a fire, but as cations such as calcium, magnesium, and potassium are lost from the soil by leaching or plant uptake, the soil pH may be expected to decrease. Conversely where a soil has a large cation exchange capacity, it is likely to be buffered against change in pH.

Soil nutrient elements

One of the main effects of burning is the production of ash containing many elements but principally calcium, magnesium, potassium, and phosphorus. The amount of increase in these elements is dependent on the amount of fuel and the intensity of the fire but, in general, fire results in some increase in these elements in a readily available form. It is the change in form from being incorporated in organic forms (except for potassium) to an inorganic state as ash that is the most significant result of the fire. Fire often causes increases in bulk density either due to removal of organic matter or associated with a reduction in total porosity. Therefore comparisons of amounts of nutrient elements in burned, compared with adjacent unburned soils or on a 'before and after' burning basis should be made on a soil volume rather than a weight basis. Following the fire, loss of ash may occur as a result of both wind and water movement.

The changes in nitrogen form and amount are difficult to determine as resulting directly from the fire because activities of the microflora involved in nitrogen transformations are also usually altered by changes in organic matter substrate, soil pH, and amounts of readily available nutrients. Knight (1966) studied the changes in total nitrogen in organic L–H layers from old growth western hemlock and Douglas-fir stands. He found that very little loss of nitrogen occurred up to temperatures of 200°c but that, from 300°c and higher, loss of nitrogen became increasingly greater. He noted that, although absolute loss occurred, the concentration of nitrogen in the residual material increased. Frequency and intensity of fire probably affects the amount and form of nitrogen. Neal et al. (1965) found that after Douglas-fir slash was burnt the amount of ammonia (NH_4^+) nitrogen increased significantly compared with the unburned condition, but after six months it decreased. Wells (1971) considered that after a series of burns the reduced annual uptake of nitrogen by seedlings was a result of its being less available rather than because of a reduction in total amount. After a period following fire, poor growth of higher plants may be attributable to a low supply of nitrogen and may in part be due to immobilization as was found by Gagnon (1963) in Quebec where a lichen (*Lecidea granulosa*) was considered responsible.

SOIL ORGANISMS

Soil fauna

Animals, especially meso and microfauna which are normally found in the surface organic layers and have limited mobility into the lower mineral soil, are particularly susceptible to destruction by fire. When only a part of the organic layer is destroyed and temperatures do not rise excessively, residual populations may be left for recolonization. Commonly, forest fires do not burn uniformly and generally islands or patches of relatively unburned soil may be zones from which animals may spread out following the fire. In northern Idaho, Fellin and Kennedy (1972) found that over a three-year period there was a general increase both in individuals and taxa of arthropods in mineral soil following fire. Three years after the fire there was a marked resurgence of a carabid (*Amara erratica*) in both mineral soil and surface organic layers and this carabid was a principal seed-destroying insect (Kennedy and Fellin, 1969). In South Carolina where burning had been carried on for 20 years there was no significant difference in numbers of mites and springtails between periodically burned (every five years) and unburned soils; where annual burning occurred, the numbers of these organisms were significantly reduced (Metz and Farrier, 1971). Annual burning can have variable results, as indicated in Table 12.2, where the effect of annual fire in a jarrah forest in Western Australia on soil faunal components is shown. Although the mean numbers are less in the annually burned soil than in the protected one, the range is somewhat greater. The proportion of mites and springtails in the total population is about the same and while there is some reduction of millipedes, centipedes, and larvae due to burning, other organisms show a pronounced increase in proportion. In contrast, Heyward and Tissot (1936) found that numbers of microfauna in soils in the longleaf pine region were reduced in the upper soil organic and mineral layers after fire. Differences in results such as these may reflect not so much the direct effects of the burning as the degree to

TABLE 12.2
Abundance of soil fauna in protected (unburned) and annually burned soils of jarrah forests of Western Australia (from McNamara, 1955)

	Estimated numbers of fauna, 10^6 per hectare	
	Protected	Burned annually
Mean	16.75	10.06
Range	4.48 – 20.6	2.12 – 25.38
	Proportion of various faunal groups (%)	
Mites and springtails	86.30	82.50
Millipedes and centipedes	1.52	1.20
Larvae	4.52	3.30
Others	7.66	13.00

which the upper soil layers are subsequently exposed to direct sunlight. It may be that reduced populations of microfauna reflect the desiccation of the surface layers as a result of exposure more than any other factor.

Soil microflora

The general pattern following fire is a decrease in microorganisms followed in many situations by an increase. The precise nature of change in microorganism population differs from one forest condition to another. In jack pine forests, Ahlgren and Ahlgren (1965) found that activity of the micropopulation decreased immediately after a fire but rose abruptly to a very high level after the first rainfall following burning. The second season after burning the number of organisms was lower than in unburned soil, but streptomycetes increased markedly in the third season. In western North American soils, Wright and Tarrant (1957) found the bacteria and actinomycetes increased in the severely burned as compared with lightly burned or unburned soils and that the ratio of bacteria: actinomycetes was increased as a result of such burning. When populations were sampled at monthly intervals in soil from beneath 25-year-old Douglas-fir and an adjacent burned area, Wright and Bollen (1961) found that, although the general level of the soil micropopulation was lower in the burned area, the numbers of actinomycetes and bacteria for the two areas were not substantially different. Changes in the soil micropopulation were primarily considered to reflect changes in seasonal moisture condition and soil pH. Jorgensen and Hodges (1971) found in South Carolina that burning had no effect in reducing fungi and that one organism – a fungus (*Gliocladium* spp.) – occurred in greater abundance as a result of fire. Burning was not found to influence rates of nitrogen mineralization.

Clear evidence of increase in asymbiotic dinitrogen-fixation following fire is not available but that it may occur has been inferred from the increase in soil pH values resulting from burning.

Ecological effects

Two immediate results from burning are increase in soil fertility and exposure to erosional forces. The improvement in soil fertility usually results in rapid and luxuriant vegetative growth and in many situations this development reflects an interaction between soil fertility, increased exposure to light, and the greater supply of soil moisture that often follows a fire. Rapid revegetation of the burned area largely modifies the exposure to erosion. The patterns of vegetational succession following fire are many and Ahlgren and Ahlgren (1960) review much of the literature on this topic. Certain aspects of revegetation may be important in soil processes. In several forest regions, species known to host symbiotically dinitrogen-fixing microorganisms may often be common in initial vegetation development. *Myrica* spp., *Alnus* spp., and *Ceanothus* spp. are woody plants known to flourish after fire; they are also host plants for dinitrogen-fixing microorganisms and, as a result of their increased abundance, the total nitrogen level of a soil may be increased. Other species of plants such as *Epilobium* and *Rubus* spp. are also common, as well as many lesser herbaceous species and it is a common observation that the litter of this vegetation is readily decomposed and results in more rapid cycling of nutrients in the upper soil layers. These vegetation changes may last only a few years, but during this period they may be significant in increasing the rate of nutrient supply rather than the total nutrient level for young forest vegetation in its period of rapid juvenile growth.

In certain forest areas, particularly those bordering on prairie or savannah, fire, especially periodic fire, may be a major factor in bringing about invasion of the forest lesser vegetation by grasses. The prolonged presence of grasses results in different forms of organic matter distribution in the soil. The annual death of fibrous grass roots within the mineral soil contrasts with the surface litter accumulation characteristic of forest soils. If vegetation changes such as the differences between grassland and forest are sustained for long periods of time, the morphology of the soils will contrast markedly. Severson and Arnemann (1973) found that the thickness of the A1 decreased from prairie to pine-hardwoods and that of the A2 horizon increased. They considered that this relationship was directly related to production and turnover rates of organic matter associated

Figure 12.3 Effect of periodic burning in a sugar maple stand

(a) unburned control

(b) area receiving light prescribed burns (1958, 1960), (1961, 1962)
Note absence of maple regeneration and more open stand. Photos courtesy Ontario Ministry of National Resources

with the different forms of vegetation, although in their study the causes for the changes in vegetation were not stated.

Periodic burning by ground fires can, without affecting larger overstory trees, change the internal stand structure. In hardwood stands of sugar maple, dense woody understory growth can be removed with the creation of a more open condition above the soil (Figure 12.3). This openness may be desirable for a number of reasons, and it will make the area easier of access for recreation, allow more air movement, and reduce availability of many browse species.

When conifer or mixed conifer hardwood forests burn, many of the tree boles of the forest remain standing. These stems (chicots) provide a degree of shade to the soil surface which diminishes as increasingly they are blown over by wind. As each chicot blows down, the major intact roots will be pulled out of the soil, usually carrying with them a considerable amount of soil. The fallen bole and its roots are finally decomposed. The windthrow process is responsible for both the creation of microtopographic variation (pit and mound topography) and for some degree of mixing of the upper horizons of the soil profile (Lutz and Griswold, 1939 Stephens, 1956).

Following fire there is often a major change in movements of water, reflected by differences in streamflow. Berndt (1971) noted three early changes in a part of Washington after a natural fire occurred in an area where stream gauging stations were in operation. Flow rate decreased during the duration of the fire as a result of vaporization of water by the intense heat; prior to the fire the diurnal streamflow showed a maximum at 08.00 hours and a minimum at 1900 hours, but after the fire daily oscillations were not evident and flow was uniform; during the two-week period following the fire, the streamflow gradually increased. The last two changes undoubtedly reflect the virtual elimination of transpirational water use. A year later Helvey (1972) noted that water yield increases averaged about 9 cm greater than before the fire and this increase apparently came largely from spring snowmelt and during the summer months. DeByle and Packer (1972) studied soil and plant nutrient losses from burned forest clearcuts in the Rocky Mountain region. The effect of logging and fire was to cause a temporary increase in the erosion of soil and loss of nutrients. Within four years, revegetation of the area had restored conditions to those prior to logging. Logging and burning did not affect soil bulk density, porosity, or organic matter content at depths greater than 2.5 cm. Soil erosion from snowmelt is illustrated in Figure 12.4. Erosion from summer storms contrasted with this since it was greatest the year following burning (approximately 134 kg/ha of solids) but showed marked reduction in the second year and by the third and fourth years was similar to losses from the unlogged and unburned forest - losses which were neg-

Figure 12.4 Erosion from snowmelt overland flow on Miller Creek Study Area during a four-year period following logging and fire. From N.V. DeByle and P.E. Packer in *Watersheds in Transition* (1972), by permission American Water Resources Association

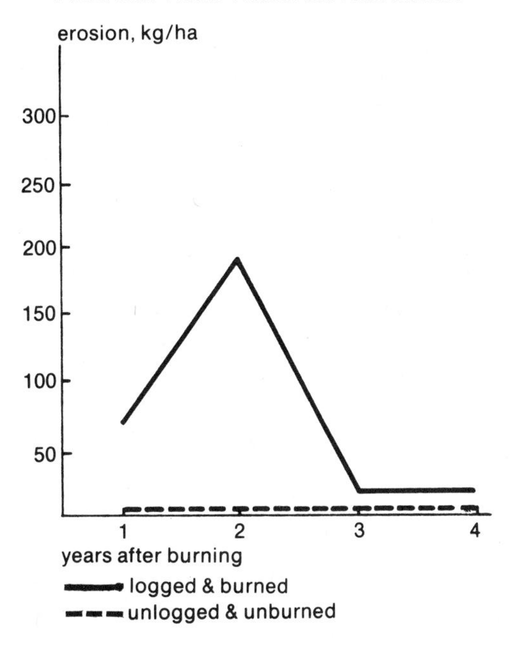

ligible. Average losses of nutrient elements in surface runoff and sediment from control and logged-burned plots is shown in Table 12.3. The area has slopes ranging from 9 to 35 per cent, averaging 24 per cent, and the soils were developed in a gravelly loam till with a superficial covering of loessial silt loam. The table shows that the loss of nutrients after burning occurs primarily in the first two years and that the total quantities involved are small. It was estimated that the losses represented the following percentages of the total 'available' amount in the upper 30 cm of soil: phosphorus, 0.5%; potassium, 1.1%; calcium, 0.6%; magnesium, 1.5%; and sodium 2.1%. While this example cannot be applied directly elsewhere, it does serve to illustrate the order of magnitude of nutrient losses for an area with steep slopes and large spring snowmelt, both factors which might be expected to intensify losses. The rapidity of revegetation, however, has obviously minimized the losses.

TABLE 12.3
Average losses (kg/ha) of nutrient elements in surface runoff and sediment from control (C) and logged-burned (T) plots – Miller Creek, Montana (from DeByle and Packer, in Watersheds in Transition, 1972, by permission American Water Resources Association)

	Years after burning									
	1		2		3		4		Total	
Element	C	T	C	T	C	T	C	T	C	T
Phosphorus	–	0.39	0.03	0.28	0.01	0.03	0.01	0.02	0.05	0.72
Potassium	0.15	1.73	0.73	1.36	0.13	0.53	0.21	0.37	1.22	3.99
Calcium	0.01	6.89	0.20	4.81	–	1.34	0.04	1.56	0.25	14.60
Magnesium	0.03	1.50	0.16	1.77	0.11	0.37	0.01	0.27	0.31	3.91
Sodium	0.19	0.76	0.29	0.79	0.10	0.43	0.03	0.16	0.61	2.14
Total	0.38	11.27	1.41	9.01	0.35	2.70	0.30	2.38	2.44	25.36

13 The hydrologic cycle

The fate of water, from the time of precipitation until it is returned to the atmosphere and is again ready to be precipitated, has been termed the hydrologic cycle (Soil Science Society of America, 1973). The properties of the soil plant system are of particular interest since they provide mechanisms which regulate interception, flow and storage of water in the cycle. It is also this system that is subject to the greatest modifications as a result of man's activities.

Water is essential for life processes and, depending upon the particular concern of investigators or users of water, often only a portion of the cycle is studied. While large comprehensive studies are not always feasible, it should be recognized that conclusions drawn from more narrowly based studies may only be valid for a limited area and portion of the water cycle. Any change that will increase outflow from one area means that other areas will have greater inflows; the converse is also true. It is usual to quantify the hydrologic cycle by preparing a water budget. Some of the components may be measured directly, as is precipitation, whereas others may be estimated from changes in amounts of other parameters. An example of the latter is discussed by Hewlett and Hibbert (1967) who mention that frequently a rapid rise in streamflow is equated with inability of water to infiltrate into the soil and therefore with runoff. Yet, in fact, it more likely reflects an increase in subsurface flow of water that has infiltrated, at least in forest soils.

Physical properties of soil affecting pore volume, sizes and distribution of pores are obviously important in determining rates of movement and soil water storage capacity. Yet forest soils normally support vegetation and because of this the properties of the vegetation resulting from its form and position with respect to the major input - precipitation - assume an important role. The soil properties affect the water cycle by exerting an influence on the kind of plant growth and its development, but the flow of water in the evapotranspiration stream through

Figure 13.1 Diagram illustrating the main components of the water cycle. C_i - canopy interception. D - deep drainage, EvT - evapotranspiration, F_i - forest floor interception, I - infiltration, P - precipitation, R_s - surface runoff, R_{ss} - subsurface flow, S - stemflow, T - throughfall, W_s - soil water storage

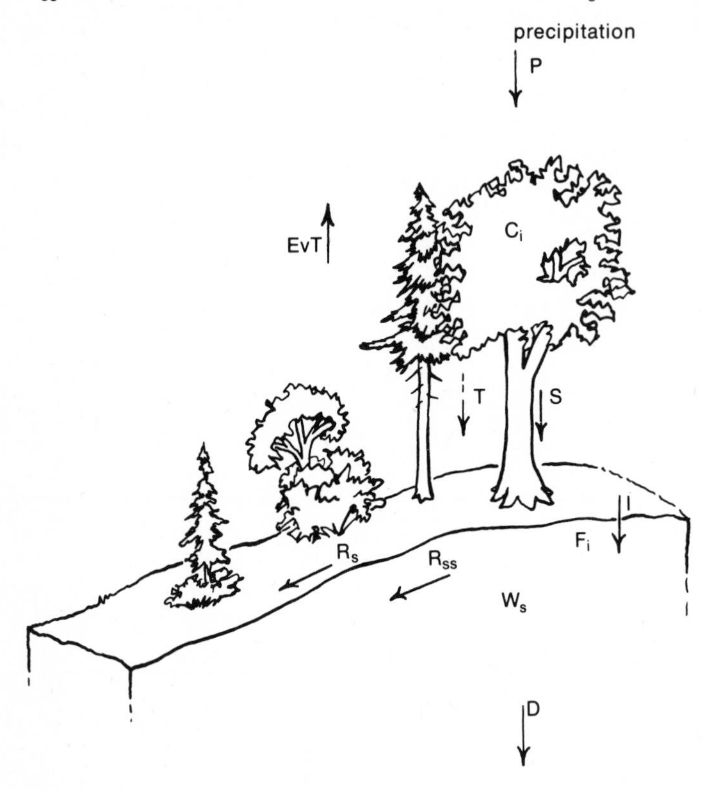

the vegetation is perhaps the most important process to affect soil moisture status. Much of the discussion in this chapter will deal with soil-plant relationships and their general role in the water cycle. For detailed considerations of plant-soil water relationships with particular emphasis on the physiological aspects, the reader is referred to Kramer (1969) and Slatyer (1967).

The water cycle

GENERAL CONSIDERATIONS

The water cycle can be portrayed in different ways, but Figure 13.1 shows a simplified cycle in which the movements of water in the soil-plant portion of the cycle are emphasized. Inputs of water into the soil are from precipitation and ground water. The amount or proportion of the precipitation (P) moving into the soil proper is dependent on:

Canopy interception (C_i) - the amount of water retained by the canopy of forest vegetation and evaporated back into the atmosphere.

Throughfall (T) - the precipitation which passes through the vegetation canopy and reaches the surface of the forest floor.

Stemflow (S). A certain amount of the water intercepted by the forest canopy is not evaporated back into the atmosphere but moves along branches to stems and down the stems to the soil and root system.

Forest floor interception (F_i). The water which enters the forest floor may be retained in part by it and evaporated back into the atmosphere.

Runoff ($R = R_s + R_{ss}$) - this is the water which moves as both surface runoff (R_s) and subsurface flow (R_{ss}). The water which moves downward into the soil or infiltrates (I) can be represented as the difference between precipitation and the losses due to crown interception, forest floor interception, and runoff:

$$I = P - (C_i + F_i + R).$$

The maximum rate at which water can enter the soil under a particular set of conditions is termed the *infiltration rate*, but the actual rate at any given time is termed the *infiltration velocity*. The rate of downward movement of water in soil that is saturated or nearly saturated is often referred to as its *percolation rate.*

Some of the water which infiltrates into soil will be retained within the soil pore system; a second portion may percolate in subsurface soil and flow to a point where it contributes to streamflow; and a third portion moves to deeper layers and may add to deep ground water stores. The upper level or surface of the ground water or zone of saturated soil is termed the *water-table.* In forest soils where the water-table or zone of moisture influence immediately above the table - the *capillary fringe* - is within the zone of rooting or the solum, it can exert considerable influence on both vegetation growth and soil development.

The major loss of water from the soil-plant system is due to *evapotranspiration*, i.e. the water moving to the atmosphere by evaporation from the soil surface and transpiration from plants. Evapotranspiration involves the change of state of water from a liquid to a gas (vapour) and this requires energy. The latent heat of vaporization of water at 25°C is about 580 calories per gram and the source of energy for evapotranspiration is radiant sunshine. This basic fact has led to the development of several techniques whereby estimates of evapotranspiration losses may be obtained primarily from climatic data. The determination of 'potential evapotranspiration' (Thornthwaite, 1948; Thornthwaite and Mather, 1957) and 'potential transpiration' (Penman, 1948, 1956) are examples of this approach. Penman (1963) and Taylor (1972) present full accounts of the methods of estimating evapotranspiration, the former with specific attention to vegetation and hydrology and the latter with emphasis in relation to the physics of irrigated and non-irrigated soils. It is important to recognize that not only will such factors as season, latitude, aspect, slope, and cloudiness affect the amount of energy received at the soil or plant surface, but also that local weather and the nature of the heat absorbing surface can be important. While evapotranspiration involves the transformation of energy and the movement of water vapour from the evaporating surface, it is dependent upon a movement of water through the soil-plant system to the surface where it is evaporated. The movement of soil water both to the soil surface but more importantly to the absorbing root surfaces of the vegetation is thus of great significance in determining actual water losses from evapotranspiration. No matter how great the energy available for evapotranspiration at a surface, there must be a flow of water of equal amount to the surface if the potential is to be achieved.

MEASUREMENTS

In addition to the use of climatic and other data for the calculation of potential and actual evapotranspiration losses and the various procedures which may be used to measure soil water (Chapter 4), there are other methods that have been used to assess water flow in plant-soil systems on both small and large scales. These employ lysimeters and stream-flow monitoring of watersheds.

Lysimeters

These are devices for measuring percolation and soil water losses from a volume of soil under a specific set of conditions. The common form is a container in which either an undisturbed block of soil or soil of desired properties may be placed. Water which moves through the soil and out the bottom is caught and measured. Drains may be placed about the surface perimeter to collect runoff for measurement and the water within the block may be monitored using conventional techniques. Lysimeters may be installed so that the total system can be weighed. Inputs from precipitation and irrigation are measured and the effects of different vegetation and treatments may be evaluated over both short- and long-term periods of time. Large lysimeters are costly to install and there are some limitations to their use. The fact that the water must move out of the soil-air interface of the drain at the bottom of the block restricts water movement unless the soil layer immediately above the interface is saturated. While this may be of little consequence for shallow-rooted plants in deep soil blocks, it can be significant for deeper rooting plants. Soil heterogeneity within the block can affect results and the relatively small surface area of the block and consequent likelihood of 'border effects' and root confinement are difficulties which have often been observed. These limitations can be particularly severe when lysimeters are used for growing trees (Patric, 1961).

Another form of lysimeter used in studies of forest soils is the *tension plate lysimeter* (Cole, 1958). The device consists of an alundum filter plate installed in soil in the field. A negative tension (suction) is maintained against the bottom of the plate approximating the matric tension at the soil's field capacity. Water passing through the plate can be collected, measured, and analysed (Cole, Gessel, and Held, 1961). This lysimeter has found application especially in determining the movement of nutrient elements in forest soils in which soil matric potentials remain fairly high throughout the year.

Watershed streamflow

In a catchment area, weirs or other devices for monitoring streamflow have been used to obtain information on the water budget for these areas. The assumption is made that streamflow (S) represents the difference between input (precipitation P) and outputs or losses due to evapotranspiration (EvT), interception (I), by both vegetation (C_i) and forest floor (F_i), change in water storage (ΔW_s) and losses due to deep drainage (D):

$$S = P - (\mathrm{EvT} + I + D) \pm \Delta W_s.$$

In hydrology, streamflow is usually equated with *runoff*, including both surface runoff and subsurface flow. The distinction is critical (Lull and Reinhart, 1972) since in forest soils where the mineral soil is not exposed, surface runoff or overland flow is negligible and changes in streamflow reflect differences primarily in subsurface flow. Rates of sedimentation can also be made in conjunction with streamflow studies.

A recent development in the technique for measuring transpiration and tree biomass is the use of tritiated water.* Kline et al. (1972) present data for jack pine and red pine plantations which were found to be comparable to estimates from other methods.

EFFECTS OF SOIL PROPERTIES

Physical properties of texture and structure, together with the nature and amount of the forest floor have the greatest effect on water infiltration, storage, and movement.

Infiltration

It is now generally considered that with a forest floor of appreciable organic accumulation the infiltration rate of virtually all forest soils is greater than the

* Water containing radioactive tritium (H_3).

Figure 13.2 Diagram illustrating change in infiltration rate for a loam, with time, at both high and low initial soil water contents.

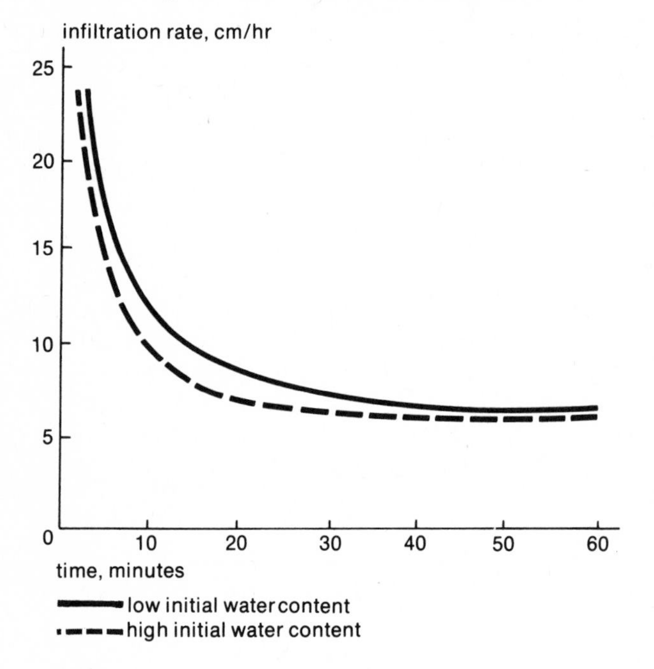

maximum rate of rainfall intensity. The forest floor absorbs the energy of raindrops and prevents the crusting and sealing of soil pores which occurs when mineral soils, especially of fine textures such as fine sands, silts, and clays, are exposed directly to rainfall. Infiltration rates for mineral soils decrease rapidly with time (Figure 13.2) and, although its rate is governed to a large extent by the soil water diffusivity (the flux of water per unit gradient of water content in the absence of other force fields) when the soil is saturated or nearly so, it is little influenced at drier soil moisture contents (Hanks and Bowers, 1963). As might be expected, the infiltration rates will be greater the coarser the soil texture. This is illustrated by results from a laboratory study by Moldenhauer and Long (1964) for soils of different texture and with varying simulated rainfall rates. Infiltration rates were measured over a 90-minute period and the final rates at the lowest (3.43 cm/hr) and highest (6.78 cm/hr) rainfall intensities are shown in Table 13.1. The differences in infiltration rates with texture assume importance in forest soils when the vegetation and surface organic layers are removed. When this occurs, surface and gully erosion is most likely during periods

TABLE 13.1
Final infiltration rates (cm/hr) for soils of different texture at low (3.43 cm/hr) and high (6.78 cm/hr) simulated rainfall intensities (from Moldenhauer and Long, in Soil Science Society of America Proceedings, 1964, by permission Soil Science Society of America)

	Rainfall intensity - cm/hr	
Soil	3.43	6.78
Luton silty clay	0.65	0.85
Marshall silty clay loam	0.58	0.73
Ida silt	1.94	1.48
Kenyon loam	1.33	1.10
Hagener fine sand	3.43	2.55

of moderate to high intensity rainfall in soils of a wide range in textures. An example of the effects of three types of disturbance occurring in forest soils is shown in Figure 13.3. The forest was old (280-300 years) western larch and Douglas-fir growing on a silty clay loam soil with average slope of 20 per cent. The large and consistent reduction in infiltration on the tractor skid road over a five-year period is notable, as is the greater increase compared to undisturbed soil of the broadcast burn treatment in the fourth and fifth year after treatment. It is likely that this increase in infiltration resulted from the vigorous regrowth of vegetation after the fire.

The capacity of the forest floor to absorb water is large. The type and thickness of the floor will determine the amount that may be held. Mader and Lull (1968) found that 6.4-cm-thick forest floors under eastern white pine stands had maximum water storage capacities of about 1 cm and the mean available storage capacity during the May to September period was 0.6 cm. Forest floors in central Washington under various stands, primarily coniferous, had maximum water holding capacities ranging from 1.9 to 3.2 cm and Bernard (1963) estimated for a range of conditions in New Jersey that the capacity was approximately 1.3 cm. Thus values of the order of magnitude of 1 to 3 cm would appear to represent the water storage capacity of many forest floors of moderate thickness. Golding and Stanton (1972) found no real difference in water holding capacities of forest floors under lodgepole pine and spruce-fir which were 0.18–0.19 cm per cm thickness of floor.

The interception losses of water by the forest floor, that is water held and then evaporated into the atmosphere under full vegetative canopy, are not large.

Figure 13.3 Infiltration on three soil surfaces as a percentage of infiltration on undisturbed soil. Undisturbed, natural forest floor; Scarified, forest floor removed or disturbed excessively by tractor in process of piling slash and soil scarification; Broadcast burned, areas or 'spots' burned; Tractor skid road, surface of roads used in tractor skidding of logs; it does not include jammer roads. From D. Tackle (1962), by permission United States Department of Agriculture, Forest Service

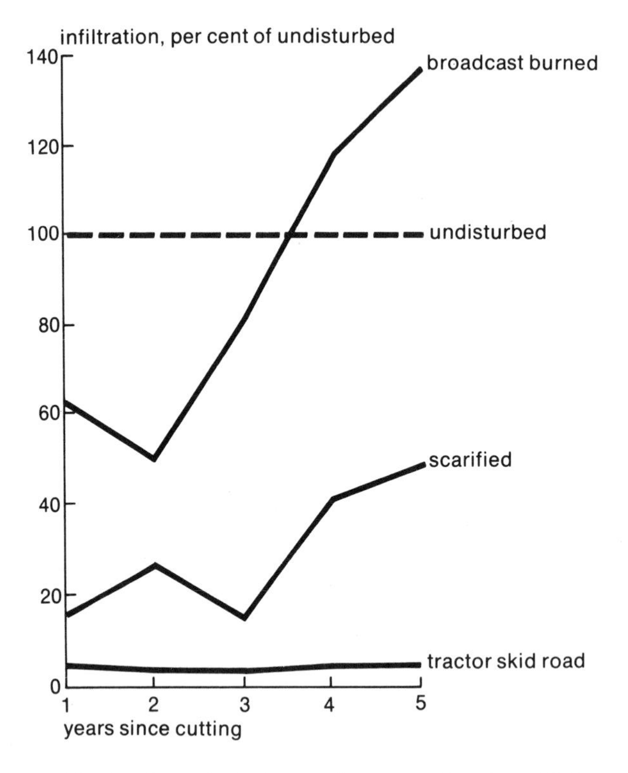

For hardwoods in the southern Appalachians, Helvey and Patric (1965) estimated that it represented only 2 to 5 per cent of the annual rainfall. In a detailed study of the effect of forest floors in ponderosa and Monterey pine forests in California, Rowe (1955) recorded annual evaporation losses from the litter surface in openings of from 3.8 to 6.6 cm for litter thickness that ranged from 2.5 to 9.1 cm. The mean annual value for 6.8 cm litter depth was 4.8 cm or 4 per cent of the mean annual precipitation which is within the range given hardwoods (Helvey and Patric, 1965). Comparisons between the effects of forest floor and bare surface are shown in Table 13.2, illustrating the negligible surface runoff where a

TABLE 13.2
Surface runoff, and amounts of water percolated and evaporated in relation to bare and forest floor covered soil in ponderosa pine forest (from Rowe, in Journal of Forestry, 1955, by permission Society of American Foresters)

Soil condition	Annual precipitation (cm)	Surface runoff (cm)	Percolation (cm)	Evaporation (cm)
Bare	93.5	33.8	25.1	34.6
forest floor		0.8	72.5	20.1

forest floor was present and the almost threefold increase in percolation water. The type of forest floor can alter the rate at which water may move into the mineral soil; Heiberg (1942) noted that infiltration rates into soil under mull were two and a half times greater than under a mor humus. Wright (1955) noted that in the afforestation of dune sands the thick felty mor which developed under Scots pine not only had a high water-holding capacity but inhibited movement of water through to the mineral soil. In contrast, under Corsican pine and birch an A1 horizon developed which, although it also held water, still allowed downward movement into the mineral soil.

The mechanisms controlling movement of water into mineral soil beneath different types of forest floor are complex. Physically the system of pore spaces and their continuity are important. Where a mor exists as a mat with no intermixing of organic and soil mineral particles, the pathways for water movement do not exist to the degree that they do where humus (H) and A1 (Ah) horizons occur and provide a connecting pore system between the upper part of the forest floor and the mineral soil beneath. The water-holding capacity can also vary and, although it increases with forest floor thickness, it is probably not the major cause of greatly different infiltration rates in different types of forest floor. Water-repellent conditions have been found to exist, especially in coarse-textured soils as a result of the production of hydrophobic compounds produced by surface organic materials. It has already been noted that such materials may move downward following fire (Chapter 12), but initial studies by Meeuwig (1971) indicate that water-repellency can develop under coniferous litter, particularly where soil faunal activity or roots from understory plants are not present.

The influence of temperature on infiltration and percolation is probably of only major significance in forest soils in which freezing or near-freezing temperatures occur. The occurrence of frost in soils has already been discussed (Chapter 3) and it has been noted that the amount of frost, particularly concrete frost, is usually greater in soil under conifers than under deciduous species. Although major floods can occur when heavy rains fall on frozen open ground, there does not seem to be evidence for major surface runoff in forest soils under vegetation even when concrete frost is present. As Lull and Reinhart (1972) state: 'If significant amounts of overland flow commonly occurred over concretely frozen soil in the forest, litter and humus would be disturbed. There are no known reports of this.'

Infiltration rates into frozen ground, as might be expected, are usually much less than into unfrozen soil. Both soil texture and type of frost are important. Stoeckeler and Weitzman (1960) found infiltration rates of 0.23 and 1.19 cm/hr into silt loam and loamy sand soils respectively with concrete frost. In unfrozen loamy sands the infiltration rate was 33.6 cm/hr compared with the same soil partly frozen, 10.1 cm/hr, and with porous concrete frost, 5.6 cm/hr.

It has, however, been pointed out by Klock (1972) that hydraulic conductivity is significantly affected over certain ranges in temperature because of associated changes in the viscosity of water. In mountainous areas of the Pacific northwestern part of North America, floods and landslides are frequently associated with snowmelt caused by rain and/or chinook winds.* Not only may the infiltration rates be reduced by low soil temperatures, but internal drainage or percolation will be slow. As an example of the reduction which may occur in infiltration rate using Klock's procedure, if the infiltration rate were 2.42 cm/hr at 15°c, it would be reduced by 32 per cent to 1.69 cm/hr at 2.5°c.

Storage

If the infiltration rates of forest soils exceed maximum rainfall intensity, it is the soil properties affecting moisture storage which will affect subsurface flow. Soil water storage consists of *retention* and *detention* capacities. Retention storage is water that is held in the soil pore system by matric potential forces and commonly is the water held below field capacity moisture status. Water in detention is that which moves in the large pores under the influence of gravity. The amount and size class together with structure determines the relative proportions of retention and detention capacities. The coarser the texture, the lower the retention capacity in relation to detention water, and conversely with finer-textured soils the amount of retention storage is proportionately greater in relation to detention storage capacity. In many forest soils there is normally a decrease in the ratio of detention: retention storage capacity with depth which reflects the greater macropore space in the upper soil layers associated with rooting intensity and higher level of biological activity. The change with depth is shown for a hardwood soil in Table 13.3.

* Warm winter winds occurring particularly east of the Rocky Mountains

TABLE 13.3
Change in water storage – retention and detention – with depth in a forest soil under hardwoods (from Hoover, in Soil Science Society of America Proceedings, 1950, by permission Soil Science Society of America)

	Water storage capacity (cm)	
Depth (cm)	Retention	Detention
0 - 15	4.37	4.90
15 - 30	4.80	2.79
30 - 45	5.10	1.57
45 - 60	5.87	1.02
Total	20.14	10.28

Obviously the water content of the soil is important in determining storage capacity for additions to the soil. Thus, if a soil is at field capacity or above, any water which infiltrates must move through the soil and add to the subsurface flow. During the growing season when soil moisture levels may be low due to evapotranspiration losses, additions of the magnitude which previously moved through the soil as subsurface flow may become part of the retention storage.

Soil depth controls soil water storage capacity. Shallow soils over impervious bedrock will retain less water than similar, deeper soils. Where fractured bedrock occurs, a simple relationship of decreasing water storage capacity may not be valid.

The content of stones (coarse fragments) over 2 mm in diameter can greatly reduce water storage capacities. Dyrness (1969) in a study of hydrologic properties of forest soils in Oregon determined that variation in stone content accounted for 80 to 87 per cent of the variation observed in retention and detention water storage capacity values respectively.

In absolute terms the water storage of forest soils will show a great range. Lull and Reinhart (1972) estimate that for soils of 10 and 25 cm water storage capacity in West Virginia, maximum storage opportunities during the April to November growing period ranged from a low of 3 to a high of 11.2 cm over 11 years. For deep soils total storage capacities of up to 80 or 100 cm are not uncommon.

EFFECTS OF VEGETATION

Interception and stemflow

The amount of precipitation which moves to the soil surface as stemflow (S) and throughfall (T) is equal to precipitation (P) less canopy interception (C_i):

TABLE 13.4
Equations for the estimation of throughfall, stemflow, litter interception, net rainfall loss, and interception loss for hardwoods of eastern United States (from Helvey and Patric, in Water Resources Research 1:193-206, 1965, copyright by American Geophysical Union), P = precipitation (seasonal) and n = number of storms per season.
N.B. Values and precipitation in inches

Value	Growing season	Dormant season
Throughfall (T)	$= 0.901\,P - 0.031\,n$	$= 0.914\,P - 0.015\,n$
Stemflow (S)	$= 0.041\,P - 0.005\,n$	$= 0.062\,P - 0.005\,n$
Litter interception (F_i)	$= 0.025\,P$	$= 0.035\,P$
Canopy interception (C_i)	$= 0.083\,P + 0.036\,n$	$= 0.059\,P + 0.020$

$$S + T = P - C_i.$$

The amount of interception will depend on the intensity of the precipitation, its form, and the nature and extent of the canopy. Canopy interception as a percentage of the precipitation decreases from maxima of 50 to 90 per cent when precipitation is less than 20 mm to between 10 to 30 per cent when precipitation exceeds 100 mm. Kittredge (1948) concluded that the interception storage varied from 0.5 mm to 2.5 mm per shower. For conifers, those species exhibiting high interception of rainfall will also show the same pattern with snowfall. Eastern hemlock, for example, which can intercept 33 per cent of rainfall compared with 19 per cent interception by red pine, also has a high snowfall interception. This characteristic makes hemlock stands preferred yarding areas in the northern limits of white-tailed deer in eastern North America. Generally hardwoods are considered to intercept approximately 10 and conifers 25 per cent of the annual snowfall which is subsequently vaporized. Where streamflow is primarily snow-fed, especially in certain alpine areas, the snow interception may be very important in decreasing water yields (Goodell, 1959). White pine will intercept as much as 17 per cent of a 5-cm rainfall and for Pacific Northwest conifer species interception appears to be of the order of 25 per cent of annual rainfall (Patric, 1966; Rothacher, 1963). Interception by deciduous species is greatest when they are in full leaf. Helvey and Patric (1965) prepared equations for the calculation of interception losses as well as other water budget values for eastern hardwoods in the United States (Table 13.4).

Stemflow is generally considered negligible for many species, especially those with rough bark; for some smooth bark species it may be a significant proportion of the precipitation. Mahendrappa (1974) found that during two growing seasons the proportion of the total rainfall contributed by stemflow during the period ranged from 0.7 to 6.1 per cent. Voigt (1960) noted that for American beech stemflow was considerably higher than for red pine and eastern hemlock and when the percentage stemflow of total rainfall was expressed on the basis of absorption area (the restricted soil area closely adjoining the bole of the tree and extending a radius of 30 cm from the bole), it was 276 per cent compared with 12 per cent for red pine and 95 per cent for hemlock.

Although most studies of interception and stemflow have focused upon the dominant tree canopy, Clements (1971) evaluated summer rainfall in a stand of largetooth aspen with three understory layers of red maple, hazel, and bracken. Interception by all canopies amounted to 22 to 46 per cent of monthly gross rainfall and more than half of this was intercepted by understory canopies. Stemflow constituted 8 to 17 per cent of the rainfall and of this bracken contributed the greatest proportion.

Stemflow and canopy interception have certain significance in the soil water budget. Water intercepted by the canopy and evaporated from the leaf surface cannot add to the soil moisture storage or contribute to subsurface flow. Where significant understory canopy interception exists, immediate gains in throughfall can be achieved by understory removal. The movement of stemflow water along main roots and its subsequent movement laterally into the soil are biologically important (Reynolds and Henderson, 1967) since it channels a certain amount of water to major rooting zones. An obvious difference in distribution of stemflow water for species which are tap or heart rooted, compared with those which typically have flat roots, can be expected.

Although it may only be important locally in regions that are mountainous or near oceans, fog drip from trees may increase the precipitation reaching the ground by two or three times the precipitation on open ground (Kittredge, 1948). In Hawaii, Ekern (1964) reported the precipitation including fog drip under a Norfolk Island pine to be 993 cm compared to 378 cm in the open. Although not as dramatic as fog drip, dew commonly occurs in many forests and both it and canopy interception moisture may result in some reduction in transpiration losses. Experimentally, Nicolson et al. (1968) determined for young eastern white pine and white spruce trees of about one metre height that intercepted water reduced transpiration for the species by 12 to 13 per cent of the total water intercepted. In other words more than 85 per cent of the intercepted water had no effect on transpiration losses. Intercepted water in tree canopies evaporates up to four times more rapidly than transpirational water (Rutter, 1967). Working with young ponderosa pine seedlings Stone and Fowells (1955) determined that dew could be of importance in increasing survival of planted trees.

Figure 13.4 Precipitation, estimated potential and actual evapotranspiration, and soil moisture storage for deep and shallow soils in a forest area in Ontario. Shading indicates when more than 50 per cent of available water has been used and is equivalent to severe drought conditions. From G. Pierpoint and J.L. Farrar (1962), by permission

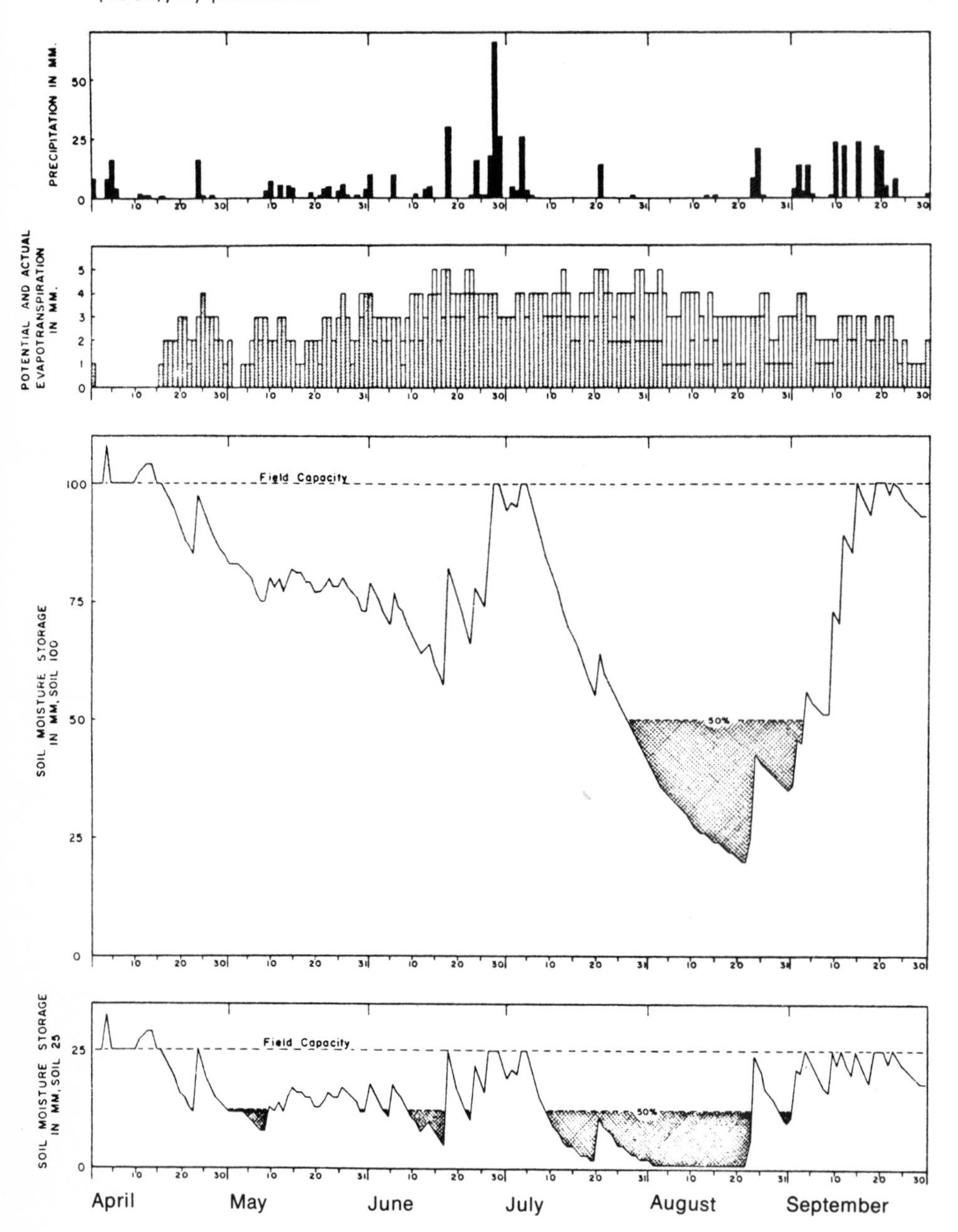

Transpiration

Water which is absorbed by plants and subsequently evaporated from their leaf surfaces is the transpiration component in the water cycle. The amount of heat energy received at the leaf surface will control the rate at which water evaporates and is lost by the plant-soil system. Direct measurement of transpiration is only possible using potted plants in lysimeters and this has obvious limitations for use with forest trees. Alternatively, formulae based primarily on climatic data can be used to calculate the energy budget available for transpiration and examples of this approach have been used successfully by Thornthwaite and Penman in particular. In forests where streamflow of specific catchments is monitored and transpiration can be altered by manipulation of the vegetation, it is possible to determine the magnitude of these changes from streamflow and other measurements. A general discussion of the theory and problems of measuring forest evapotranspiration by Federer (1970) has emphasized the necessity of combining micrometeorological methodology with data from soil physics and plant physiology.

An example of the use of a heat budget formula (Thornthwaite) to obtain a quantitative approximation of the water balance in a forest area is given by Pierpoint and Farrar (1962). Their estimates of runoff (surface and subsurface flow) were found to agree closely with measured streamflow for the same area. In the forest studied the soils derived from glacial till occur in varying depths over bedrock and two soil moisture storage capacities were assumed - 100 mm for the deeper soils and 25 mm to represent shallow soils (Figure 13.4). This approach is particularly useful in providing an overall view of the soil water budget.

The transport of water in the soil-plant-atmosphere system is dependent on characteristics other than climatic and soil properties. Depth of penetration of roots into soil, the density of rooting per unit of soil volume exploited, the internal resistance of the plant to water flow, and the critical value of leaf water potential related to stomatal closure are all important in determining moisture loss (Cowan, 1965).

An illustration of the evapotranspiration and other water values for bare soil, herbaceous vegetation, and aspen-herbaceous cover is given in Table 13.5 for an area in Utah. It can be noted that evaporation from bare soil was much less effective than transpiration in removing water from the soil. In this study it was found that the aspen roots occupied the soil to a depth of 1.8 m, whereas the roots of the herbaceous vegetation extended to 1.2 m.

Transpiration takes place from leaf surfaces and therefore it can be expected that factors which alter the leaf surface area may change the pattern and amount of soil moisture depletion. The difference that may occur between an evergreen and deciduous species is illustrated in Figure 13.5. Virtually no transpiration takes place from the oak stand until full leaf development in June, whereas the

TABLE 13.5
Summary of total precipitation, water losses, and water available for streamflow from bare soil, soil with herbaceous cover, and soil with aspen – herbaceous cover – mean annual values for a 3-year period (from Croft and Monninger, in Transactions, American Geophysical Union 34:563-74, 1953, copyright by American Geophysical Union)

Water factor	Bare soil	Herbaceous	Aspen-herbaceous
Precipitation (cm)			
Winter	115.4	115.4	115.4
Summer	18.6	18.6	18.6
Total	134.0	134.0	134.0
Water losses (cm)			
Summer interception	0	2.0	2.9
Snow evaporation	7.6	7.0	6.4
Summer evapotranspiration	28.5	37.7	45.0
Winter evapotranspiration	0	0	2.5
Total	36.1	46.7	56.8
Water available for streamflow			
Surface runoff	1.0	0.05	0.03
Subsurface runoff	96.9	87.4	77.2
Total	97.9	87.45	77.23

red pine transpired water in April and May. The removal of soil water for transpiration by forest trees takes place initially in relation to root concentration from a uniformly moist soil and later absorption tends towards more equal extraction with depth. This pattern of water removal for soil depths up to 1.5 m is illustrated in Figure 13.6. A depletion pattern with depth for deciduous trees growing in deep soil materials is shown in Figure 13.7; chestnut and scarlet oaks were among the main tree species and their roots extended to a depth of 5.5 m. In layered soils it is common to observe differential root development at depth particularly when bands of finer-textured materials with high moisture contents occur within a matrix of coarser-textured soils. Although the amount of water which may be absorbed by relatively few roots in such heterogenous conditions is difficult to assess, the greater tree growth usually associated with it would indicate that it is significant (White and Wood, 1958).

For equal volumes of rooted soil during the growing season when transpiration is taking place, the amount of water transpired appears to be relatively independent of species. Zahner (1955) estimated that, during a six-week summer period, water depletion was about 48mm per day from the upper 1.2 m of soil

Figure 13.5 Patterns of soil moisture depletion for oak and red pine growing on deep sands in Michigan. From D.H. Urie (1959), by permission United States Department of Agriculture, Forest Service

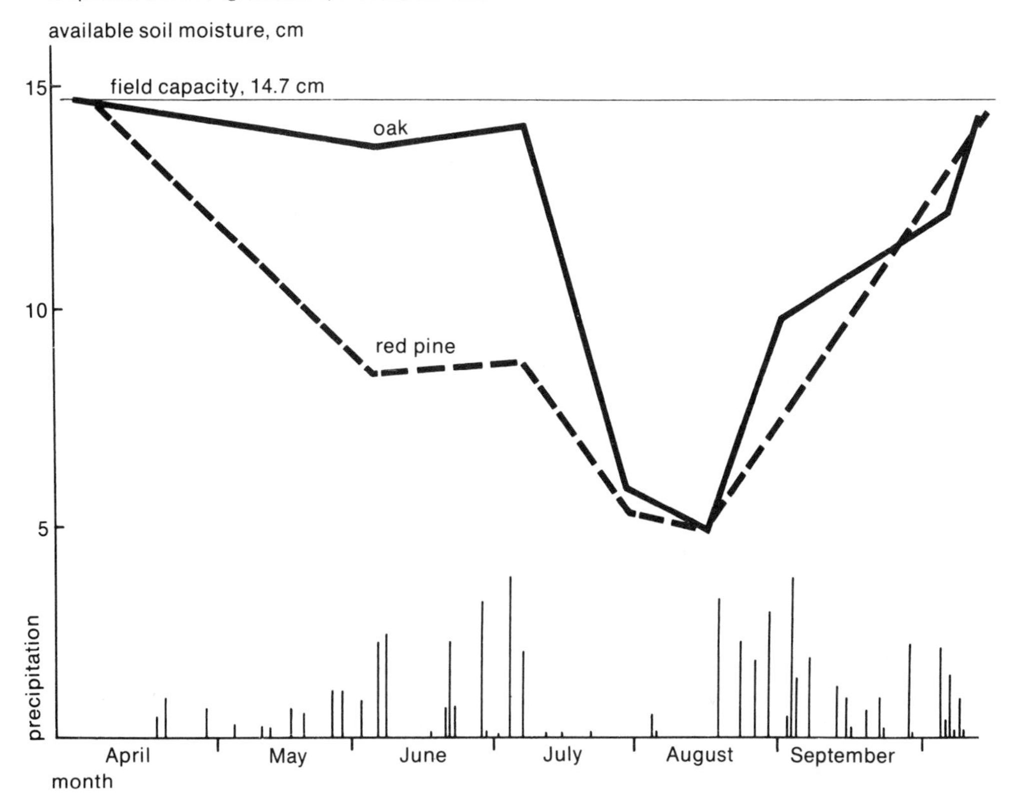

from both a 35-year-old oak stand and a 20-year-old loblolly-shortleaf pine cover. In Africa, Pereira and Hosegood (1962) could find no difference in water use by pines, cypress and bamboo forests and concluded that the substitution of exotic conifers would have no adverse effect on the water budget. The density of vegetation in some conditions can affect amount of soil water use; for example, in a ponderosa pine forest growing on a coarse loamy sand pumice soil, thinning and understory vegetation removal resulted in dramatic differences in water use; at a pine density of 2,471 trees per hectare, the water use was 1.6 times that at a density of 153 trees per hectare and use was 45 per cent greater when the understory vegetation was present than when it was removed (Barrett and Youngberg, 1965).

Figure 13.6 Initial six weeks of water loss from four depths in a silt loam under pine and hardwood cover. From R. Zahner in *Forest Science* (1955), by permission Society of American Foresters

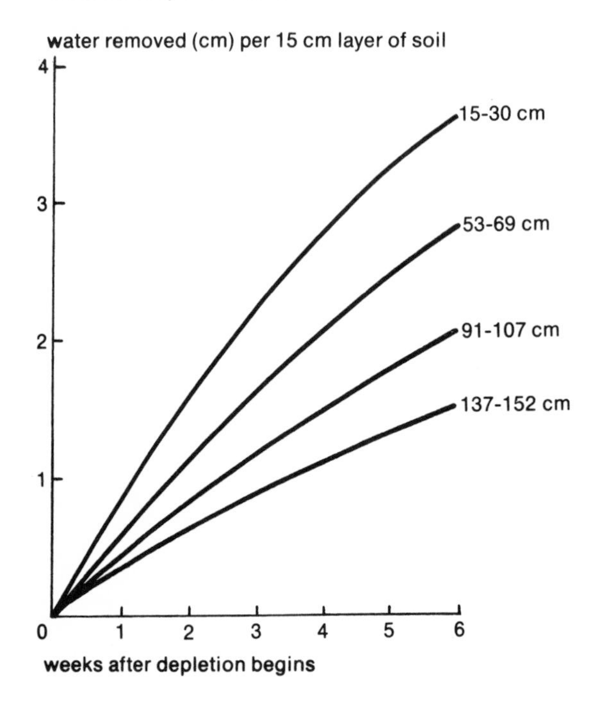

Figure 13.7 Soil water absorption rates (cm) for deciduous species growing on a deep mountain soil. Rates shown are total and for 0–1.8, 1.8–3.7, and 3.7–5.5 m depths. From J.H. Patric, J.E. Douglass, and J.D. Hewlett in *Soil Science Society of America Proceedings* (1965), by permission Soil Science Society of America

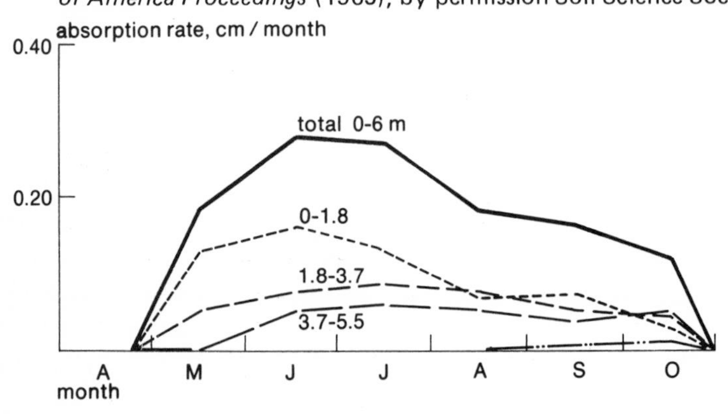

From a summary by Ovington (1962) of annual water circulation in a number of forest stands in both lysimeter and catchment installations it is evident that evapotranspiration losses approximate 50 to 70 per cent of the total precipitation. Conifers generally are in the upper part, 60 to 70 per cent, whereas deciduous species are usually but not necessarily somewhat lower.

FOREST MANAGEMENT EFFECTS

In forest management the interrelations between the water budget and vegetation are of direct consequence to the objectives of management. In practical terms these objectives relate primarily to plant growth, water yield (including water-table location), and erosion. Although management may be concerned with only one of these objectives, the consequences of actions may impinge on the achievements of other objectives. A forest managed to produce the maximum current increment of wood fibre may be manipulated silviculturally to do this and at the same time result in marked changes in water yields and erosion.

Plant growth

During the growing season in many forest areas, supply of soil water is the most important factor limiting growth. The occurrence of natural forest types often reflects differences in water supply; changes in forest tree species' composition and growth with differences in soil moisture supply are commonly observed and recorded. Height growth in seedlings and young trees is often inhibited by reduced soil moisture supply. Stransky and Wilson (1964) found for loblolly and shortleaf pines that terminal elongation was inhibited with soil moisture matric suction of no more than two and stopped completely at three and a half atmospheres. Reduction in diameter growth in relation to change in soil moisture supply was measured by Moehring and Ralston (1967), who noted that, regardless of the amount of soil water, growth was curtailed when rates of evapotranspiration losses were rapid and consequently the soil moisture content at which growth ceased was variable. This same pattern of growth response was observed in a study by Bassett (1964), who found for shortleaf and loblolly pine that diameter growth ceased when soil matric suctions in the upper 30 cm of soil were greater than three to four atmospheres during July and August. It is commonly assumed that water available for plant growth is represented by the amount held at soil water potentials between field capacity and wilting (a range of one tenth or one third to 15 atmospheres). It is increasingly apparent that the amount and rate of supply – the flux of soil water – are more critical and that moisture stress together with associated physiological changes and growth reduction can frequently occur at matric suctions well below 15 atmospheres.

Manipulation of stand density is one of the most obvious techniques by which water use can be altered. Usually thinning can achieve this: for example, the increased water supply resulting from heavy thinning in lodgepole pine averaged 11 cm per year over a five-year period (Dahms, 1973). In one of the early studies of growth response due to increase in soil moisture resulting from thinning, Zahner (1955) showed that the increase in diameter growth resulted from both an increased rate and duration of time over which diameter growth took place. Following the reduction in density by thinning, there was an associated increase in lateral root extension and increased moisture uptake was partly attributable to the greater absorption surface areas of the residual trees. The increase in soil moisture absorption which can occur as root extension takes place is very important in providing a plant with water (Kramer and Coile, 1940).

There are often qualitative as well as quantitative differences in plant growth when soil moisture supply changes as a result of a management practice. Understory vegetation will increase in abundance, frequency, and number of species. That this is not entirely a response to light and temperature but rather to soil moisture was demonstrated experimentally in North America by Toumey and Kienholz (1931) in trenching studies.

Water yield

A removal of vegetation will normally result in an increase in water yield. In the northeastern United States Lull and Reinhart (1967) estimated that removal of all vegetation on a well-stocked forested watershed would increase water yield to 10 to 30 cm the first year. The increase in water yield as measured by streamflow comes primarily in the mid to late growing period. Federer (1973) noted in New Hampshire that forest transpiration speeded the recession of streamflow. During the dormant season streamflows continued for days at rates of 1 mm per day, whereas during the growing season the recession rate declined within a few days to less than 0.05 mm per day. The transition between these regimes occurred in June and late September–October, coinciding with the time of transpiration start and finish. In other areas, depending on the types of forest and climatic conditions, the pattern and amount of water yield may differ. In a Pacific Northwestern area in Oregon, clear cutting the coniferous forest resulted in an increase of annual water yield of 45 cm, most of which occurred in the October to March period (Rothacher, 1970). Where snowmelt is a significant part of the streamflow, forest cover will prolong or delay the melting period and, when the forest is removed, the streamflow pattern will change accordingly. The different effects that deciduous and evergreen forests have in terms of snow interception and soil snow cover have already been discussed and removal of hardwood vegetation would not normally be expected to change snowmelt patterns to the extent that removal of evergreens would.

Figure 13.8 Changes in soil moisture storage for barren, clearcut, and forested soils in West Virginia. From C.A. Troendle (1970), by permission United States Department of Agriculture, Forest Service

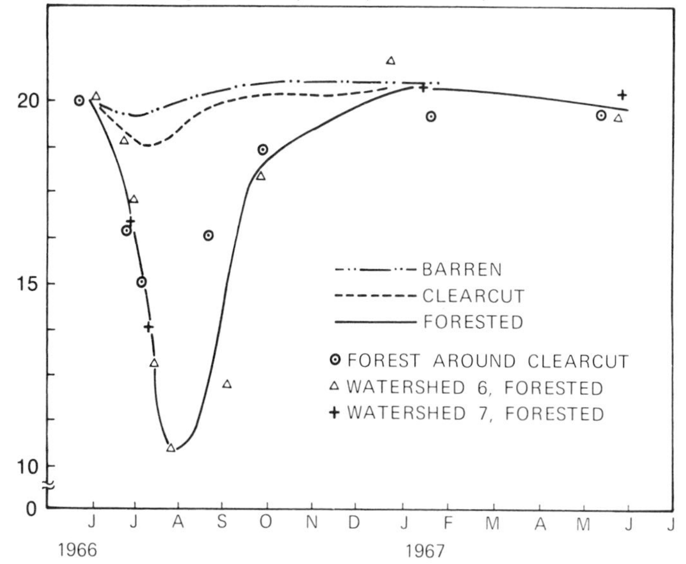

The increase in water yield resulting from drastic reduction in transpiration occurs because the soil water content remains high and as a result, when additional water infiltrates following rainfall, the storage capacity is slight, compared with what it would be if water were being removed by transpiration. Figure 13.8 shows the differences in water storage for barren, clearcut, and forested soils. When soil water content is high, the percolation rates will be high because of the increase in hydraulic conductivities. Cole (1966), using tension lysimeters, provided evidence of this increase in flow-rate within the soil and its response over a five-day period to precipitation inputs (Figure 13.9). Removal of vegetation resulted not only in a shortening of the delay to peak flow but it also increased the flow rate. Partial removal of vegetation can increase streamflow and Douglass (1967) indicated that the water saved from evapotranspiration is proportional to the severity of forest cutting. Using data from a series of northeastern watersheds, Douglass and Swank (1972) calculated a regression for estimating first-year streamflow increases as a function of reduction in stand basal area:

$$Y = -1.39 + 0.13\,X$$

where Y is the first-year increase in streamflow (inches) and X is the reduction in forest stand basal area (per cent).

Figure 13.9 Flow of soil moisture at 91 cm depth under forested and clearcut conditions. From D.W. Cole in *Proceedings Society of American Foresters* (1966), by permission Society of American Foresters

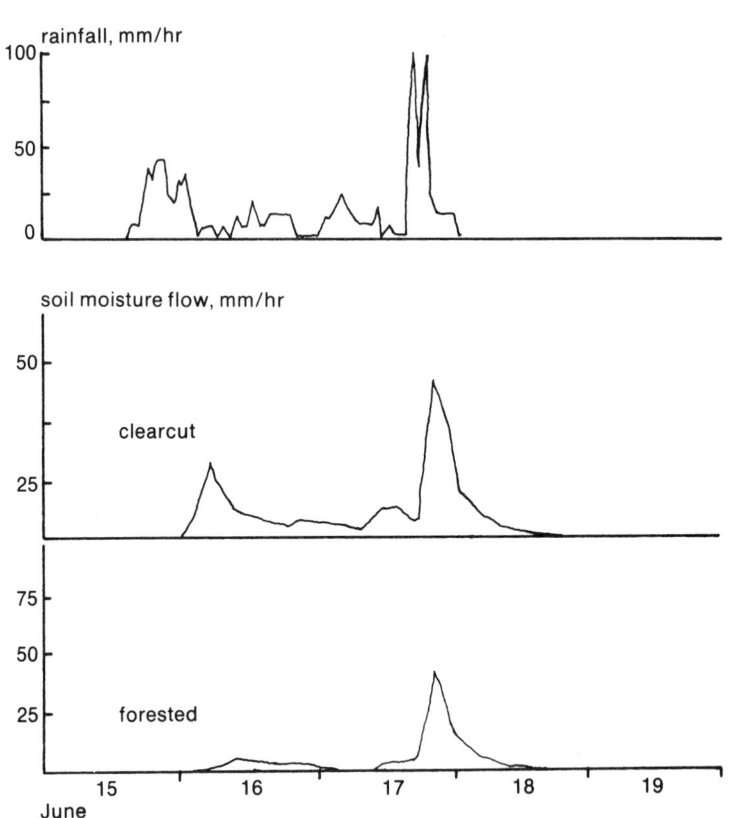

Under normal conditions revegetation takes place and, as might be expected, there will be a concomitant decrease in streamflow (Table 13.6). Although the foliage mass 13 years after the clear cut treatment is almost the same as prior to treatment, streamflow is still significantly greater, and this could be due to an incomplete root exploitation of the soil, especially at depth, together with a much smaller total standing biomass of vegetation, which represents a considerable storage capacity for water. Hibbert (1967) presenting data for the same area showed that after a 23-year period of regrowth following clear cutting, a second clear cut resulted in a return to the same increase in streamflow as resulted from the original clear cut.

TABLE 13.6
Relative change in streamflow with changes in forest basal area and foliage mass for a clearcut area - Coweeta (from Meginnis, 1959)

Time	Stand basal area (m^2)	Annual foliage mass (kg X 10^3)	Streamflow increased over yield prior to treatment (cm)
Pretreatment	10.3	3.26	0
1st year after cut	0	0	38
13th year after cut	4.8	3.17	13
40th year after cut (estimated)	9.3	3.26	0

Removal of understory vegetation can sometimes be reflected by streamflow increases. Johnson and Kovner (1956) found that removal of mountain-laurel and rhododendron increased annual streamflow by 5 cm for six years following treatment. This suggests that the increase in soil moisture resulting from understory removal is not necessarily used by the residual vegetation. Further evidence is provided by Lambert et al. (1971) from a seven-year-old red pine plantation growing in glacial outwash. When competing weeds were killed by herbicide treatment, they found that of the 2.7 cm precipitation occurring in the June–August period, evapotranspiration was 1.6 cm where weeds were present and only 0.9 cm where weeds were removed; virtually all the difference in evapotranspiration was water lost from the root zone which resulted in a 61 per cent increase in drainage water.

Water table

When roots from transpiring plants have access to water at a water-table, it follows that if they are removed the general level of the ground water should rise. Evidence that this does occur has been provided from a number of studies such as that of Trousdell and Hoover (1955), who demonstrated that the rise in the water-table as a result of removing vegetation would occur over short distances, as for example when the strip cut was only 20 to 40 m wide. A striking example of the effect of transpiration losses over a period of one day is given in Figure 13.10 and the removal of water by transpiration in the afternoon period is clear. The yearly variation in water-table movement was studied by Holstener-Jørgensen (1961) who found that the lowering of the table was the same from year to year, although the character or pattern of lowering varied. He was able to estimate the differences in evapotranspiration for stands of different species and to determine that, during the winter months, Norway spruce consumes water, com-

Figure 13.10 A comparison of rates of water level change in shallow wells under forest and clearing. From P.W. Fletcher and R.E. McDermott in *Soil Science Society of America Proceedings* (1957), by permission Soil Science Society of America

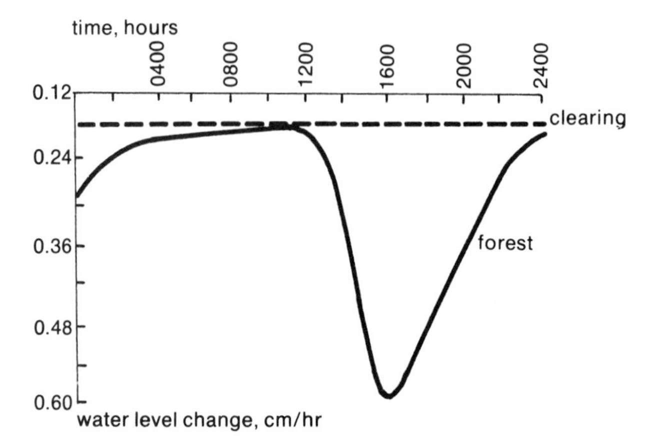

pared with lack of consumption by European beech. An important conclusion from this study was that the position of highest stable water-table is an important factor. During the growing season the water table was lowered less in European beech stands where the highest stable water-table is higher than where it is deep and the water-table was lowered less in young than in old stands. This response was not valid for other species such as oak, but there was a general increase in ability to lower the water table with increase in rooting intensity. In a later study (Holstener-Jørgensen et al., 1968) from detailed measurements about single trees showed that water consumption was greatest near the trees, so that during the day the water table was lower beneath the tree than 2 to 6 m distance away, but at night there was ground water movement to restore the table under the tree.

Within an area where water-tables occur periodically, the patterns of fluctuation give rise to differences in soil morphology and this, in turn, reflects changes in specific soil properties. Lyford (1964) observed the movement of organic substances in water even in the winter and suggested that where disturbances such as fire, blowdown, decay, insect injury, and cutting affected the vegetation, certain soil-forming processes might occur during short periods of a few years. In areas where the soil is normally frozen, winter thaws have caused increases in ground water levels (Sartz, 1967), and this can affect water consumption and root activity in the spring.

Erosion

Erosion processes based on water movement may be sheet and gully, streamflow, and mass earth movement. Erosion is natural; the management of forest soils should be concerned with ensuring that the natural rates are not increased.

Sheet erosion and associated gully erosion do not occur to any significant degree if the forest floor is intact. Under such a condition the infiltration rates are usually much greater than maximum rainfall intensity. Where the surface organic layers are removed or destroyed either by accident or in the process of management, the possibility of sheet and gully erosion must be considered, and appropriate counter measures taken. A modicum of forethought can often prevent the destruction resulting from inappropriate management activities. Finer-textured soils such as silts and clays are particularly susceptible to sheet erosion when surface organic layers are removed. With the increase in scarification of forest soils for regeneration purposes, the probability of greater sheet and gully erosion also increases, unless the same principles that have been effective in agricultural tillage practices are followed. These consist of cultivating in such a manner as in contour plowing so that the rate of overland flow is as slow as possible and avoiding baring or cultivating the natural depressions where gully erosion could begin. Smith and Stamey (1965) provide certain guidelines of tolerable erosion from soils with vegetative cover and, allowing for some increase as a result of man's activities, suggested tolerable limits of annual sediment yield to be 224 to 1,345 kg/ha. Megahan (1972) has summarized the problems of sheet and gully erosion in forestry activities by emphasizing that probably 90 per cent or more of the increase in erosion is a result of roads. The disturbance is due not only to the road itself and drainage channels but frequently results from the areas of soil that are bared or moved carelessly during the construction of the road. Roads may be carefully located and built to minimize disturbance when they are completed, but it is during their construction that much of the accelerated erosion is initiated.

Streamflow

Streamflow erosion always occurs, but concern is generally with the large increases associated with storms and flood conditions. The influence of the forest in modifying stormflows varies considerably with the size of storm, the season, and the watershed area in relation to stream channel storage capacity. Lull and Reinhart (1972), in a comprehensive review of forests and floods in the eastern United States, concluded that sustained yield programs of forest management offer no flood threat when attention is directed to maintaining appropriate age class distribution and degree of forest cover and that the potential for reducing flood damage in forestry was greatest in the reforestation of non-forested lands which produce substantial overland flow. As an example of the amounts of sediment that may be associated with different soil cover in Wisconsin, they cite a range of suspended sediment (ppm): clean, tilled soil, 238,000; heavily grazed woodland, 55,900; alfalfa meadow, 19,800; logged forest, 3,600; old field, 300; undisturbed forest, 100.

Mass earth movement

Soil mass movements are the dominant natural process of erosion and slope reduction in the steep, mountainous regions of western North America (Swanston, 1971). Mass movements are related to two general types of conditions. In the first, movement occurs when shallow soil materials occur over impermeable layers and in the second the flows are deep seated and massive. In both instances the movement stems from instability resulting from the shear strength of the material being less than the gravitational stress. Water content and slope, length and degree are the major contributing factors. As soil water content increases, the cohesion between particles, particularly finer-textured ones, becomes less, the weight of unit volume of soil increases and, depending upon slope angle and length, the ability of the soil to remain in place will be overcome by the force of gravity. Where permafrost exists, mass movement or flows can occur on slopes of five per cent or less and soil materials which in other climatic areas would never be subject to soil flow. The ability of root systems of trees to increase shear strength and stability has been mentioned in Chapter 11.

Irrigation and drainage

Irrigation and drainage are management practices which affect the water cycle by adding and removing water, respectively. Both practices are undertaken under certain conditions to increase growth of natural or more usually introduced vegetation. When irrigation is used to supplement natural rainfall in order to increase plant growth, additions are usually made to the root zone during periods when soil moisture supply is low and efficiently undertaken irrigation ensures that a very high proportion of the added water is absorbed by the crop roots. The additional water does not, therefore, move in subsurface flow and contribute to streamflow, but is recycled rapidly into the atmosphere by evapotranspiration.

Another type of irrigation is the recycling of waste water in which treated municipal sewage effluent is spray-irrigated on forest soils. The longest period, experimentally, during which such a practice has been undertaken is 11 years and rates ranged from 2.5 to 15 cm a week which were considerably less than the infiltration capacities of the soils supporting white spruce, red pine, and mixed deciduous forest. The amounts of water added over periods of 23 to 52 weeks were quite large and contributed to the surface flow. Most attention has focused on the effect of these additions on water quality.

Drainage of forest lands has somewhat different results depending upon the type of soil to be drained. Where mineral soils are drained, the main concern is with minimizing of sedimentation and erosion, primarily at ditch sides. The effectiveness of drainage depends mainly on the texture of the soil and the intensity of draining. The *drainage capacity* of a soil can be expressed as the difference between the total pore space and the pore space occupied by water at the field capacity expressed as per cent. The drainage capacity of sands and sandy loams is considerably greater than that of clays.

When organic soils are drained, it should be recognized that as a result of improved aeration there is increased microbiological activity. These soils are warmer than if saturated and the improved aeration, nutrient supply, and temperature regime will increase the activities of heterotrophic organisms which use the organic soil as a substrate. Records of the subsidence of organic soils in the Florida Everglades since 1914 show the average rate to be 32 mm per year - twice the rate for subsidence in the central United States.

Irrigation and drainage are undertaken on a very small proportion of North American forest soils. Locally the major concern is with disruptive changes in the water cycle, primarily related to ground water levels and streamflow.

14
Nutrient cycling

Living organisms require a supply of nutrient elements and much of it is dependent on a cycling of these elements in the biosphere. Carbon and nitrogen are two elements whose prime source in the cycle is the atmosphere, whereas others such as calcium, phosphorus, and potassium enter the cycle from weathering of geologic materials. These latter are termed 'mineral elements' because their source is the minerals of the earth's mantle.

A consideration of nutrient cycling is somewhat analogous to a discussion of economics. The size of the area which is taken into account can be of great importance. The economic factors of major concern to a person working in a business or factory in a small town are income and cost of living and both are capable of clear, quantitative expression as wages and expenses. To this person global-scale economic factors may seem remote, but periodically they impinge on his own situation, usually in a disruptive manner. In a parallel, the nitrogen economy of a forest soil may be in a certain state in terms of inputs and outputs so that they are essentially in balance, but through natural or management activities – fire, insect attack, windstorm, harvesting, grazing – the economy is disturbed either directly or indirectly. Thus the assessment of nutrient cycling must take into account the actual or most probable changes that can occur when what may appear to be a 'balanced' cycle or economy is disturbed.

The most common approach to the study of nutrients is to prepare a balance sheet or budget. In order to do this, the amounts of an element in the larger components of the system are determined. These represent the pools of nutrients. We speak of the pool of total nitrogen or potassium in the soil, biomass, or forest floor. The size of the pool may be stable or it may increase or decrease, and these changes or lack of change reflect the differences in the inputs and outputs for the particular pool. The changes can be expressed quantitatively as the amount of an element and its rate of transfer from one pool to another; this is the *flux*.

The major sources of input of elements into the forest-soil ecosystem are atmospheric, geologic, and biologic. Elements from the atmosphere enter either as gases, in solution or as particulate matter. From the geologic source they may be present as solids or in dissolved form and they may enter the system as seepage flow or as material deposited by water, volcanic action, or other depositional force. The biologic sources of input are in most situations related to man, for example, the introduction of fertilizers. If these are the three sources of input, they are also the sources or means of loss. Certain elements may be volatilized and enter the atmosphere (nitrogen and carbon); others may be removed in erosional processes (geologic losses); removals by animals or man, the main biologic agencies causing loss, are obvious.

The inputs and outputs related to these three sources or agencies are most important in the overall cycling of elements, but within a particular forest ecosystem the smaller internal cycles, for example the cycle of an element from soil to vegetation to soil, may be of critical importance for the growth of that particular vegetation. If there is a blockage in the part of the cycle from the soil back to the plant, as for example an immobilization in the forest floor, the development of the vegetation may be restricted. Yet, in terms of total budget, the amount within the system remains the same. Here again the analogy with economics is appropriate. On a national basis the economy may be in fine shape, but within certain regions surplus or shortage of raw materials or labour supply can depress the economic life of the community, and such a condition is usually only alleviated by a removal of the inhibiting factor. For a forest manager it is often the local or internal cycles and associated processes which are of immediate concern, yet in a broader ecological sense these cannot be considered as separate from the overall budget of the system.

There are many sources of data relating to nutrient cycling in the literature and Rodin and Brazilevich (1967) present a comprehensive review of previous studies.

General considerations

Although many studies of particular aspects of nutrient cycling have been made and provide much useful information, there is a need for studies which enable the total inputs and outputs to be determined. In order to do this, it is necessary to study cycling on a watershed basis. As Bormann et al. (1968) pointed out, the watershed should be underlain by tight bedrock or other impermeable base so that unmonitored losses are negligible; hence a balance may be determined from the difference between inputs and outputs for the main sources – meteorlogic, geologic, and biologic. Watersheds that are instrumented for hydrologic studies are often also most appropriate for work on nutrient cycling.

An approach used primarily in the study of cycling in Douglas-fir soils of the Pacific Northwest utilizes a tension plate lysimeter (Cole et al., 1961) and the fluxes associated with soil water movement are determined. These data may then be related to analyses of nutrient pools in the soil and biomass as well as measures of meteorologic inputs. In order to obtain estimates of nutrient quantities in the various pools, some form of sampling must be undertaken. In Appendix 1 the matter of size and variation in sampling forest soils is discussed. Somewhat similar problems exist when the pools in the biomass are sampled and studies such as that of Ovington et al. (1968) provide guidance for tree biomass estimation. Many studies, by their nature, can provide information only about nutrient cycling over limited periods of time. Frequently, for example, inputs or changes in pool size are measured over a few weeks or one growing season and yet for forest ecosystems this timespan is usually a very small proportion of the period of stand development. Another approach puts emphasis on the study of systems which are thought to be in a state of dynamic equilibrium as in natural forests of all-aged, self-regenerating species which have not been disturbed. These are, however, infrequent and in the main unrepresentative of the forest conditions which occur generally.

Nutrient inputs and outputs

ATMOSPHERIC

The basic source of carbon and nitrogen is the atmosphere. It is generally considered that the supply of carbon dioxide is non-limiting to natural vegetation, although under controlled growing conditions growth responses have been obtained when carbon dioxide concentrations have been increased.

The amount of nitrogen and other elements added in precipitation varies considerably, but is generally less in the cool temperate regions and greatest in tropical regions. Table 14.1 gives values for precipitation inputs for nitrogen and other elements from studies in various forest regions. Proximity to industrial areas or large fires may increase additions, particularly of nitrogen. In coastal areas, significant additions of sodium, calcium, and magnesium may be supplied from ocean spray. Additions of nitrogen from the atmosphere are obtained by symbiotic and non-symbiotic fixation (Chapter 6), and these amounts may be considerably greater than the quantity added in precipitation. Table 6.2 shows values of the order of 1 to 55 kg/ha/yr by non-symbiotic fixation and symbiotic fixation probably accounts for amounts of the same order if not greater in some instances. In a study of the nitrogen budget of a tropical rain forest in Puerto Rico, Edmisten (1970) found the same rain input of 14 kg/ha/yr of nitrogen as Nye (1961) did for Nigeria (Table 14.1) and an output of 29 kg/ha/yr. These

TABLE 14.1
Inputs of elements in precipitation for various forest regions.
All values for elements in kg/ha/yr

Precipitation (cm)	N	P	K	Ca	Mg	Area and source
185	14.0	0.41	17.5	12.7	11.3	Nigeria (Nye, 1961)
–	–	–	12.5	14.0	3.3	Malaysia (Kenworthy, 1971)
215 - 251	0.90 -1.08	0.27	0.11 -0.27	2.33 -7.65	0.72 -1.32	Oregon (Fredriksen, 1972)
–	–	–	2.4	3.3	2.1	New York (Woodwell & Whittaker, 1968)
95.3	4.94	0.09	4.0	5.6	0.8	Central Ontario (Foster, 1974)
173.9	9.13	0.35	3.91	12.54	5.36	Great Britain (Carlisle et al., 1967)
–	0.8 -4.9	–	1 - 4	6 - 9	–	Sweden (Emanuelson et al., 1954)
–	–	0.28	5.6	3.4	–	New Zealand (Will, 1959)
98.2	–	–	2.01	2.74	5.36	Australia (Attiwell, 1966)

amounts were small in relation to the amount of nitrogen cycling and exchanging with the atmospheric nitrogen which was estimated at more than 100 kg/ha/yr and resulted in a major net gain to the system. Much of this nitrogen was probably added to the system, not in the soil but through fixation by organisms on the above ground portion of the vegetation, primarily leaves. Nitrogen-fixation of this type is termed *phyllosphere* fixation. The introduction of higher plants with dinitrogen symbionts is therefore a common practice to bring about an acceleration of nitrogen addition from the atmosphere. Gadgil (1971) demonstrated that by introducing a legume on coastal dunes in New Zealand large enough quantities of nitrogen would be fixed over a four-year period that Monterey pine could then be planted. The amounts of nitrogen which may be fixed by natural stands of alder have ranged from 15 to over 100 kg/ha/yr. Van Cleve et al. (1971) provide data for accretion over a 20-year period by alder stands of 156 kg/ha/yr. It is also known that the introduction of many conifer species enhances the fixation by microorganisms in the soil (Richards and Voigt, 1965), but the nature of the mechanism is not known, although estimates of 56 kg/ha/yr have been made for the amount fixed.

The chemical composition of the portion of the precipitation reaching the forest floor as throughfall and stemflow is somewhat different from that of pre-

TABLE 14.2
Chemical composition of precipitation and throughfall for an old (450-yr) Douglas-fir forest (Abee and Lavender, 1972) and a moist tropical forest (Nye, 1961). All values in kg/ha/yr

	N	P	K	Ca	Mg
Douglas-fir Forest					
Precipitation	1.40	0.23	0.11	2.09	1.27
Throughfall	3.35	2.74	21.72	4.42	2.12
Gain from canopy	1.95	2.51	21.61	2.33	0.85
Moist Tropical Forest					
Precipitation	14.01	0.42	17.49	12.67	11.32
Throughfall	26.45	4.10	237.51	41.58	29.14
Gain from canopy	12.44	3.68	220.02	28.91	17.82

cipitation in the open. Some elements may be absorbed by the vegetation; others are washed from the vegetation. In general, there is an increase in the amounts of elements, particularly potassium. An illustration of the quantities involved is given in Table 14.2 for an old Douglas-fir forest and a moist tropical forest. Stemflow will also contribute elements to the soil, but these are usually much less in amount than that contained in throughfall. Mahendrappa (1974) monitored stemflow during a growing season on four hardwood and six conifer species in eastern Canada and found that potassium in stemflow amounted to as much as 4.5 kg/ha but that all other elements were less than 1 kg/ha.

Although the amounts of elements contained in throughfall and stemflow are sometimes referred to as inputs, this is not so because much of what they contain has been added by the vegetation, and this portion is part of the internal cycling of nutrients.

Output losses from meteorologic sources are represented by the movement of gases as, for example, when nitrogen is volatilized by microbial activity in the soil or by removal of particulate matter in wind storms. Few data are available for volatilization losses from natural forest soils, but such losses for nitrogen from fertilized forest soils are well documented (Overrein, 1968, 1969; Bernier et al., 1972). When nitrogen is applied in ammonium or nitrate forms, volatilization losses are negligible whereas, when applied as urea, significant losses can occur. The amount of loss depends on the quality of urea applied and moisture and humus conditions, but losses may range from 5 to 30 per cent of the amount applied. Some of the nitrogen volatilized from the soil surface as ammonia may be absorbed by the vegetation and the atmospheric loss reduced. Loss of windblown particles from forest soils is normally very small because of the forest floor cover and only when it is removed are losses likely to occur. Inputs of particulate matter from road dust may result in additions of elements, particularly calcium to the soil in zones bordering the road; the amounts are not large (up to 6 kg/ha) for calcium but much less for sodium, potassium, and phosphorus and extend only a few meters' distance (Tamm and Troedsson, 1955).

GEOLOGIC

The inputs and outputs of geologic source are for the most part related to gains and losses due to erosion and infrequently to major geologic forces such as volcanic activity or major uplift. In relatively stable landscapes the inputs from geologic source are usually very small and the outputs are represented by the element content in streamflow.

Inputs from erosion gains of alluvial or colluvial deposits or other deposition will depend very much on the chemical composition and amount deposited and generalities are not possible. In a study of accumulation of nitrogen in Alaska, Van Cleve et al. (1971) determined for recent (time zero) flood deposits of silt and sand bars that the total nitrogen to a 69-cm depth was 397 kg/ha or an average of 5.75 kg/ha for each centimetre depth of deposition.

The nutrient budgets, in terms of output related to input, for the Brookhaven forest of oak-pine in New York, a Douglas-fir forest in Oregon, and a European beech forest are given in Table 14.3. The greatest loss in the Douglas-fir forest is of silicon, but losses of sodium, calcium, and to a lesser degree magnesium occur. The losses of nitrogen and phosphorus are negligible and in all forests losses of potassium are small. Undoubtedly the losses of silicon, sodium, calcium, and magnesium are derived to a large degree from chemical weathering of minerals; certainly estimates of primary mineral decay in the Brookhaven soil indicate quantities released that are mostly in excess of the amounts lost in output. The values for the European beech in the Solling district in Germany reflect the higher levels of nitrogen, sodium, and calcium in the atmosphere, which might be expected in a densely populated industrial area. The outputs are similar to or less than those for the Brookhaven forest. Table 14.3 illustrates that, in relation to the total quantity of nutrient elements in the forest-soil system, the amounts of input and loss in absolute terms are small under normal natural conditions. The geologic losses reflect the composition and weathering rates for particular areas and kinds of rocks and minerals and therefore losses vary greatly from one forest soil to another. Values of an annual loss (1964) from a pumice soil supporting Monterey pine in New Zealand are given (kg/ha/yr) as: nitrogen, 4.82; phosphorus, 0.22; potassium, 12.3; sodium, 10.1; calcium, 26.9; and magnesium 2.6 (Will, 1968). These were determined using a lysimeter and, while there are

TABLE 14.3
Nutrient budget for dissolved elements. Amounts expressed as kg/ha/yr.
Brookhaven data from Woodwell and Whittaker (1968), Douglas-fir from Fredriksen (1972), and European beech (Ulrich and Mayer, 1972)

	N	P	Na	K	Ca	Mg	Si
Brookhaven Forest							
Input (precipitation and dust)	–	–	3.3	2.4	3.3	2.1	–
Output (to ground water)			19.4	3.3	8.0	6.1	
	–	–	-27.1	-4.6	-11.3	-8.6	–
Douglas-fir							
Input (precipitation)	0.90	0.27	2.34	0.11	2.33	1.32	trace
Output (streamflow)	0.38	0.52	25.72	2.25	50.32	12.44	99.3
European beech, Germany							
Input (precipitation)	23.9	0.48	7.3	2.0	12.4	1.79	–
Output (to ground water)	6.2	0.01	8.8	1.6	14.1	2.40	–

differences between these values especially for nitrogen, the general order of magnitude is remarkably similar to the values for output in Table 14.3.

BIOLOGIC

Man is the agent most responsible for losses and gains of nutrients. The actions of animals in terms of consumption by grazing and browsing and deposition as faeces, although they may be of some local importance in certain forests, are not usually of a magnitude to be included in the nutrient budget. The removal of large portions of a forest biomass by man in exploitation or management often constitutes major nutrient losses. In Table 14.4 data are given for nutrient removal which may occur after clearfelling and bolewood utilization of a range of forest species. These data show some consistent patterns such as the small quantity of phosphorus removed compared with large amounts of nitrogen, potassium, and calcium. Whereas much of the mineral nutrients contained in the biomass is derived from the weathering of soil minerals, which may be viewed as an internal transfer of capital from one form (inorganic) to another (organic), the nitrogen must ultimately be derived from the atmosphere. If the annual nutrient additions in precipitation (Table 14.1) are multiplied by the ages of the stands given in Table 14.4, it will be seen that for most elements the precipitation input is about equal to or may exceed the amount removed in the bolewood harvest. Outputs in streamflow are most likely derived primarily from mineral weathering (Fredriksen, 1972).

TABLE 14.4
Estimates of amounts of nutrients removed by clearfelling and removal of boles (nutrient amounts in stemwood - kg/ha)

Species and age (years)	N	P	K	Ca	Mg	Reference
Jack pine						
65	84.5	4.8	52.2	90.6	12.1	Morrison, 1973
Mixed northern hardwoods						
45 - 50	120.3	12.1	60.2	129.6	13.0	Boyle and Ek, 1972
Black spruce						
65 (to ca. 8 cm top diameter)	42.7	11.7	25.0	98.3	8.1	Weetman and Webber, 1972
Scots pine						
55	161	14	98	210	42	Ovington, 1962
Monterey pine						
26	127.8	17.9	156.9	105.4	–	Will, 1968
Oak						
115 - 160	386	17.5	219	769	58	Duvigneaud and S. Denaeyer-De Smet, 1970
Douglas-fir						
36 (wood and bark)	125	19	96	117	–	Cole et al., 1968
Loblolly pine						
40	138	11.4	96	121	–	Switzer and Nelson, 1973

The level of loss associated with forest harvesting can be altered greatly in two ways. The first is the intensity of utilization and the second is the rotation period for harvesting. An increase in intensity of utilization for black spruce full tree harvesting as compared with bolewood removal was estimated by Weetman and Webber (1972) and resulted in a three- to four-fold removal of nutrients. In mixed northern hardwoods, Boyle and Ek (1972) found that inclusion of branchwood removal with estimates for bolewood harvest would increase nutrient removal about twofold. In neither instance was the increase in this removal from fertile soils considered likely to result in reduced growth from deficiency of nutrient supply in the next rotation.

The effect both of a change in degree of utilization and a shortening of the rotation for loblolly pine is shown in Table 14.5. The increases range from 14 to 61 per cent with rotation and utilization changes, depending upon the element. An increasing intensity of cultural treatments in forest management, particularly in the regeneration period combined with shortened rotation periods and greater intensity of utilization further increases nutrient losses. Switzer and Nelson (1973) have summarized the reasons as follows:

TABLE 14.5
Effect of shortening rotation length and intensifying utilization on the mean annual nutrient demands of natural loblolly pine stands on good sites in the southeastern United States (from Switzer and Nelson, 1973). All values kg ha/yr

	Partial utilization[1]		Complete utilization[2]	
Nutrient	20-yr rotation	40-yr rotation	20-yr rotation	40-yr rotation
N	14.6	11.6	19.2	14.2
P	0.84	0.69	1.35	1.01
K	8.4	7.3	10.6	8.6
Ca	8.9	7.6	10.2	8.7

1 Stemwood and stembark removed.
2 Entire above ground portion of tree removed.

1 / Shortening the rotation means that a greater proportion of the rotation time is taken up in regeneration. Under intensive culture, associated with short rotations, increased disturbance of the soil will occur, resulting in further erosion losses from both increased frequency of harvesting and regeneration activities.

2 / Output leaching losses increase under shorter rotations. With greater frequency of regeneration periods there will be increased decomposition of the forest floor and losses of mineralized nutrients by outflow. The loss is also enhanced by increased streamflow associated with cut-over areas.

3 / In short rotations a greater proportion of the forest stand development period is taken up with the development of foliage which is the most nutrient-rich portion of the biomass produced. Thus two full canopies must be produced by a 20-year rotation compared with one for a 40-year stand.

The shortening of a rotation by itself need not result in major increases in nutrient removal if the reduction occurs in the later years of the stand long after full canopy development and root occupation of the soil have occurred. Thus in a boreal conifer forest the reduction of rotation age from 100 to 80 years on fertile soils is unlikely to result in losses that may significantly affect the overall nutrient budget. Any increase in the intensity of utilization of the forest biomass by removal of roots of trees as well as total above-ground parts and the utilization of lesser vegetation such as woody shrubs will inevitably increase the magnitude of nutrient removals from the forest soil-plant system. Another aspect of nutrient removal is the differential uptake of nutrient elements that may occur, not only between species but also within species as a result of genetic variation in growth and nutrient uptake (Smith and Goddard, 1973). Such variation in nutrient uptake has been demonstrated for a number of forest tree species: cottonwood (Curlin, 1967), slash pine (Walker and Hatcher, 1965; Pritchett and Goddard, 1967), Norway spruce (Fober and Giertych, 1971), and Douglas-fir (van den Driessche, 1973).

The major biologic inputs to forest soils are additions of fertilizer materials. Although such treatments are made only on an extremely small proportion of the total forested area of the world, the areas are increasing. Particularly where there is an intensification of forest management, silvicultural practices will often involve fertilization. In northern coniferous forests the major element applied is nitrogen, usually as an ammonium nitrate or urea at rates of 100 to 200 kg/ha nitrogen. In other areas such as Europe, the southeastern United States, Australia, and New Zealand, phosphatic fertilizers are often used at the time of plantation establishment. The use of potassium fertilizers is limited in many forest regions and in North America has been effective primarily in eastern North America in reforesting coarse-textured soils that had been cleared of the native forest and used for agriculture for several decades. Although such fertilizer additions are usually considered as absolute gains, they should be regarded as net additions. When nitrogen is added as urea, some of the nitrogen may be volatilized. Further, the indirect effect of adding materials, particularly anions such as sulphate (SO_4^{2-}) or chloride (Cl^-) can stimulate cation leaching losses under certain conditions and thus an addition of one element will result in loss of another.

The application of treated effluent waste to forest soils has been almost exclusively on an experimental basis, but may eventually be applied on an operational scale. Three potential hazards exist with such treatments: disease, water pollution, and heavy metal toxicity.

Internal cycling in the forest and soil

So far the main consideration of nutrient cycling has been with the overall aspects of input and output. In forest management, prime concern is with the cycling of nutrients within the forest-soil system, since it is by manipulation of both forest and soil that the major dislocations occur.

THE FOREST

The sizes of the nutrient pools within a Douglas-fir stand are given in Table 14.6. The large size of the soil pools for nitrogen and phosphorus is quite evident and it should be noted that the soil phosphorus is large compared with the values for potassium and calcium, because only exchangeable amounts for these two elements were determined. It can also be seen, if potassium and the other elements in the forest component are compared, that the amount of potassium represents a much larger proportion of the total than do the amounts of the other elements in the system.

TABLE 14.6
Distribution of N, P, K, and Ca between major components of a 36 year-old Douglas-fir forest (from Cole et al., in Primary Productivity and Mineral Cycling in Natural Ecosystems, 1968, by permission University of Maine Press)

Ecosystem component or pool	N		P		K		Ca	
	kg/ha	%	kg/ha	%	kg/ha	%	kg/ha	%
Forest	320	9.7	66	1.7	220	44.6	333	27.3
Subordinate vegetation	6	0.2	1	0.1	7	1.4	9	0.7
Forest floor	175	5.3	26	0.6	32	6.5	137	11.2
Soil (0 - 60 cm)	2,809[1]	84.8	3,878[1]	97.6	234[2]	47.5	741[2]	60.8
Total	3,310	100	3,971	100	493	100	1,220	100

1 Total.
2 Exchangeable with pH 7 N ammonium acetate.

TABLE 14.7
Annual transfers (fluxes) of nitrogen, phosphorus, potassium, and calcium (kg/ha/yr) between components of second-growth Douglas-fir forest (from Cole et al., in Primary Productivity and Mineral Cycling in Natural Ecosystems, 1968, by permission University of Maine Press)

Transfers (fluxes)	N	P	K	Ca
Input from precipitation	1.1	trace	0.8	2.8
Uptake by forest	38.8	7.23	29.4	24.4
Return to forest floor (litter)	16.4	0.60	15.8	18.5
Leached from forest floor to mineral soil	4.8	0.95	10.5	17.4
Output leached beyond depth of rooting	0.6	0.02	1.0	4.5

The annual transfers between components of this system are given in Table 14.7. The small amounts of nutrients which are inputs or outputs, compared with the transfers within the system, are apparent. Cole et al. (1968) conclude that under these conditions there is an annual depletion of nutrient capital within

Figure 14.1 Distribution and annual transfer of nitrogen in 36-year-old Douglas-fir. Areas of circles represent size of nitrogen pools and width of arrows the amount of transfer. From D.W. Cole, S.P. Gessel, and S.F. Dice in *Primary Productivity and Mineral Cycling in Natural Ecosystems* (1968), by permission University of Maine Press

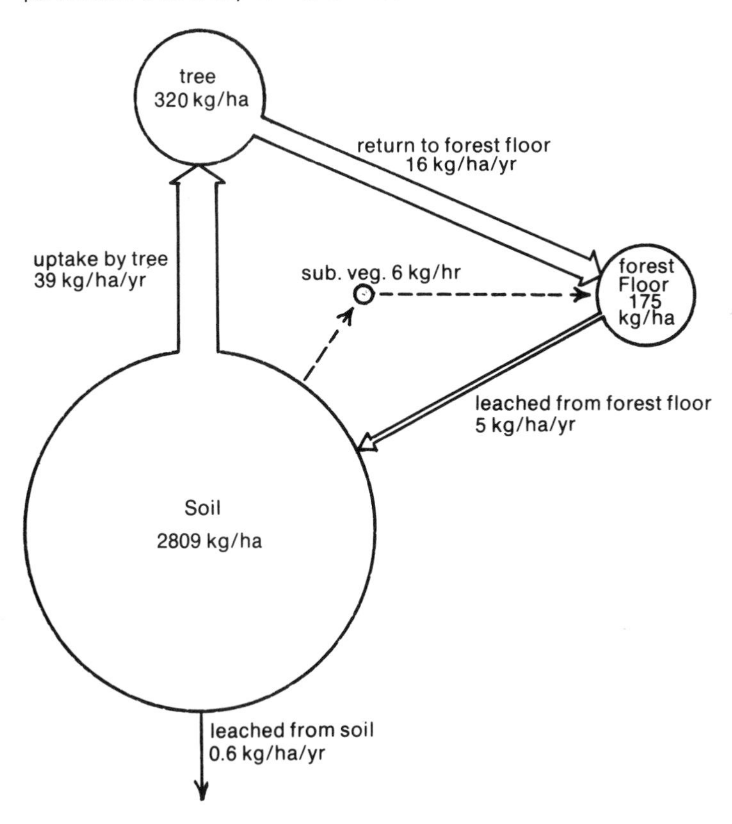

the soil as follows: nitrogen, 34.6; phosphorus, 6.3; potassium, 19.9; and calcium 11.5 kg/ha. For nitrogen and phosphorus these depletions are not great in relation to amounts in the soil pool and for potassium they are large in relation to soil exchangeable quantities, but probably not if the total amounts were known. Changes in sizes of nutrient pools will alter as a stand develops. Diagrammatically the pool sizes and transfers for nitrogen for this Douglas-fir system are shown in Figure 14.1. As the forest vegetation increases in dimensions, the tree biomass and nitrogen pool within it will increase, although for nitrogen the increase will be relatively less than for nutrients which are immobilized and

TABLE 14.8
Total amounts of nutrients (kg/ha) in vegetation and soil, and annual uptake and return (kg/ha/yr) from two deciduous forests in Belgium (from Duvigneaud and Denaeyer-De Smet, 1970)

Virelles-Blaimont: mixed oak, dominants 70 – 75 years old;
Biomass 156 × 10^3 kg/ha; soil weight 1360 × 10^3 kg/ha

Component	Nutrients (kg/ha)				
	N	P	K	Ca	Mg
Vegetation	533	44	342	1,248	102
Soil total	4,500	900	26,800	133,000	6,500
Soil exchangeable	–	–	157	13,600	151
Total annual uptake	92	6.9	69	201	18.6
Annual return – litter throughfall and stemflow	62	4.7	53	127	13

Wève-Wavreille: mature oak-ash; oaks 115 – 160 years old;
biomass, 380 × 10^3 kg/ha; soil weight 6318 × 10^3 kg/ha

Component	Nutrients (kg/ha)				
	N	P	K	Ca	Mg
Vegetation	1,260	95	624	1,648	156
Soil total	13,800	2,200	185,000	33,300	50,100
Soil exchangeable	–	–	767	13,865	1,007
Total annual uptake	123	9.4	99	129	24
Annual return – litter throughfall and stemflow	79	5.4	78	87	19

retained, such as calcium, rather than returned or recycled. Potassium is an element which, as has been demonstrated by its large contribution to throughfall, is returned in rather large amounts to the soil from tree canopies.

In contrast to the sizes of nutrient pools and annual transfers for a young conifer stand of Douglas-fir it is of interest to consider values for a deciduous forest. The amounts of nutrients both in total vegetation and the soil as well as annual uptake and annual return for two hardwood forests are given in Table 14.8. The larger amounts of nutrients in the Wève-Wavreille forest vegetation reflect the larger biomass and those of the soil pool the deep soil of Wève-Wavreille compared with the shallower soil supporting the Virelles-Blaimont stand. The magnitudes of annual uptake and return are generally similar, although some differences do exist. The uptake and return for calcium is greater for the younger stand. The large quantities of total and exchangeable potassium, calcium, and magnesium in relation to annual uptake or even more importantly annual uptake–annual return serve to emphasize that the forest-soil system is particularly efficient in recycling elements. If the amounts of nutrients in the soil pools are expressed on a unit weight basis (i.e. the soil amounts are divided by the total soil weight), the Virelles-Blaimont forest has greater nitrogen and phosphorus, about the same exchangeable potassium, slightly less exchangeable magnesium, but almost fivefold the amount of exchangeable calcium. This supply of calcium results in greater uptake in the younger forest.

It is interesting to note that the amounts of nutrients in the younger deciduous stand (Virelles-Blaimont), with the exception of calcium, are not too dissimilar to those for the vegetation of the Douglas-fir (Table 14.6). Duvigneaud and Denaeyer-De Smet (1970), in comparing their data for the two oak stands (Table 14.8), consider that the nutrient quantities are similar to those for oak forests studied by Remezov and others, but are greater than are found for beech, spruce, and pine forests. The annual transfers for a European beech forest of approximately the same size as the Wève-Wavreille oak stand show some similarities but also major differences (Ulrich and Mayer, 1972). The total annual uptakes are generally greater for the oak but the greatest difference is for magnesium which is eightfold, calcium, fourfold, and potassium twice in the oak stand what it is in the beech forest. The litter return, as might be expected, reflects these differences in plant uptake.

In most temperate forests the annual return of nutrients in the litter is a major component in the annual cycle. In tropical forests there is evidence that not only may the throughfall additions be of considerable magnitude (Nye, 1961) but that in certain tropical forests the transfers of nutrients in the litter may be negligible compared with the transfers by throughflow and stemflow (Jordan, 1970).

An example of differences in nutrient cycling for potassium between adjacent oak and pine forests growing under similar conditions is shown in Figure 14.2. Although the uptake by the oak is greater than the pine, there is in the pine stand a secondary cycling of potassium in the ground flora which is greater than for the oak or pine trees. The plant responsible for this transfer is bracken. This example illustrates that in a particular forest-soil system lesser vegetation may play a major role in the cycling of one or more nutrients. A forest management practice designed to modify only the dominant forest cover but resulting in a large change in lesser vegetation may have considerable effect on the nutrient cycle. Equally an alteration on the lower canopy without direct change in the forest cover may alter nutrient relations. In the example shown (Figure 14.2), if the bracken were killed or removed, there would be a sizable increase in the pool of soil potassium.

Figure 14.2 Annual transfers of potassium by oak and pine growing in adjacent woodlands under similar conditions. Stand ages - 47 years. Width of arrows indicates magnitude of flow. From J.D. Ovington, *Woodlands* (1965), by permission The English Universities Press

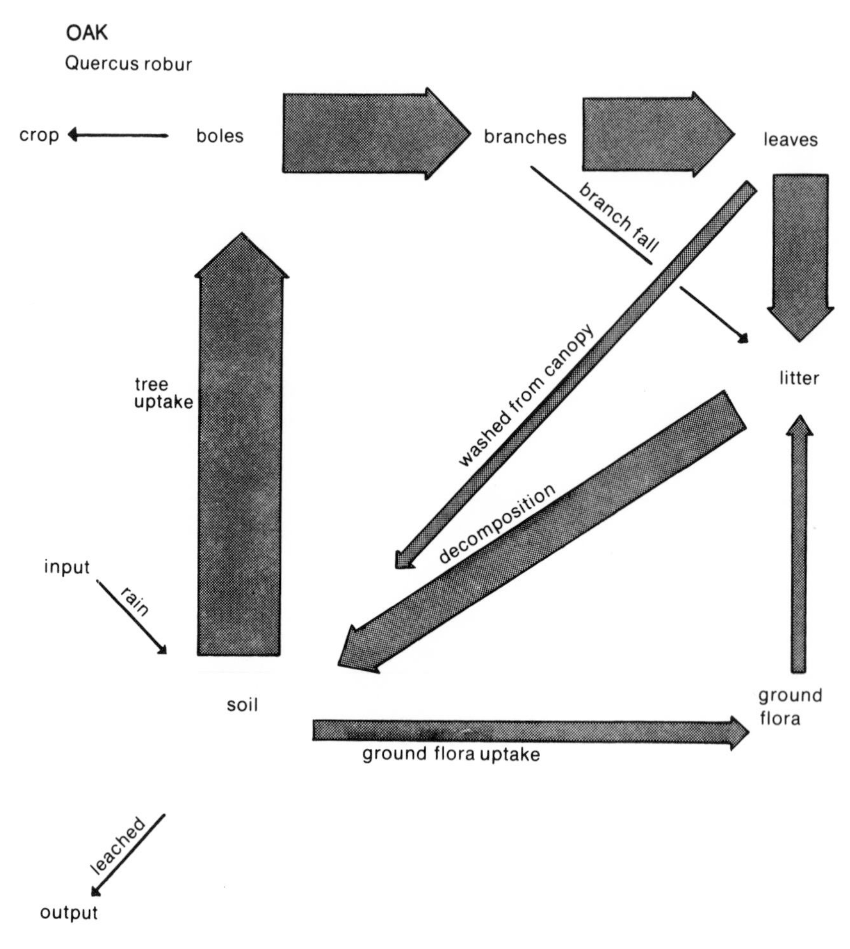

Another factor that merits attention in nutrient cycling is the stage of stand development. As a forest grows and develops, the rate of nutrient uptake varies. In the earlier stages of development the crown canopy and foliage comprise a larger proportion of the total tree biomass than in later years. This is reflected in Figure 14.3 by the accumulation of nitrogen in loblolly pine stands. The total accumulation reaches a maximum in the foliage mass before 20 years and the

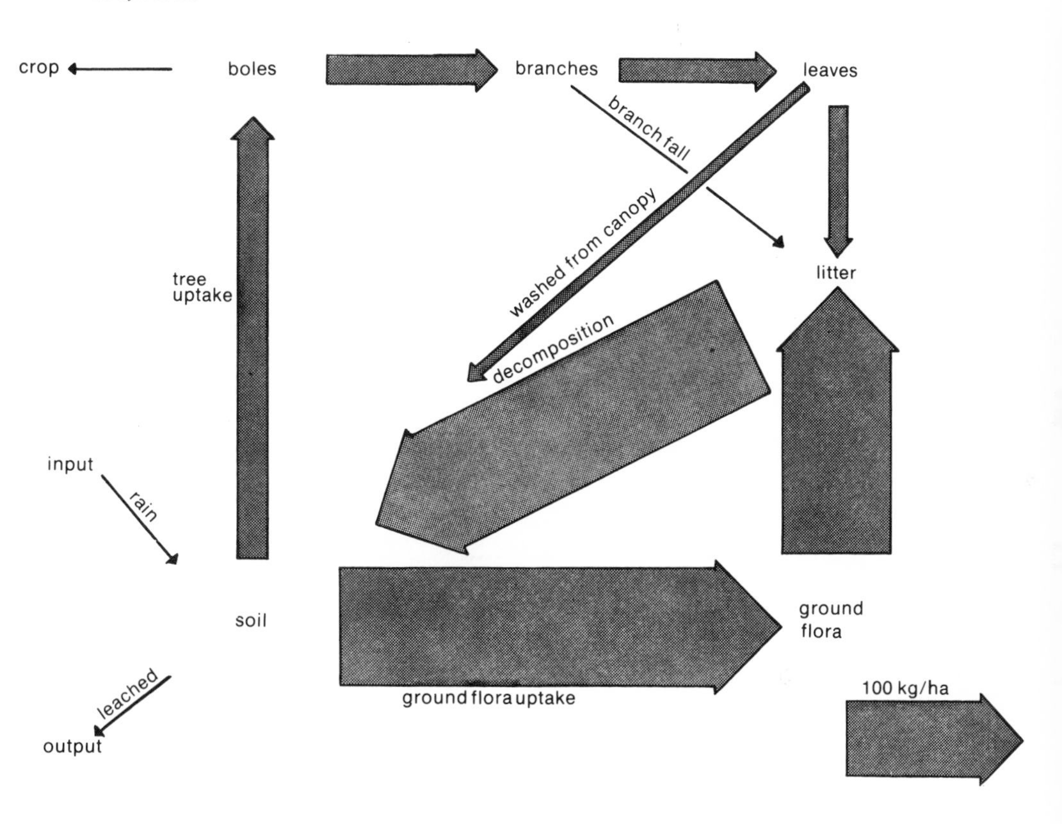

rate of accumulation declines even before this. By the time the stand is 40 years of age, total accumulation is virtually at a maximum, in other words the size of the stand nitrogen pool has become nearly constant. The mean annual accumulation rates for major nutrients in loblolly pine stands are given in Table 14.9. The quantities are greatest for nitrogen and potassium and the rates highest in the early period of growth culminating for all elements by 30 years; the maxima occur in different periods of development, that for nitrogen in the first decade, phosphorus and potassium in the second, and calcium and magnesium in the third.

Figure 14.3 Accumulation of nitrogen by loblolly pine stands of different ages on good sites. From G.L. Switzer, L.E. Nelson, and W.H. Smith, in *Forest Fertilization - Theory and Practice* (1968), by permission Tennessee Valley Authority

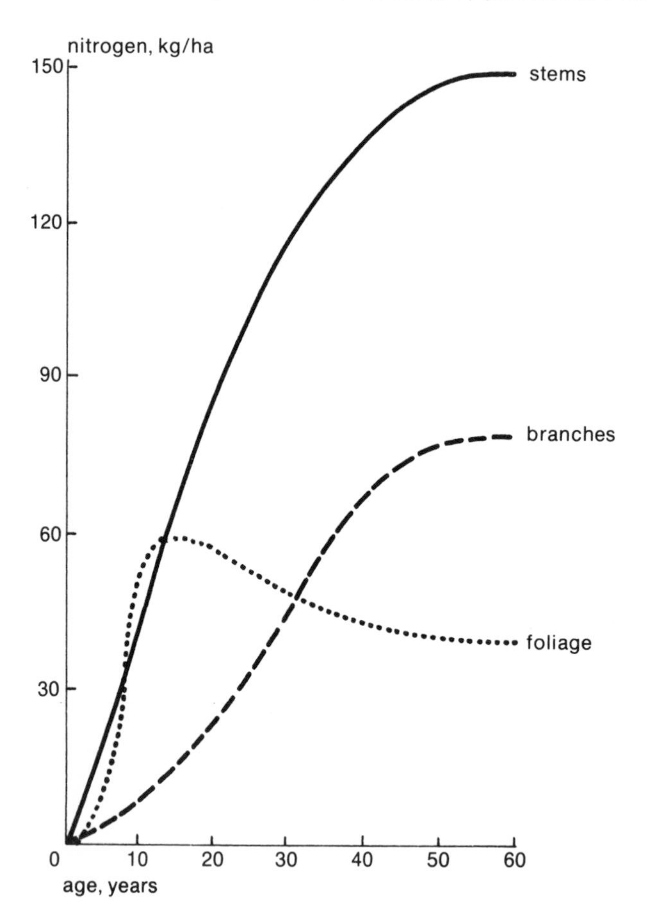

The progressive changes in the sizes of the nutrient pools during the first twenty years of loblolly pine plantation development is evident from Table 14.10. As the pine have developed in the first ten years, the herbaceous vegetation has diminished until at age 10 there is no nutrient pool in this component. During the same decade the nutrient pool in the forest floor increases greatly, and at 10 years the proportions that the nutrients in the forest floor constitute of the total amount in the tree-herbaceous-forest floor system are: nitrogen, 46; phosphorus 42; potassium 20; calcium 64, and magnesium 50 per cent. These

TABLE 14.9
Mean annual accumulation by decades for nutrients in loblolly pine stands – kg/ha/yr (from Switzer et al., 1968)

Decade	N	P	K	Ca	Mg
0 - 10	8.3	0.8	4.1	2.6	0.9
10 - 20	8.0	1.8	5.6	4.9	1.5
20 - 30	4.9	0.5	3.5	5.2	1.7
30 - 40	3.0	0.1	1.7	3.8	1.4
40 - 50	0.8	0.04	0.8	1.9	0.8
50 - 60	0	0.02	0.4	0.8	0.3

TABLE 14.10
Amounts of nutrient elements in trees, herbaceous vegetation, and forest floor at 5-year intervals in the development of a loblolly pine plantation (from Switzer and Nelson, in Soil Science Society of America Proceedings, 1972, by permission Soil Science Society of America). Amounts – kg/ha

Age years	Forest component	N	P	K	Ca	Mg
0	Trees	0	0	0	0	0
	Herbaceous	75	7.8	10	23	8.2
	Forest floor	0	0	0	0	0
	Total	75	7.8	10	23	8.2
5	Trees	22	2.3	13	7	3.4
	Herbaceous	43	3.3	7	14	3.8
	Forest floor	15	1.1	5	16	2.3
	Total	80	6.7	25	37	9.5
10	Trees	85	9.5	49	33	10.5
	Herbaceous	0	0	0	0	0
	Forest floor	75	6.9	12	59	10.5
	Total	160	16.4	61	92	21
15	Trees	140	15.8	82	62	17.3
	Herbaceous	0	0	0	0	0
	Forest floor	108	8.2	14	73	14.2
	Total	248	24	96	135	31.5
20	Trees	174	19.3	99	91	24.2
	Herbaceous	0	0	0	0	0
	Forest floor	124	9.1	16	80	15.4
	Total	298	28.4	115	171	39.6

proportions decrease somewhat with further development of the stand to age 20. At this age, an average of 18 per cent of the nutrient quantities is cycling in the system ranging from a low of 7 per cent for calcium to a maximum of 28 per cent for potassium. For a late successional stand of oak-pine in New York, Woodwell and Whittaker (1968) measured not only nutrient pools in the biomass but also those in the soil, as well as inputs and losses, and concluded that the proportion of cations cycling annually in the system was probably in excess of 11 per cent of the total. In the study of loblolly plantations (Table 14.10), Switzer and Nelson (1972) concluded that in the nutrient cycling at the twentieth year the source contribution was different for each element. Forty per cent of the nitrogen came from release by decomposition (mineralization) of litter and about the same proportion from internal transfer, whereas 50 per cent of the potassium came from canopy wash and leaching (throughfall and stemflow) and 60 per cent of the phosphorus requirements were met by internal transfer. At this stage in development of the system it would appear that the relative role of the soil in providing a nutrient supply is minor compared to the quantities cycling within the components of the system.

THE SOIL

In most studies of internal cycling greater attention has been paid to the biomass, partly because of ease of sampling and partly because it is more readily altered by man. There are, however, certain aspects of nutrient transfer within the soil which can be of significance in relation to forest development. The nutrient content of soil mineral materials is usually considered as a capital amount in the overall nutrient budget. Quite obviously, depending upon geologic origin of the materials, the capital will vary. For this reason many earlier studies emphasized recognition of the mineralogical composition of a forest soil as a basis for assessment of forest soil fertility. The higher the proportion of minerals with greater base content of such elements as calcium, potassium, and magnesium, the greater the level of inherent fertility. This approach is limited, because the form and availability of the nutrient, both in terms of absorption by organisms and leaching losses, are dominant factors in the utilization of the soils' nutrient capital. Rooting depth and intensity, the patterns of soil moisture movement, and temperature regime can modify most dramatically the ability of a soil to supply nutrients to a particular biological system.

In a study of the change in soil chemical properties in afforested dune sands, Wright (1956) found that there was a redistribution of nutrients following fixation of dune sand movement. In the young age classes with shallow root systems the removal of nutrients can equal or exceed the supply, but as tree crowns develop and canopy closure occurs there is not only an accumulation in the litter and

Figure 14.4 Inputs, outputs, and net fluxes for nitrogen and potassium from the forest floor under Douglas-fir during a twelve-month period. From D.W. Cole and T.M. Ballard in *Tree Growth and Forest Soils* (1970), by permission Oregon State University Press

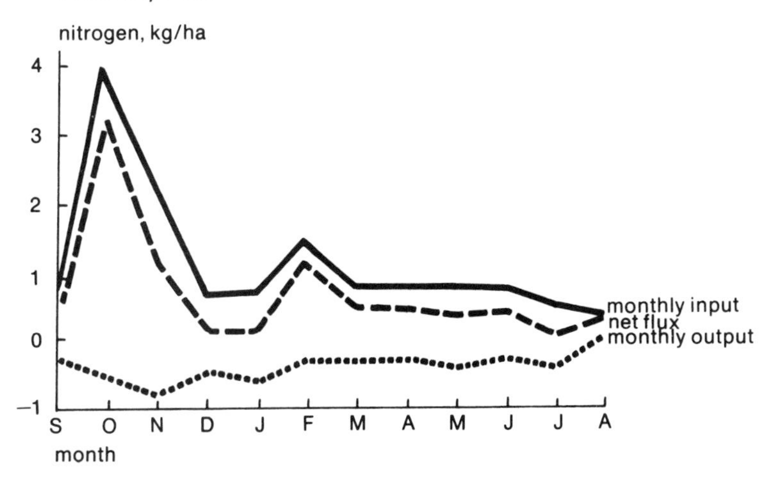

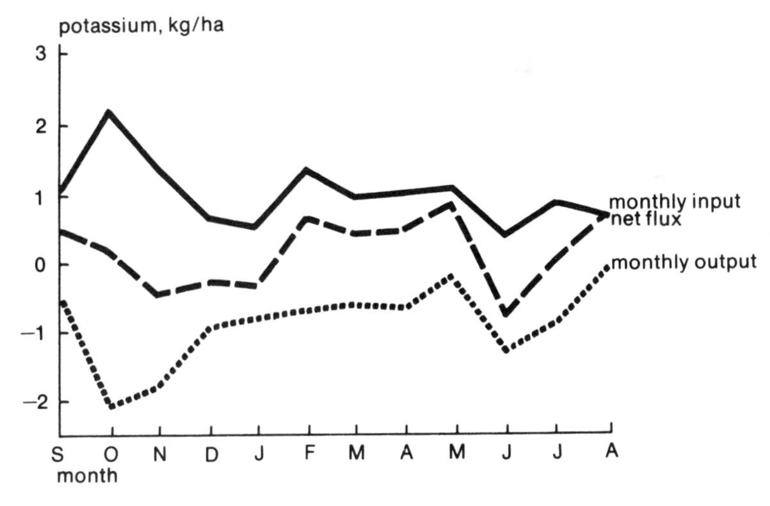

developing forest floor but leaching and downward movement from the litter into the mineral soil.

A more detailed study using tension plate lysimeters traced the movement of elements in the forest floor under Douglas-fir. Cole and Ballard (1970) provided

Figure 14.5 Precipitation and amounts of calcium leaching through soil profile during 1970–71 growing season in gravelly sandy loam under Douglas-fir. From C.G. Grier and D.W. Cole in *Research in Coniferous Forest Ecosystems* (1972), by permission United States Department of Agriculture, Forest Service

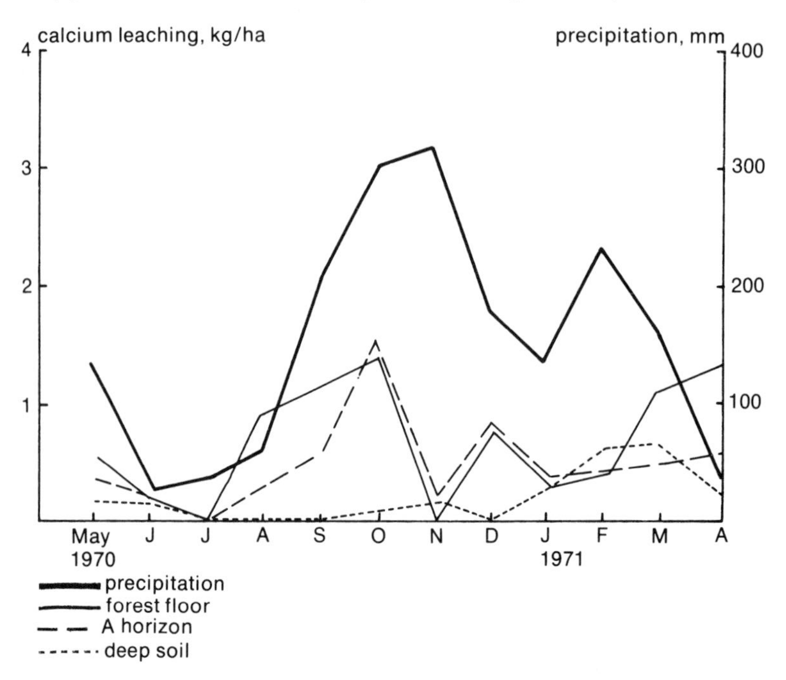

data on net fluxes over a twelve-month period which are given for nitrogen and potassium in Figure 14.4. The peak of input for nitrogen is derived from the litterfall in autumn and the accretion in net flux is evident whereas for potassium, although there is an autumnal peak, it is much less in relation to the monthly inputs for the remainder of the year. The net flux for potassium shows a balance between input and output. The output is from the forest floor to the underlying mineral soil. In the same study area, Grier and Cole (1972) compared the nutrient capital of the soil (0 to 60 cm depth) in 1965 and 1971 and could detect no significant change. There were, however, transfers within the soil, especially for calcium (Figure 14.5), where the movements through the soil profile are illustrated and are generally related to the amount of precipitation. The minimal amounts leaching to deep soil in relation to that leaching through the forest floor and A horizon are apparent.

One of the soil properties exerting a major control on the movement of cations in the soil is the level of anions (McColl, 1972). The most abundant anion is normally bicarbonate (HCO_3^-) which originates to a large degree as an indirect product of respiration and therefore is in large concentrations in the forest floor. In the forest floor of a Douglas-fir stand McColl measured a carbon dioxide concentration of 1.3 per cent in a wetting front in June. Anions of organic acids produced as decomposition takes place will also increase cation movement.

Nutrient cycling in relation to certain ecological and silvicultural considerations

As a forest stand grows and develops, the major increases in nutrient pools are in the biomass and forest floor. The rates at which these increases are greatest is in the initial years of development, at least for conifers. The nutrient capital in the mineral soil is usually large in relation to both the total in the soil-plant system and especially the annual amount cycling within the system.

If in a forest the dynamic balance is such that pool sizes and transfer rates remain essentially constant, the nutrient requirements of the system will be met by internal cycling and the small net losses of geologic output over combined natural inputs will be met primarily by a drain on the soils' nutrient capital. Following the pathological maturity of a forest or a major change in its structure, there will be associated changes in nutrient pool sizes, depending upon circumstances in transfer rate.

For natural forests the occurrence of major forest disturbances related to fire, wind, insects, and disease would appear to be the rule rather than the exception. The impact of these disturbances will vary but frequently is greatest in those forests where the dominants are declining in vigour. Each disturbance vector will have different results, but a few examples can be considered. The effects of fire and resulting changes in nutrient pools have been described in Chapter 12. In a moist tropical forest, Greenland and Kowal (1960) estimated that, with the exception of nitrogen, about 50 per cent of the nutrients in the vegetation would be released if the forest were cleared and burnt. These quantities would amount to (kg/ha): phosphorus, 67; potassium, 449; calcium, 1,345; and magnesium, 196. Such amounts are large in relation to the supply of 'available' nutrients in the upper 30 cm of the soil which were (kg/ha): available phosphorus, 11; exchangeable potassium, 650; exchangeable calcium, 2,578; and exchangeable magnesium, 370. The ecological rationale for shifting cultivation in relation to the conservation of soil fertility by forests is based on this periodic change in nutrient pool size. In forest soils where not only the above-ground biomass nutrient pool is reduced and added to the soil but where mor humus layers of appreciable thickness occur, some portion of the forest floor pool will also be available to developing vegetation. Where forest floor pools are not developing towards an equilibrium state in the first few years but increase continually with consequent increase in the immobile nutrient pool, periodic fires may well be the primary natural factor maintaining soil fertility. It also seems clear that, if the frequency of fire occurrence is high, the nutrient pools in the biomass and forest floor will be drastically reduced. Further, the leaching, erosional, and other losses associated with each fire will constitute a larger proportion of the total nutrient supply and the nutrient-supplying capacity of the soil will be reduced. When fire occurrence is infrequent or is regulated so that only a small portion of the biomass and forest floor pools is removed, then nutrient losses will be insignificant and the nutrient-supplying capacity of the soil can, in fact, be enhanced because of greater nutrient availability.

In contrast to fire, the effect of major wind disturbance is to enlarge the nutrient pool of the forest floor by blowing down a portion of the tree biomass. Any increase in nutrient transfer from this enlarged amount of nutrients depends on microorganism activity and related decomposition.

The effect of cutting in forests on cycling and particularly on changes in output can vary greatly. In Corsican pines growing on sand dunes, Wright (1957) found that, where tree growth was good, thinning did not change the nutritional states of the residual trees, but where trees of poorer growth were thinned heavily there was a delayed removal of water from the root zone and an increase in levels of nitrogen, potassium, and magnesium in the residual trees. Quite possibly the increase in moisture supply also contributed to increase in uptake.

Cole and Gessel (1965) followed the movement of nutrients at two depths in a gravelly sandy loam under clear cutting and fertilizer additions which were made in the fall of the year. The release of nutrients over a 10-month period from September to August is given in Table 14.11. Both clear cutting and application of nitrogen resulted in greater transfers from the forest floor into the mineral soil beneath than occurred in the control. Although clear cutting resulted in some increase in movement below 90 cm and therefore may be considered as an output loss, the amounts were small and only the calcium could be considered appreciable. Addition of urea fertilizer did not result in any appreciable movement from the 90 cm depth of any element, but ammonium sulphate brought about the same loss as when the stand was clear cut and increased the potassium loss. This cation output was undoubtedly brought about by the addition of the sulphate anion. No nitrate was detectable in the soil solution. Cole and Gessel noted that the soil pH was 5.2 to 5.9. Greatest release of cations was in the fall and early winter following treatment.

TABLE 14.11
Release of nutrients from beneath forest floor (2.5 cm) and at 90 cm during 10 month period - September to August (from Cole and Gessel, in Forest-Soil Relationships in North America, 1965, by permission Oregon State University Press)

Treatment	Depth (cm)	N (kg/ha)	P	K	Ca
Control	2.5	3.9	0.84	7.44	11.68
	90	0.54	0.03	0.91	4.07
Clearcut	2.5	10.7	2.31	16.12	20.85
	90	0.98	0.11	1.07	8.75
N @ 224 kg/ha	2.5	173.5	5.55	13.37	9.09
as urea	90	0.69	0.08	0.97	5.68
N @ 224 kg/ha as	2.5	201.5	4.51	13.73	22.17
ammonium sulphate	90	1.1	0.17	2.60	8.51

In contrast, a clear cutting followed by an unusual herbicide treatment applied for two consecutive seasons resulted in very large increases in nitrate nitrogen and cation output in streamflow at Hubbard Brook (Likens et al., 1970). For a three-year period the mean annual losses from the treated area were (kg/ha/yr): nitrate nitrogen, 175; potassium, 43; calcium, 103; and magnesium 23. It would appear, however, that where clear cutting occurs, much of the increase in cation output will reflect the increase in amount of subsurface water movement resulting from reduced transpiration. Where nitrification rates are increased there may be a corresponding loss of nitrate in streamflow. When considering losses which may be associated with any form of treatment, the proportion that the treatment area occupies in the watershed area should be kept in mind. Although there may be increased losses for a short period of years from several hundred hectares, the magnitude in a watershed of several thousand hectares may be extremely small and streamflow concentrations subject to considerable dilution before and after leaving the watershed.

Applications of nutrient elements in fertilizer form can be more effective when related to the stage of stand development and knowledge of the nutrient cycling characteristics of the system. For young trees grown in nurseries under intensive management the seasonal patterns of nutrient absorption are different. Nitrogen uptake rates for white spruce seedlings parallels growth rate, whereas phosphorus shows a bimodal pattern (Armson, 1965). In a plantation of loblolly pine, Smith et al. (1971) detected three different nitrogen accumulation rates during the growing period, an increase in the spring which levelled off in summer and increased again to a maximum in late summer and early fall. During this same period the dry matter increase was almost linear with time, and current foliage accumulated nitrogen over much of the period by transfer from other organs. Although soil moisture may be a factor in determining uptake patterns, nevertheless the application of a fertilizer at a time when there is a maximum possible uptake or in a form that may be retained in available form within the rooting zone, is desirable.

In relation to stand development the most effective time for addition of nutrients, biologically, will be in the earlier stages of development, when rates of growth and uptake are approaching a maximum. The particular time will be somewhat different depending upon species and conditions but should be when the greatest amount can enter into the cycle; this is especially true for the more mobile elements within the system - nitrogen, phosphorus, and potassium, and less critical for calcium. Thus in the initial stand development the trees do not usually have the capacity to absorb efficiently because their root systems are not dense enough. In addition, their small biomass and canopy represent limited nutrient pool size to absorb the added nutrient. In the later stages of growth approaching maturity, when the cycle is essentially one of relative equilibrium, pool sizes are large but the biological capacity of the trees to respond is often correspondingly less.

Dramatic results are frequently obtained when a deficient nutrient is added to a soil which has been depleted as a result of its previous use. The application of potassium to coarse-textured acid soils in New York which had been previously cleared and cultivated is a classic example of how limited amounts of fertilizer applied in the early stages of stand development result in growth increases for red pine (Stone and Leaf, 1967). The growth response persisted for many years and it has been postulated that where reforestation follows severe soil degradation, as by exploitive agriculture, there will be a gradual return to a higher, stable level of 'inherent' fertility. Where the specific deficient nutrients are added in fertilizer form, this return is accelerated (Heiberg et al., 1964). In effect it is suggested that in order to maintain a cycle, potassium in this example, which will ensure the growth and development of a conifer forest, a minimal pool supply must be present. In the natural forest which preceded the clearing and exploitive agriculture the major pools of potassium in these systems were the biomass and forest floor. When these were removed and the soil pool further depleted by agricultural use, the system could no longer provide adequate potassium when a demand for it was created by planting conifers, which in the early years, although small, have a large annual uptake requirement. There are other examples of additions of deficient nutrients resulting in long-term growth responses; the addition

Figure 14.6 Diagram of relative growth response to fertilizer inputs into a degraded soil (a); and non-degraded soils with (b) single fertilizer addition, and (c) multiple fertilizer additions at moderate and high levels.

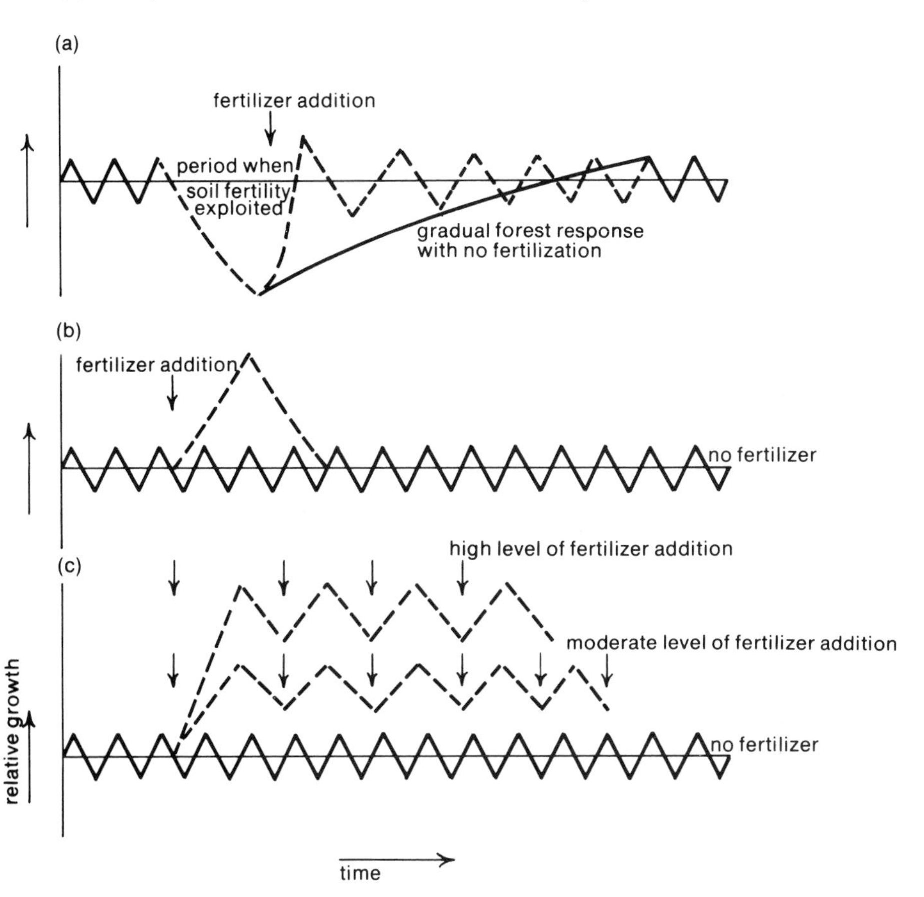

of phosphatic fertilizers to heathlands in northwest Europe, Les Landes in France, New Zealand, and the southeastern United States yield striking examples of such growth improvements.

Where the objective is to increase yields by silvicultural practices involving a choice of species, special establishment techniques, and spacing, these will often entail increasing and maintaining nutrient pools at levels above those which would naturally occur. A schematic illustration for fertilization of non-degraded and degraded soils is shown in Figure 14.6. Where soils are degraded, a single addition to the nutrient pool may be all that is required to return growth to its previous normal level. On soils that are not degraded a single treatment will most likely result in short-term growth responses and rates will revert to about normal. Where a sustained increase in growth is required as an objective of management, then the levels of input will usually have to be sustained.

Many factors affect and in turn are modified by increase in fertility and associated growth. When nutrient cycles are manipulated, directly or indirectly, by altering the levels of input and output, the greatest attention and consideration should be given to the overall effects such changes will have on the total ecosystem.

15 Forest soil development

When living organisms colonize an exposed portion of the earth's surface, the processes of soil development begin. The geologic materials in which the soil processes are at work may already have been subjected to some degree of physical and chemical weathering. In many places soil development begins on materials which have been transported from their place of origin. Windblown loessial deposits, glacial till areas, and waterlaid sediments, while retaining certain attributes of the materials from which they arose, will display new properties reflecting their mode of transportation and deposition.

Exposure to the climatic forces, primarily of temperature and precipitation, together with the establishment of vegetation, brings about the development of soil – a development chronicled by the soil profile. This development will be modified by relief and the time during which the soil processes occur.

Climate, geology, organisms, relief, and time are the five factors usually considered to determine soil development. Although it may be convenient to consider their function separately, the factors are interrelated in terms of the soil. Organisms which colonize the surface may arrive in a haphazard way, but those which develop as dominants will do so because of favourable or non-limiting climatic and geologic conditions which may be modified by relief. Species with high water requirements may be found in the water-collecting depressions in a region where precipitation normally limits their growth.

The groups of factors are not only interrelated but also, and perhaps more importantly, they themselves become altered with time. The weathering and removal of calcium minerals from one soil can change it from a medium that was originally basic to one that becomes acidic. The organisms change as various stages of succession alter the nature of the forest vegetation. Infrequent major disturbances of climatic origin such as winds and floods or in a longer time scale the changing nature of the entire climate will result in both temporary and permanent alterations in the soil. Some of the results of soil development will in turn modify or change dramatically the soil properties. The progressive development of a layer impervious to downward moving water, as when a pan develops, can change a well-drained soil into a poorly drained one.

The chances are that the soil which you see, describe, and measure today can be understood only partly in relation to its present environment but, if it is carefully examined, the pieces of evidence which it may contain can provide clues to past factors and processes which helped form and bring it to its present state.

Soil-forming processes express themselves in four basic ways: by *additions* of materials to the soil, for example in the addition of organic matter; by *losses* from the soil as when calcium and other elements may be removed from the solum; by *translocation* when, for example clay particles are moved downward, and by the *transformation* of materials such as the change in form of organic matter or iron compounds. The clues therefore about soil development are largely related to the kinds, amounts, and location of the results of these four mechanisms and it is from this evidence and knowledge of the general nature of soil forming processes that inferences may be drawn and hypotheses made concerning a particular soil. As with many scientific disciplines, the comparative studies of different soils with known similarities and dissimilarities are often helpful in the formation of hypotheses about soil development.

Factors of soil development

The concept that soil development reflects the action of climate, organisms, parent materials, relief or topography, and time was expressed by Russian soil scientists in the late nineteenth century. It was, however, Jenny (1941) who gave mathematical expression to the concept:

$$s = f(cl, o, r, p, t, \ldots)$$

where s, the soil, is dependent on climate, cl; organisms, o; relief, r; parent material, p; and time, t. The dots indicate that additional soil-forming factors may have to be included. Jenny distinguished between the external or environmental factors represented in his equation and the soil properties which could be described and measured and which would enable a soil to be defined by them. In making this distinction he particularly encouraged the comparison of soils which were similar except for one specific factor which varied. This approach gave impetus to the quantification of soil properties and the establishment of relationships, for example, the characterization of the nitrogen content of grassland soils as a function of a humidity factor for soils located along a particular isotherm.

Another aspect of a quantitative approach to soil genesis or formation is a recognition that the various processes going on in the soil are in many instances proceeding at different rates. A soil may have an organic status which is in dynamic equilibrium – the gains and losses are balanced – but may still exhibit active loss of calcium carbonate or development of a mineral horizon. Arnold (1965) suggested that progress in attempts to explain soil genesis by proceeding from quantifiable measurements of the soil and related phenomena make it possible to establish multiple working hypotheses in which soils may be considered as sequential models in a particular landscape. The differences in soil properties in terms of losses and gains of organic matter, carbonates, clay, and other components are then related to profiles in a particular sequence. It is from the morphology and soil properties arrayed spatially in a sequence that hypotheses may be formulated and subsequently tested, modified, or rejected.

This approach emphasizes that soils are geographic bodies with gradations rather than sharp boundaries between them. Simonson (1959) proposed that soil genesis consisted essentially of two overlapping steps: (1) the accumulation of parent materials and (2) horizon differentiation. The processes of additions, removals, translocation, and transformation may occur in both these steps, but are related primarily to the second one, horizon development in relatively stable landscapes.

In quantifying gains and losses related to these four processes the question arises as to what datum or base should be used. A common one is to relate gains or losses to the C horizon (not soil). It is assumed that the status of the C horizon is now similar to its condition when soil profile development first commenced. This assumption can only be verified if the same C horizon materials are exposed to soil development under similar conditions of climate, organisms, and relief as those at the present time. If the exposure has taken place at several periods of time, then the soil profiles that have developed represent a sequence in which differences reflect the influence of time, and the soils and their properties then constitute a *chronosequence*. Other sequences may occur – *toposequence* (relief), *biosequence* (organisms), *lithosequence* (parent materials). More often soils do not vary in terms of one external factor but in terms of several and it is then much more difficult to determine the precise influence of each factor separately. In many regions the not-soil beneath the solum is not a C but a IIC horizon; the solum has developed in a material to the extent that no C horizon remains. One example is the development of a profile in shallow glacial till deposits over bedrock, where the solum extends to the bedrock surface (R). Another example evident in many forest regions of northeastern North America occurs where a lodgment till, deposited during glacial advance, is covered by an ablation till during the retreat of the ice. Then in the dry cold interval between ice retreat and plant colonization there has been a superficial covering, perhaps a few centimetres thick, of windblown particles. Development of the soil profile proceeds and may now exhibit in its morphology an A horizon primarily in modified windblown fine sand, a B horizon in ablation till, and a not-soil C horizon which is lodgment till. If the properties of the horizons are to be compared, the problem of a basis for comparison arises. There is no internal standard for comparison of rates of gains and losses within the profile. For many forest soils this type of heterogeneity in the soil profile is common.

Another procedure uses, as a basis for comparison, values for soil properties in horizons that appear genetically similar in their morphology. The numerical values for specific soil properties can then be treated as samples from a population and subjected to appropriate statistical analyses. This approach provides data that may form the basis for the numerical classification of soils described in Chapter 9.

Processes of soil development

ADDITIONS

There are several principal means by which additions are made to the soil. Atmospheric additions of water and certain elements have been discussed in Chapter 14. In local areas atmospheric components such as sulphur dioxide or heavy metals originating from industrial activity may exert a powerful influence on the soil, primarily by modifying and frequently killing vegetation so that the soil is exposed to erosional forces. Additions of certain organic compounds which may be washed from the forest canopy, together with similar components from litter, can play a role in soil weathering processes.

Wind and water are the major general agents which add solid material to the soil, although in local areas volcanic and colluvial deposits may predominate. In extensive areas of forest, wind deposition is normally negligible. Where the soil was previously cleared and cultivated or was occupied by unstable non-forest vegetation, the soil profile may show horizontal bands of windblown deposits often delineated by the accumulations of organic matter associated with the shorter periods of stability (Figure 10.9). In areas where arable fields and small forest stands exist contiguously, the forest acts as a windbreak and may trap considerable quantities of solid material. In a Norway spruce windbreak in Denmark, Holstener-Jørgensen (1960) measured amounts of between 221 to 5,771 kg/ha of windblown solids from an adjacent field over a range of 5 to 80 m distance within the stand during March to July.

Alluvial deposits on forest soils occur infrequently, usually as the result of large catastrophic storms and this results in the burying of the previous profile (Figure 15.1). The depositions of material may be infrequent and often annual. Strang (1973) describes a multilayered white spruce root system in a river valley

Figure 15.1 Forest soil profile buried by a thick alluvial deposit. Photo courtesy D. Burger

Figure 15.2 Accumulation of calcium carbonate about red pine roots growing in calcareous sand. Photo by K.A. Armson

where alluvial silt deposits were probably annual. The thickness of deposit from the lowermost root to the present surface was 190 cm and, on the basis of ages and depths of roots, he estimated a mean annual net addition of alluvial silt at 0.5 to 1.3 cm. Estimates based on the radiocarbon dating of organic detritus 7.75 m below the present surface were for a net annual accumulation of 0.3 cm.

The most obvious and significant addition to forest soils is organic matter. A major distinction between forest soils and those supporting other forms of vegetation is in the amount, form, and location of the annual organic additions. In forest soils the major addition is to the soil surface as litter, whereas in grassland soils a major contribution is the annual mortality of grass roots. The surface additions not only modify the surface soil layer physically and chemically but also exert a major influence as the substrate for the soil fauna and flora. Compounds present in the litter at the time of litterfall or produced as decomposition takes place affect the weathering of the soil. In certain situations where decomposition of organic material is slow, it accumulates and histosols (organic soils) may develop.

LOSSES

Erosion and leaching are the two processes that result in losses from forest soils. Wind erosion losses are usually considered negligible in forest soils supporting vegetation and surface erosion by water is generally related to exposure of surface mineral soil layers as a result of a major disturbance. There are numerous examples in eastern North America of forest soils which were cleared and cultivated for agricultural use, with surface soil materials being removed by sheet and gully erosion over this period. Where revegetation, either natural or man-made, has since taken place, differences in profile morphology will be evident, resulting from the prior agricultural use; where the forest cover develops and the upper soil layers show signs of forest soil development processes, evidence of the previous state may be obscured or non-existent. The thickness of the upper mineral horizon (A) and consequently the solum depth may be markedly less in more elevated locations than in lower slope positions which may have accumulated the erosional material. The evidence of stone piles or vestiges of fence lines indicating previous agricultural use should serve to focus attention on signs of erosional losses and gains in the soil.

Leaching losses are those of materials which leave the soil in solution. They are represented by the flux of elements moving from the solum to deeper layers and by the materials in solution from runoff (surface and subsurface flow) which are measured in streamflow. Examples of some of the quantities are given in Table 14.3.

TRANSLOCATION

Water and ice

The movement of materials either in solution or as solids within the solum is called translocation. The form in which a material can be moved by a particular agent is obviously important and a consideration of the translocation of some particular soil component – organic matter, iron, phosphorus, etc. – is clearly linked to a knowledge and understanding of form changes. It is convenient, however, to arbitrarily discuss some features of the agents of translocation with specific reference to forest soils.

Water will move materials in solution and in suspension. The movement out from one part of the profile is termed *eluviation* while the movement and deposition in another part of the profile is *illuviation.* A distinction is often made in the movement of certain materials; Duchaufour (1973) refers to the movement in suspension – 'l'entraînement mécanique' of clay and iron as *lessivage.* In North America (Buol et al., 1973), lessivage refers to the movement of fine soil particles (mainly clay) in suspension from one part of the soil to another portion where they accumulate as clay skins.

The movement of water will vary, depending upon the season and particularly the balance between precipitation and evapotranspiration. The depth, therefore, to which water moves materials in solution or in suspension will differ greatly during the year for a given soil. The general depth of the solum and zones of deposition within it of particular materials such as clay, calcium carbonate, organic matter, free iron, indicate the pattern of extent of movement. The nature of the soil materials - texture in particular - will affect the depth, with thickness of the solum increasing with increase in the coarseness of texture over certain ranges in the same area. When the uniformity by which water enters and moves through a soil is altered, variations in the horizonation of materials which have been translocated may be found. Microtopographic irregularities account for some of the variation, but the tree vegetation itself can modify the water movement pattern. Stemflow concentrates the water entry into the soil and results in both physical and chemical changes reflecting an increase in movement of water and solutes (Gersper and Holowaychuk, 1970a,b; 1971). Further, the downward extension of roots, particularly from trees with tap and heart root form, will channel water movement and increase the irregularity of profile horizon development. Roots affect the movement of water not only downward but also in directions towards the roots as water is absorbed; in some soils there will be accumulation of water-translocated substances, as for example calcium carbonate about the tree roots (Figure 15.2). Under certain conditions of excess of evapotranspiration over precipitation in soils containing soluble salts of sodium and calcium sulphate, the net upward movement of water during a period of the year may result in the salts being deposited in the upper soil horizon, where they can influence not only the chemical but also physical and biological properties. Solonetzic soils are an example of non-forest soils in which this movement is a diagnostic feature. Lateral movement of water through soil can occur down slopes, especially where a barrier or impeding layer prevents vertical movement. Water movement of this type has been recognized as an asset for tree growth.

The periodic vertical movement of the water-table in the region of the solum results in movement of some materials. The reduction of iron and other elements can make them mobile and they may then move in the ground water. The process of mobilization of materials under such reducing conditions is both a transformation and movement and is termed *gleyzation.*

Although it is uncommon in forest soils, the alternate drying and wetting of soils having a large content of 2:1 lattice clay minerals results in large volumetric changes within the soil as the clay minerals shrink and swell in response to the changes in moisture content. Soil particles will fall into the large cracks which occur when the soil is dry or be washed into them with the first rain. As the soil wets, it increases in volume and, because of the increase in particles which have

Figure 15.3 Cross-section of a hummock showing horizon disturbance and movement resulting from cryoturbation. Photo courtesy S.C. Zoltai.

fallen into cracks when dry, the soil aggregates push upwards to form characteristic small ridges and mounds termed *gilgai* relief. The result of such action is a mechanical mixing of the upper mineral horizons. This process would be drastically modified, if not prevented, by the presence of a tree canopy and forest floor.

The formation of ice lenses can result in movement and physical disturbance, particularly of the surface centimetres of mineral soil. This small-scale mixing occurs during periods of freeze-thaw cycles and an account of its action has been given by Schramm (1958). Where litter and forest floor organic materials cover the soil surface, such movement is usually negligible but, where mineral soil is exposed, the result of such freeze-thaw cycles is an accumulation of smaller-sized coarse fragments at the surface to form a stone line from these 'frost-heaved' particles.

In cold climates where permafrost is present in the soil, seasonal differences in temperature can result in phase changes from ice to water and back in the 'active layer' - the zone above permafrost. These changes in form result in physical disturbance of the soil - *cryoturbation*; there is a mixing of mineral and organic material and typically the soil surface becomes hummocky (Figure 15.3).

Figure 15.4 Changes in soil profile as a result of earthworm activity in a New Brunswick podzol: 1958, before earthworm activity; 1961, after earthworm activity. From K.K. Langmaid (1964), with permission from Canadian Journal of Soil Science

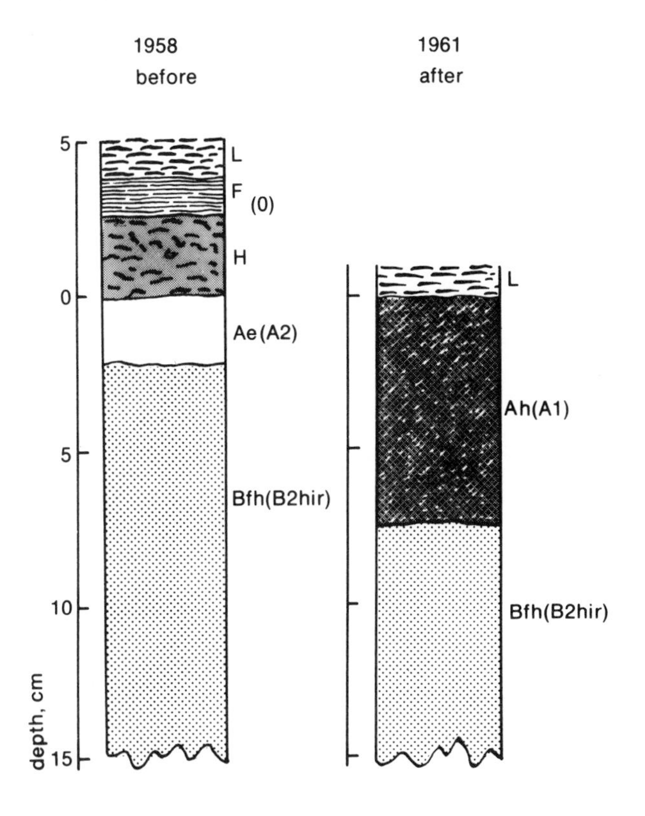

Biotic

The activities of certain fauna in moving soil have already been mentioned in Chapter 6. An illustration of the effect of earthworm activity in moving soil materials has been given by Langmaid (1964) - Figure 15.4. The movement and incorporation of organic and inorganic materials to form an Ah (A1) horizon during a three year period is most striking. In a study of soil-forming processes in the humid tropics, Nye (1954, 1955) considered that two genetic horizons developed, a lower sedentary one and an upper horizon which could be divided into two zones, an upper one of one to three centimetres thickness formed from casts of the earthworm, *Hippopera nigeriae*, and a lower thicker zone resulting from activity of the termite *Macrotermes nigeriensis.* The combined thickness of the

Figure 15.5 Soil profile disturbance resulting from tree windthrow in balsam fir-spruce forest, Québec. Light-coloured zones are Ae (A2) horizon from a spodosol (orthic ferro-humic podzol). Photo by K.A. Armson

upper horizon resulting from the faunal movement of soil could be up to 30 cm. This horizon of finer-textured soil was subject to soil creep on slopes. Nye (1955) estimated that the annual deposition of soil in worm casts amounted to 5×10^3 kg/ha annually.

Another mechanism of soil movement peculiar to forest soils is the movement of soil resulting from windthrow of trees. This process has been described by Lutz and Griswold (1939) and Lutz (1940). In Wisconsin (Milfred et al., 1967), Ontario (Armson and Fessenden, 1973), and the northeastern United States (Stephens, 1956), the proportion of the forest soil surface showing evidence of such movement varied from 19 to 35 per cent in the stands studied. After a study of the windthrow of deciduous species in central New York, Mueller and Cline (1959) concluded that much of the upper 60 cm of forest soil had been disturbed during the previous 500 years. The distortion of profile morphology associated with windthrow disturbance is shown in Figure 15.5. Windthrow provides a continuing exposure of fresh mineral surface for weathering in the upper part of the profile and return from a lower horizon (B) material which has been previously translocated and immobilized. It therefore minimizes the intensity of horizon (A and B) development and thus paradoxically, while providing increased variability, also results in greater homogeneity by minimizing the horizontal zonation. The exposed mineral soil surfaces are often essential as seedbeds for many tree species and the pit and mound or cradle knoll topography commonly associated with windthrows can be a factor in holding back snowmelt waters.

The general process of soil movement by mixing is termed *pedoturbation*. In many areas man is a common agent in bringing about soil mixing. In forest areas where shifting cultivation is employed the effects are essentially ephemeral, but in the temperate forest regions the effects of soil mixing by plow and spade may persist for many decades.

TRANSFORMATION

Soil development is accompanied by many physical and chemical changes in soil materials. For simplicity, the transformations of certain components and elements will be treated separately. The general weathering processes of minerals and rocks were discussed in Chapter 7.

Organic matter

The quantities and chemical composition of organic additions to the forest floor were discussed in Chapter 4. As might be expected, the focus of attention on processes which transform soil organic material has been on the soil; the by-products of these processes, especially organic acids, have long been considered to play a key role in soil profile development. One of the most important changes in our knowledge of forest soil processes came about in the 1950s when, instead of looking exclusively at the soil, attention was directed to the origin of the soil's organic matter – the vegetation. The causes of mull and mor soil development have engaged the attention of soil scientists and foresters for many years; the development of one or other form have variously been attributed to quantity and quality of litter, soil pH, nitrogen status, and soil faunal and floral differences. Handley (1954) has pointed out in an exhaustive review that no one soil feature seems to be a primary determining factor and concluded that there was evidence that some attribute of the plant material itself, which in turn might be capable of modification, could be determining. Handley then demonstrated that certain mesophyll tissue was subject to a coating process by precipitation of proteinaceous compounds, analogous to tanning; this precluded a rapid breakdown and use of the cellulose substrate by the soil micropopulation. Experimentally, he showed that when model substrates (gelatins) were prepared, using aqueous extracts from leaves of mull- and mor-forming species, there was a differential breakdown; the models from the mor-forming species were more resistant. The masking of the cellulose components of the foliage by some proteinaceous coating precipitated by a material that can be removed as an aqueous extract means that, when such leaves arrive as litter at the forest floor, a major energy source substrate, cellulose, has been removed from immediate and rapid conversion by the micropopulation and also that a portion of the nitrogen is immobilized in the form of the cellulose-masking proteinaceous material. It is noted that the protein-precipitating materials when not united to protein can be acted upon by microorganisms, whereas when complexed with protein they are resistant.

Contemporaneously with Handley's studies, Bloomfield (1953a,b) reported that water-soluble compounds washed from the leaves of Scots pine and kauri were capable of mobilizing iron and aluminium from their oxide form. This ability was corroborated for extracts from fresh litter of a number of eastern North American tree species by Delong and Schnitzer (1955). The most important single group of compounds in the leaf extracts involved in iron mobilization was the polyphenols (Bloomfield, 1957). Enzymatic oxidation of the polyphenols decreases the activity of their water extracts, and species whose extracts were least affected by aging of the plant tissue were those most commonly associated with iron mobilization and transfer in the soil. Extracts of leaves from deciduous species may be capable of extracting more iron than extracts of conifers. The paradox that extracts of deciduous species, such as maple and poplar, least associated with soil iron mobilization were found to take up more iron than extracts from American beech (Schnitzer, 1959) could be explained by speculating that biochemical processes and the activity of soil organisms may act as opposing influences and that, in the process of senescence of foliage on the plant, the polyphenolic or other compounds may be rendered inactive. Other organic components such as reducing sugars are also capable of mobilizing iron, and their control in canopy drip may in fact be greater than that of polyphenols (Malcom and McCracken, 1968). Nykvist (1959a,b, 1960a,b, 1962) showed the decomposition rates of fresh litter to be generally related to their initial content of water soluble substances. It is thus possible that the reducing sugars are rapidly inactivated by soil microflora. There are other possible mechanisms for inactivation of the mobilizing substances when they pass through an acid mor than when through a more basic mull soil.

The nature of certain soil-plant interrelations and the role of polyphenols have been studied by Coulson et al. (1960a,b) and Davies et al. (1964a,b). They determined that the greatest quantity and diversity of polyphenols were found in aqueous extracts of foliage from trees growing on low base, mor humus soils and lesser amounts from those on base rich, mull soils. Extracts of humus and litter were singularly low in their ability to complex iron, but the illuvial B horizon of a podzol contained organic matter of polyphenolic origin. Imbalanced supply of nutrients - deficiences of nitrogen and phosphorus - were found to result in higher polyphenolic levels in the foliage. It was also determined that polyphenols can act under certain conditions as a masking agent in the manner demonstrated by Handley (1954) and that this process was most effective at low pH values of 3 to 5.

It appears that constituents in tree foliage can modify the availability of important substrates such as cellulose to microorganisms in the soil and thus modify the processes of decomposition. These constituents are also present in the throughfall from tree canopies and, depending upon the type of surface organic layers and related soil populations, the degree varies to which they survive unaltered and pass into the mineral soil, where they can mobilize ferric iron and under aerobic conditions complex it and transport it in a reduced state to lower soil layers. The complexing of iron and aluminium and their subsequent movement by this means is termed *cheluviation* (Swindale and Jackson, 1956). The nature of the mechanism by which the organic-iron-aluminium complex is precipitated may be related to increase in soil pH values or may be merely a simple function of water removal by evapotranspiration, a dominant process in the soil during the growing season when such throughfall would contain the organic complexing substances. Certain soil constituents may react specifically with polyphenols; for example, Kyuma and Kawaguchi (1964) indicated that allophanic material and hydrated iron oxide accelerated the oxidation of polyphenols to give dark coloured, highly polymerized humic substances. It has been estimated that under a mixed wood forest in North Carolina 20 kg/ha/yr of organic matter could be added to the soil in the throughfall (Malcolm and McCracken, 1968). Although this is a very small amount in relation to the total annual litterfall, its significance in mobilizing iron and aluminium within the soil may be very great.

In addition to the relatively small quantities of organic material involved in moving through the soil there are the much larger quantities, the bulk of which is deposited as litter annually. One extreme in transformation is the tropical forest where, although the annual additions are great, the rates of decomposition are equally great, so that organic matter does not accumulate. Bates (1960) recorded for a Nigerian forest soil which was undisturbed for 50 years that only a light litter existed and this lost its structure in 14 to 21 days. The organic carbon in the upper 5 cm of soil was 3.95 per cent, but below this it fell rapidly to 0.38 per cent and less. Another extreme occurs when, as a result of low temperature and/or reducing conditions, the decomposition rates are very low. Where it is primarily a function of low temperature, the amounts of litter are also invariably small. Douglas and Tedrow (1959) estimated that the decomposition rate in the upper 2.5 cm layer of an Arctic brown soil could be 1,270 kg/ha/yr but, since

Figure 15.6 Changes in organic carbon content with depth of soil and time. From B.A. Dickson and R.L. Crocker, *Journal of Soil Science* (1953), Oxford University Press, by permission of The Clarendon Press, Oxford

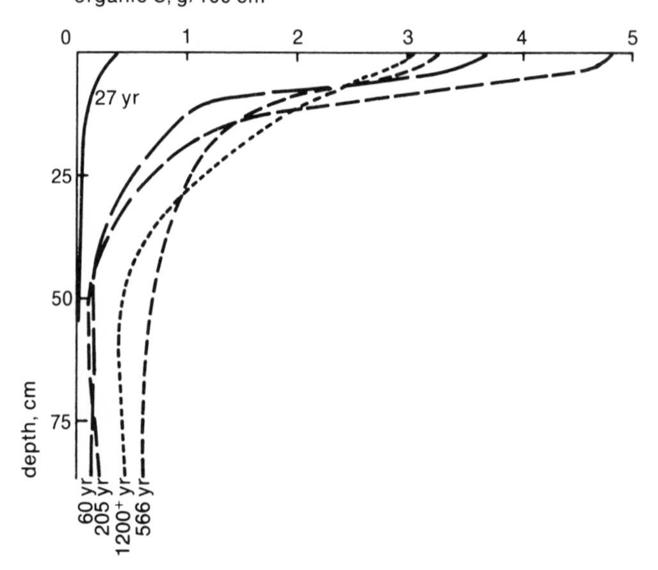

this layer was usually discontinuous, the actual amount decomposed might be about one half that amount. When reducing conditions resulting from saturation occur, relatively little decomposition takes place and peat soils develop.

Between the extremes there exists a wide range of conditions. Dickson and Crocker (1953) studied a chronosequence of forest soil development in California. The forest floor thickness reached a maximum at 205 years and declined rapidly until an equilibrium was reached some time later. The organic carbon content within the soil reached a steady state at 566 years as reflected by its distribution with soil depth (Figure 15.6). In contrast Syers et al. (1970) estimated that for windblown sands in New Zealand a steady state of organic matter was not achieved under 10,000 years. The rate of organic matter turnover in northern spodosols (podzols) under coniferous forest was studied in Sweden; estimates for the ages of humus in the B horizons ranged from 330 ± 65 to 465 ± 60 years for soils from a northerly region (Tamm and Holmen, 1967). There is obviously a continuous breakdown of humus materials going on in these lower mineral horizons. In the northern soils, 70 to 80 per cent of the soil's store of organic matter was in the B horizon, 5 to 10 per cent in the A horizon, and between 10 to 20

Figure 15.7 Distribution of organic matter in a podzol profile under undisturbed yellow birch-red spruce forest. Volumes of blocks represent weights (kg/ha) of organic matter in each horizon. From W.W. McFee and E.L. Stone, *Soil Science Society of America Proceedings* (1965), by permission Soil Science Society of America

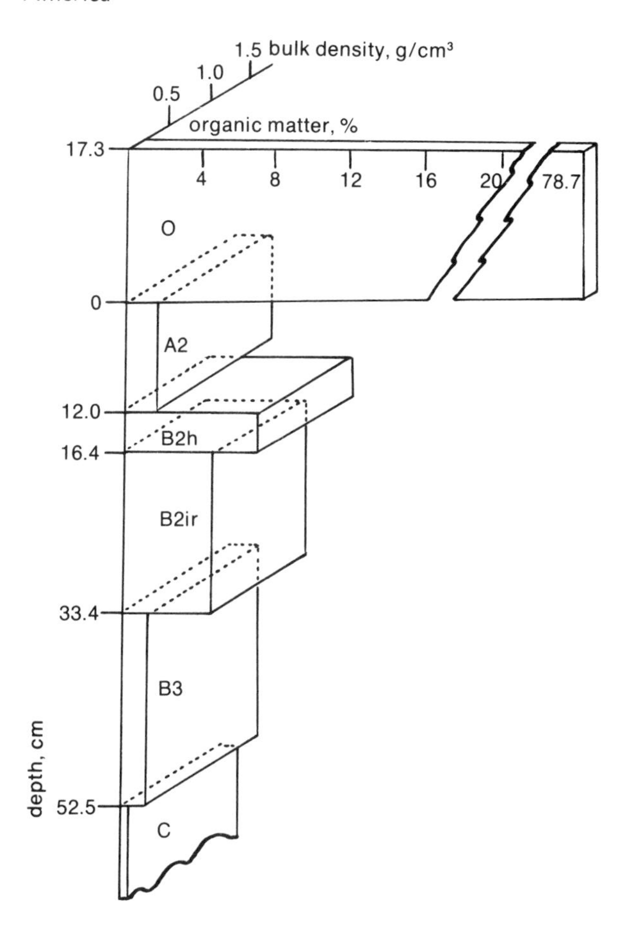

per cent in the mor layer of the forest floor. This was not true for the middle Swedish soils where only 32 to 51 per cent of the total soil organic matter was found in the B horizon and 32 to 46 per cent was in the mor humus. The distribution of organic matter in a spodosol from New York is illustrated in Figure 15.7. The maximum amount is in the forest floor, but the B horizons show secondary accumulations. In contrast, where the forest floor accumulation is minimal and mull surface horizons develop, the maximum is usually in the A1 (Ah) and no secondary maxima occur in deeper horizons.

Clay

Certain forest soils are enriched by clay in B and sometimes upper C horizons. Typically, the clay is a coating or skin about the surface of peds and lines the sides of pores. Clay skins are an example of a type of cutan.* Buol and Hole (1961) in a study of the origin of clay skins in an alfisol concluded that:

1 / Clay skins form in soils with relatively stable aggregates and root channels.

2 / There has to be an adequate supply of clay.

3 / During a period of time they may form, be disrupted, and finally removed. Clay may be moved from the upper B horizon and deposited at greater depths as in the C1 horizon.

4 / The clay is moved by percolating water (lessivage) and deposited when percolation stops. Areas of oriented clay in the upper B horizon ped interiors were originally from larger pore walls but have been incorporated by turbation (soil mixing process).

5 / Natural clay skins are formed by increments of minute lamellae.

6 / The mixing or turbation processes such as freeze-thaw and dry-wet cycles, as well as by organisms, cause a disruption of clay skins.

Often clay skins contain higher quantities of iron, manganese, and phosphorus than the matrix ped material in which they occur. In young soils derived from geological materials laid down during Pleistocene or more recent times the bulk of the clay transformations are mainly simple ones, involving a release from deposits and their redeposition in the profile, with some degree of weathering but primarily a reorientation. In older landscapes, particularly although not necessarily in present day, warm, humid climates, the geological materials may have been subjected to very long periods of weathering processes, and much of the clay in soils may have resulted from in situ weathering of clay-forming minerals. In such soils, many ultisols for example, the clay may then be redistributed by lessivage.

Calcium

In forest soils calcium carbonates are normally removed by weathering from the solum or at least the upper part of it as water containing carbon dioxide moves

* Cutans are soil microstructures formed from either the concentration of certain materials or in situ modification of certain soil components.

through the profile. Transported as a soluble bicarbonate, it may be reprecipitated in the carbonate form lower in the profile, often at the lower boundary of the solum and beneath it, associated with roots and root channels. The solution of carbonates in the upper part of the profile often results in the release of clay particles as, for example, from the weathering of calcareous shales, and it is from this type of material that the transformation of calcium and clay occur concurrently. Following a removal of calcium from the upper part of the solum and the creation of an acid rather than an alkaline soil reaction, more intense weathering may occur and, in particular, cheluviation with concomitant movement of iron, aluminium, and organic material frequently results in a solum with two sets of profiles - a *bisequum.*

Iron, aluminium, and silicon

Weathering requires moisture and its rate not only increases with increase in temperature but becomes more intense the longer the period of occurrence. During weathering, three elements which are transformed, often to a considerable extent, are iron, aluminium, and silicon. The depth to which the weathering processes can occur may be many metres. An example is shown of the development of a jarrah root system in laterite materials in Western Australia (Figure 15.8). The weathering results in a removal of silicon in solution and the formation of iron and aluminium sesquioxides at lower depths. Clay is usually present as 1:1 kaolinite. When weathering has been intense and has occurred over a long period of time, soils such as *oxisols* are formed. Oxisols include soils which were formerly termed latosols, or if they contained a sesquioxide rich zone high in iron and had certain other characteristics they were *lateritic* soils. The sesquioxide-rich material associated with lateritic soils is now termed *plinthite*; it hardens irreversibly when exposed and is usually mottled with lighter coloured material.

The formation of plinthite is related to fluctuating ground water occurrence and van Schuylenborgh (1971) considers the weathering process resulting in an oxisol with plinthite to comprise three stages: (1) A period of general weathering under conditions of free drainage. Materials such as kaolinite, gibbsite, and goethite are formed and there are losses of bases and some silica and alumina. As weathering proceeds and the amounts of clay and sesquioxide colloids increase, permeability is decreased and a temporary groundwater table is established. (2) Under the influence of the temporary groundwater table, alternate conditions of oxidation and reduction occur, and pseudo gley with mottling develops. This is the beginning of plinthization. With further weathering, drainage and evapotranspiration cannot result in complete water removal and permanent saturation prevails. (3) With the onset of permanent saturation, a normal gley is superimposed on the pseudo gley.

Figure 15.8 Root system of jarrah (Eucalyptus marginata) growing in laterite in Western Australia. Note intensive rooting in upper soil zone and depth of root development. From P.C. Kimber, *The Root System of Jarrah* (1974), courtesy Forests Department, W. Australia

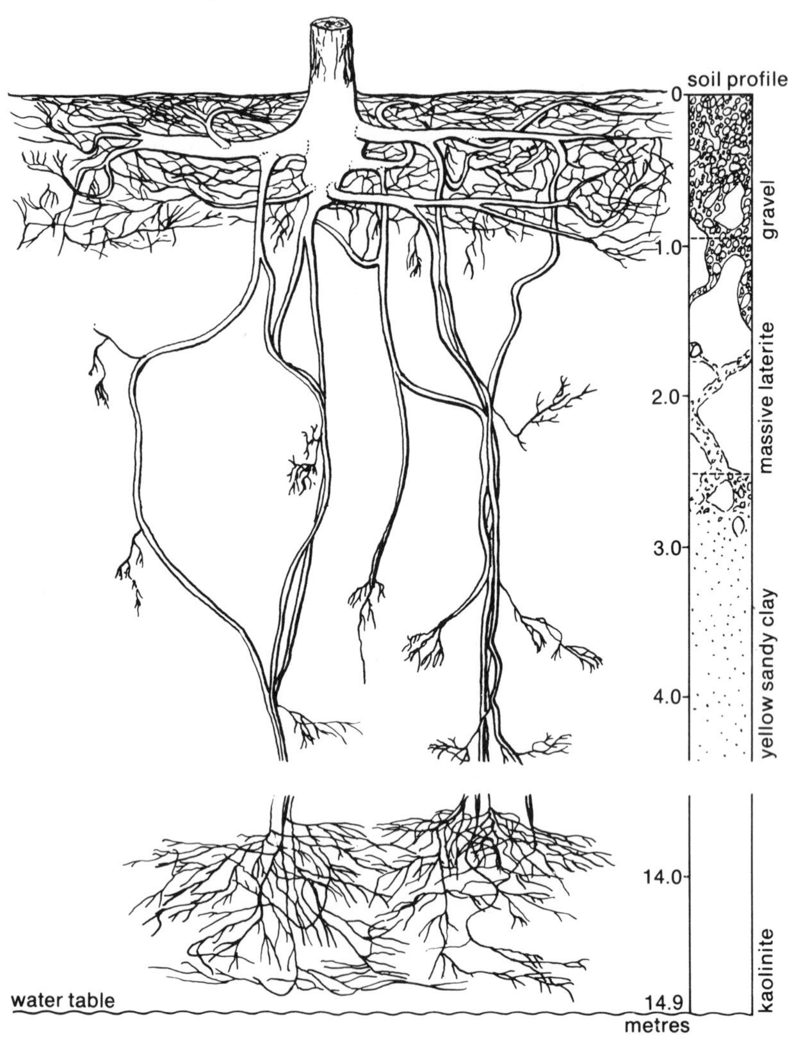

Although a large proportion of laterites have been developed in this way, de Villiers (1965) considered that others were formed following desiccation and hardening of residual sesquioxides after intense weathering.

A modal oxisol with plinthite has been described as having six zones, in vertically descending order of: top zone, plinthite, mottled clay, pallid zone, rotten rock, rock (van Schuylenborgh, 1971). These zones may be variable and some may be missing.

In regions where the weathering processes are less intense (McCaleb, 1959), removal of silica will not occur to the same degree and, although primary minerals are weathered to secondary clays, often these are of the 2:1 lattice and some of the 1:1 lattice form. The bulk of the clay is produced from in situ weathering. Iron and aluminium sesquioxide deposits give the soil material red and yellow coloration. Ultisols have properties which result from this type of weathering.

In much of the cool temperate forest area of the world the complexing and transformation of iron and aluminium are associated with organic matter, a cheluviation process already described. There does exist a considerable difference in the intensity of both weathering and redeposition, and these differences can often result in markedly contrasting profile morphology. Ball and Beaumont (1972) found little change with depth in the profile of a dystrochrept (brown earth), but in a soil that was an intergrade between inceptisol and spodosol some of the weathered aluminium had been moved downwards although there was a high proportion of iron in weathered form. In a spodosol (podzol) there is a marked accumulation of sesquioxides in the B horizon. A schematic representation of the net change in extractable iron (dithionite) and aluminium (oxalate) and dithionite iron:clay ratio for several forest soils is shown in Figure 15.9. These serve to provide generally a set of quantifiable differences which may be used diagnostically to categorize different profiles.

The weathering processes by which the three elements, iron, aluminium, and silicon are transformed are complex and interrelated. As Acquaye and Tinsley (1965) have noted, both iron and aluminium can exert a controlling influence on the solubility of silica. Ferric iron and silica changed from high to low solubility between pH 4 and 5 and aluminium and silica in the same way between pH 5 and 6; as a reflection of this there is frequently a zonation in the depositional pattern in which these components may be found in the soil profile, although some modification occurs where organic matter is also involved.

Conclusions

The foregoing discussion of soil development must be regarded as introductory; a more comprehensive account may be found in Buol et al. (1973). It should serve to indicate, however, that knowledge of processes is still drawn inferentially from measurements of various properties of soil profiles and less commonly

Figure 15.9 Schematic representation of diagnostic depth functions for dithionite extractable iron (Fe_d), oxalate extractable aluminium (Al_o) and the ratio of dithionite extractable iron to clay ($Fe_{d/clay}$). From H.P. Blume and U. Schwertmann, *Soil Science Society of America Proceedings* (1969), by permission Soil Science Society of America

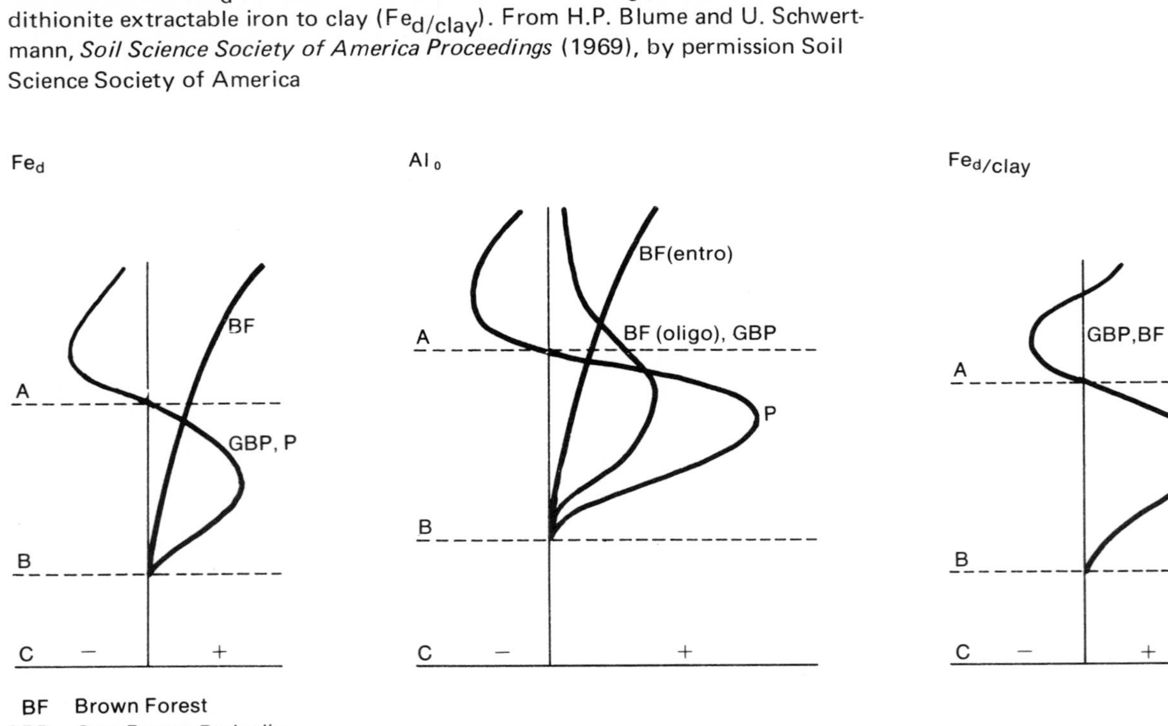

from direct measurement and observations of processes as they occur. It has been possible to demonstrate experimentally a number of soil-forming processes. Within the last decade or so there has been an increase in studies of this type. The relevance of such studies to the understanding of soil development, particularly in soils of recent development - within a few centuries or millennia - are obvious. It is when very ancient landscapes and soil materials are examined that greater difficulty is experienced because of uncertainty about past climatic, vegetational, and erosional conditions.

For many years it was common to treat as different the processes of soil development in certain regions of the world. They were given distinguishing names - *podzolization, laterization, calcification,* etc., and there was a tendency to view these processes as unique or peculiar to these regions. Such compartition can have serious repercussions scientifically. The coining of a catch phrase or jargon does not of itself give knowledge or understanding; it is, in fact, often a blanket for ignorance. The belief that soil development takes place under completely different processes in various parts of the world will often preclude the objective examination of processes that are actually occurring and restrict studies to the selection of whatever evidence seems appropriate to the maintenance of a conceptual fiction.

A soil reflects not only the present physical, chemical, and biological activities but also those previously occurring which may not have been the same or of similar combinations. As a result of past and present development the soil itself may modify the nature and intensity of subsequent soil development. When a broad understanding of the major changes which a landscape has undergone has been achieved, there is a greater likelihood that a clear knowledge of the processes of soil development will emerge.

16
Soils and changing landscapes and use

The very act of digging into a soil and examining its profile is to look at a form of historical record. The question of deciphering and understanding the nature of the record depends not only on knowledge of soil properties but also on a realization of their significance both present and past. It is for this reason that the nature of the geological surfaces in which the soil has developed are important. Not only do the geological materials comprise the 'body' of the soil in most instances but the time span during which they have been formed provides a broad scale in which the shorter period of soil development can be placed. In geologically 'old' landscapes, erosional surfaces or depositions can provide materials for 'young' soils.

If geological materials provide the 'body,' it is the organic component which quite literally brings life to the soil. The vegetation and forest species in particular provide the major source of the organic material and determine the direction and magnitude of many soil processes. In contrast to geological deposits, usually only the present vegetation is directly visible to the observer and knowledge of the past history of vegetation is dependent on specialized techniques such as pollen analysis not normally available in field examination. A knowledge of the local forest conditions and the successional changes allows for some reasoned conjecture concerning the immediate past history of vegetation; beyond this the only evidence is the soil profile itself which can provide a base from which inferences may be drawn.

The historical record in many forest regions is incomplete indeed if the effects of man, both direct and indirect, are not seen in the profile record. The soil by recording man's effects on the geological materials and more often on the vegetation becomes a monument to his actions. Sometimes the monument has been eroded and weathered, sometimes turned over and moved, quite literally. It is almost as if nature were playing capricious pranks to confound the would-be historian.

Yet, if the difficulties and obstacles are many in attempting to unravel the record in the soil, the reward is also large. By understanding the record of the past and present from the soil an insight can be gained into the place of man and his actions in the natural world, both now and in the future and he must surely gain a deeper understanding of his own relationship with the land.

Soils and landscapes

There are many different approaches possible when studies are made of soils in relation to change in landscapes. The approach depends upon the particular background, knowledge, and experience of each investigator. A geologist looks for the geological evidence; an ecologist or forester is more likely to stress the evidence from the vegetation itself; an archaeologist will be more concerned with the artifacts of human presence. Yet each will find much, if not all, of their evidence within the soil body. The presence of soils formed in response to conditions significantly different from those presently occurring at their location often presents key evidence of landscape changes. Such soils are termed *paleosols*. Paleosols are frequently eroded or otherwise modified and may be completely buried or exposed. They are of particular value when the materials in which they occur can be dated either relatively or absolutely. The following discussion considers soil and related landscape changes primarily in terms of geological, vegetational, and cultural factors.

GEOLOGIC

The general nature of relations between geomorphology and soil morphology and genesis have been described, together with specific examples by Daniels et al. (1971). In general, younger deposits may be found on top of older beds, but this simple pattern may be altered and distortions occur as a result of strong geologic forces. The age of the soil developed in a geological material is often closely related to the pattern of erosional activities. On a plateau of low relief the depth of weathering and soil age may be expected to be greater than either at the edge of the plateau where continuous or sporadic erosion may occur or on lower slopes below the plateau where soils develop in materials moved from upslope locations. Both solum thickness and intensity of profile development may be less than on the plateau itself.

In central North America where the lower limits of the main deposits from the Pleistocene glaciations occur, soils have developed during interglacial times on the depositional materials of the previous glacial advance. A generalized view of these soils has been given by Thorp et al. (1951). They noted that Sangamon interglacial soils are deeply leached and that many modern soils are formed in loess of Wisconsin age and in truncated surfaces of the A,B, and C horizons of

Sangamon origin soils. Thus, the soils vary not only in age but also in present-day properties, depending on the surfaces or materials in which they have developed. Where paleosols occur, part may be buried, but a portion will occur as a component in the continuum of present-day surface soils. Ruhe (1956) noted that the solum thickness increased from the profiles developed in late Wisconsin deposits (62 cm), late Sangamon (132 cm), to those of Yarmouth-Sangamon age (203 cm). He also found that the thickness of the B horizon and clay content increased with age. The paleosol B horizons high in 2:1 lattice clay and gray in colour were given the name *gumbotil* by geologists (Flint, 1957). Where paleosols of this type are exposed contiguously with soils developed from the most recent glacial till deposits, a considerable heterogeneity can exist in a small surface area of a few hectares. This not only creates problems in the survey and mapping of such soils but also in soil management practices, largely because of differences in physical properties of texture and drainage (Bushue et al., 1970).

Two areas that have been studied from the standpoint of soil development and geological changes are situated in Australia and the southeastern United States. They are of interest because some of the oldest soils in the world, dating back to the Pliocene, are exposed. A general pattern of soil distribution of laterite and lateritic soils in southwestern Australia has been given by Mulcahy (1960). Jessup (1960a,b) presents a detailed account of the relationships in southeastern Australia. He describes the remnants of a former landscape surface which rises 30 to 90 m above the surrounding plain. These remnants (plateaus) consist of a thick portion of the pallid zone of a laterite soil possibly formed in the Pliocene. The plateaus are capped by a layer of quartzite which varies from 60 to 500 cm in thickness. When the original laterite soil was eroded, the profile was truncated to the pallid or mottled zone. Water-laid deposits were then deposited over the eroded surface and silicification occurred, resulting in the duricrust of quartzite. This duricrust is absent from younger surfaces. Subsequently there was a period of large lake formation in the plains area in which lacustrine deposition occurred. As the lakes dried out, the entire landscape was buried beneath wind-transported deposits of at least two layers in which soils of Quaternary age developed. Periods of soil formation in these layers were associated with relatively humid climates when erosion would be minimal because of vegetative cover. This was followed by an increasingly arid climate in which vegetation decreased and the landscape became unstable.

In the southeastern United States, soils of the upper and middle coastal plain have been exposed for several million years. Daniels et al. (1970) studied various properties of these ultisols and found that progressively from the younger soils of 2 to 3 $\times$ 10^6 years to the older of 6 to 10 $\times$ 10^6 years, there was an increase in solum thickness, and proportion of soils with plinthite and gibbsite content.

Conversely, the kaolinite content and proportion of soils with eluvial bodies in the B horizon decreased with age. For each property the changes were curvilinear with time and the rates of change were not uniform. In fact, sometimes they rose to a maximum, then steadily decreased. From this it was tentatively concluded that the relationship between expression of a soil property and time is not directly tied to the age of the geomorphic surface in which the weathering occurs. The presence of high water-tables inhibited the development of some properties such as gibbsite accumulation while enhancing other properties. Previously (Daniels and Gamble, 1967), it had been noted that abrupt to gradual changes in soil drainage could cause changes in soil texture in certain ultisols. The change in texture was produced by eluviation from A2 to B2 horizons in relation to water-table regimes. Thus the area of greatest eluviation from the A2 was at the dissected edge of a flat to gently undulating surface where the water table was relatively deep, but the greatest eluviation from the B2 came about when there was a high fluctuating water-table in moderately well to poorly drained soils.

These examples indicate that the relations between soils and their specific occurrence within a landscape may not be simple and straightforward. Not infrequently, repeated observation of two related features, for example slope position and depth of solum, may lead to the relation being viewed as one of cause and effect. It is only when anomalies are encountered that the causal hypothesis is challenged. Unfortunately, anomalies are sometimes overlooked because they do challenge the accepted relationships.

VEGETATION

An obvious approach is to select an area in the landscape where vegetation changes on the same soil materials. Often, however, the difference in vegetation is associated with other landscape differences. For example, Finney et al. (1962), using transects, studied the soil differences in four Ohio valleys. They found thinner A1 and thicker A2 and B horizons in the southwest-facing slopes than on the northeast-facing slopes. The forest vegetation was also different, mixed deciduous species on the northeast-facing slopes and mixed oak on the southwest-facing slopes. The microclimatic factors and soil temperature and moisture regimes also differed as did the degree of slope. Thus not only the vegetation but also associated changes in these other attributes were responsible for the variation in soil profiles.

In regions where prairie-forest transitions occur, there have been numerous studies of soil differences related to the vegetation changes. Bailey et al. (1964) studied two gray brown podzolic soils under oak-hickory and two gray brown podzolic-brunizem intergrades under grass, all of which were developed in loess and till of similar age. They found that the profiles were more highly weathered under forest cover than under grass and had greater organic matter and bases in the A1, greater clay maxima, lower calcium:magnesium ratios, and a lower content of bases in the B2 horizon. In a detailed study of properties for soils showing a developmental range from haploboroll (chernozem) to eutroboralf (gray wooded), Severson and Arnemann (1973) noted that the abrupt change both in vegetation (forest-prairie) and soil properties contrasted with the gradual gradients in topographic and climatic factors. The study area was about 10 km wide (north-south) and 60 km long (east-west). The west half of the area was known to have been in prairie vegetation for the past 8,500 years, and the soils present are udic haploborolls. The vegetational history of the east half of the area is different. From 8,500 years to 4,000 years BP (before present) it was an oak-savannah and a portion of the area is still in this vegetation with soil classed as a udic haploboroll, but differing from the similarly classified profile under prairie in that there has been a more intense removal of carbonates and translocation of clay to the B and upper C horizon. From 4,000 to about 2,000 years BP, a mesic deciduous forest occupied the greater part of what was formerly oak-savannah. On the area where this mesic deciduous forest formation still exists the soils are classed as typic udic argiborolls and show a lesser A1 development than the oak-savannah soil, and an A2 horizon. Finally, for the past 2,000 years a pine-hardwood vegetation has existed on an area which previously supported mixed deciduous forest. The soils under the pine hardwoods are typic eutroboralfs with thin A1 horizons, intensive removal of carbonates and bases as well as clay, and considerably thicker B horizon development. A diagram outlining the horizon differences is shown in Figure 16.1.

Evidence of vegetation which may have existed centuries or thousands of years previously is usually non-existent .or difficult to obtain. Plant material which has been preserved in acid peat bogs under anaerobic conditions is one of the principal sources of such records. Pollen grains and woody fragments are the most commonly used plant parts. Under certain conditions these components may be preserved in well-drained soils. Dimbleby (1961) has used pollen analysis in acid mineral soils to establish the nature of previous vegetation. The technique has proven particularly effective in studies of the development of British heathlands and associated soils. In acid soils pollen may be preserved for periods of up to 2,000 years and it is by establishing the frequency of occurrence of pollen at different depths that a relative abundance and dating are obtained. It is assumed that pollen washed to deeper soil layers is older than pollen above it. When pollen layers can be dated in relation to other evidence – the presence of archaeological artifacts, radiocarbon dating or known erosional or depositional surfaces – the history of the vegetation becomes clear. An example of pollen distributions

Figure 16.1 Diagrams of soil profiles associated with different vegetation and developed on medium-textured calcareous till in a forest-prairie ecotone in Minnesota. Based on data of Severson and Arnemann (1973)

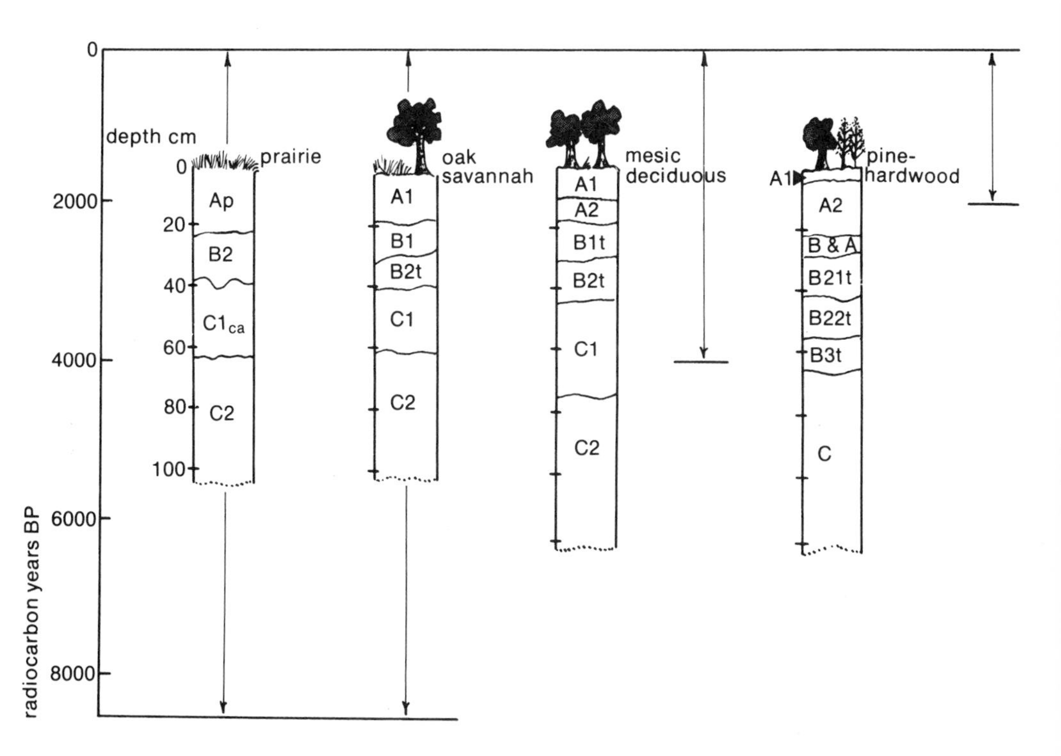

given by Dimbleby (1962) is shown in Figure 16.2a,b. In the pollen analysis of the heathland soil, a spodosol with a well-developed A2, three zones are apparent. In the upper 20 cm the development of heathland from hazel (*Corylus* spp.) is clearly shown by the increasing preponderance of Calluna pollen. From 20 to 35 cm there is a zone in which there is an abrupt change where the proportion of grass pollen shows a large increase at the top (20 cm) followed by a varied pattern, until at 35 cm there is clearly a pollen spectrum representing forest conditions. In close proximity to the heath area the pollen profile for a soil under oak (*Quercus robur*) was determined (Figure 16.2b). The profile was described as an acid brown forest soil with a mor humus layer; no A2 layer was discernible. In the lower part of the soil the pollen reflects a range of smaller quantities of tree species and larger proportions of Calluna and grasses. The pollen spectra from the bottom to the top show increasing abundance of trees and other

Figure 16.2 Pollen profiles from two soils near Burley, New Forest, England: (a) podzol soil supporting heath vegetation; (b) acid brown forest soil with oak forest. For each genus or family, histograms on left side indicate numbers of pollen grains per gram of soil and histograms on right side their per cent. Reproduced from: G.W. Dimbleby, The Development of British Heathlands and their Soils, © 1962 Oxford University Press, by permission of The Clarendon Press, Oxford

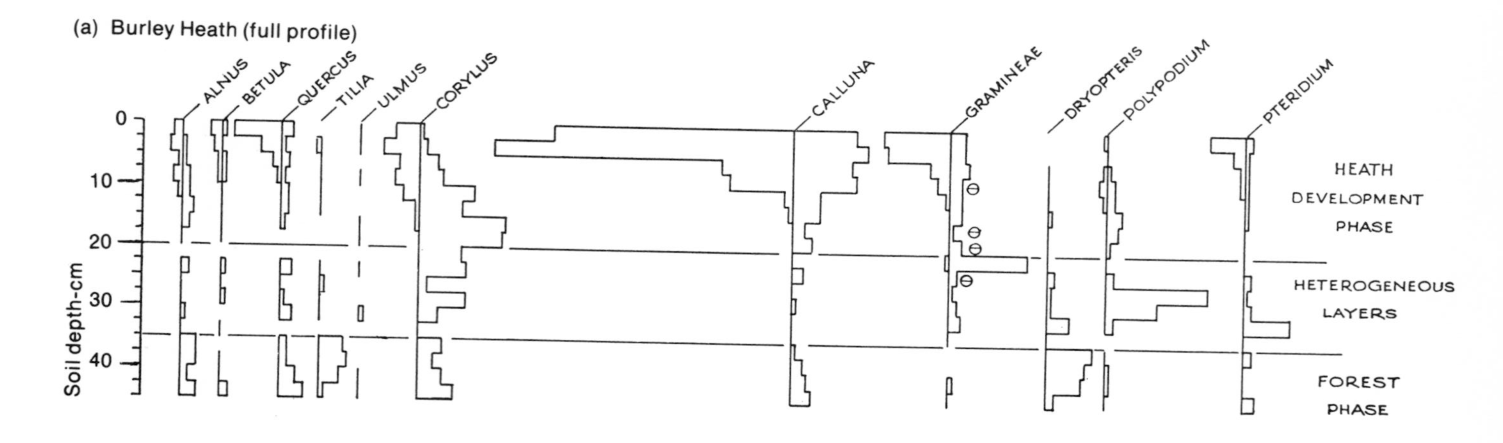

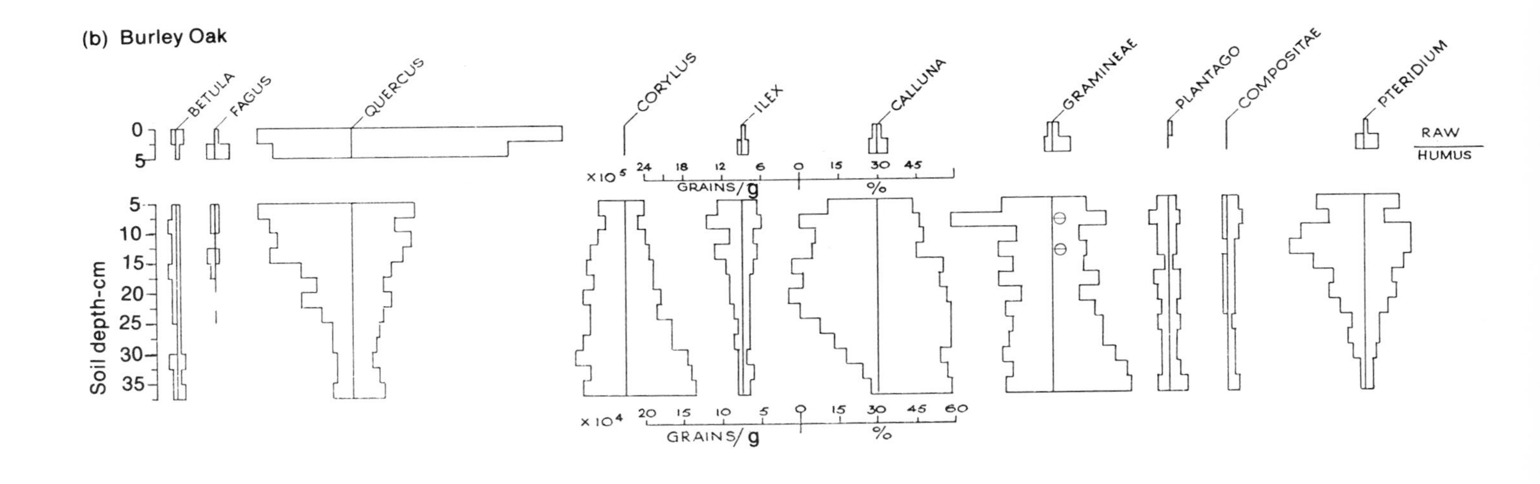

(b) Burley Oak
BETULA
FAGUS
QUERCUS
CORYLUS
ILEX
CALLUNA
GRAMINEAE
PLANTAGO
COMPOSITAE
PTERIDIUM
RAW
HUMUS
Soil depth-cm
0
5
10
15
20
25
30
35
×10⁵
24
18
12
6
0
15
30
45
GRAINS/g
%
×10⁴
20
15
10
5
0
15
30
45
60
GRAINS/g
%

woodland flora (holly and bracken). These two examples serve to illustrate the changes in vegetation that can be documented in the same area in geological deposits (Barton sands) in which contrasting soil profiles have developed. Dimbleby (1962) presents convincing evidence, especially when the pollen profiles and soil development are linked to archaeological evidence, primarily from Bronze Age times, that the spodosol development has arisen largely as a result of 'man's assault on the landscape' during that period. Fire and grazing, combined if not separately, would reduce the tree component and in turn most likely result in decrease in fertility due to reduced nutrient cycling. The dominance of shallow rooting heath vegetation such as Calluna(ling) accelerates the decline in fertility. The heathlands and their soils are a legacy which the twentieth century has inherited, in all probability from the actions of man some 2,000 to 3,000 years ago. In some northern areas, changes in vegetation may have come about in response to climatic shifts, but the evidence in many other heathland areas certainly indicates that the changes were wrought by Neolithic and Bronze Age people (Gimingham, 1972).

Other evidence of previous vegetation is related to the absorption of silica by plants. Much of the silica occurs in the plant as plant opal or *phytoliths*, which vary in size, shape, and abundance. When the organic part of the plant decomposes in the soil, the phytoliths remain. Phytoliths are subject to weathering and those from tree species such as sugar maple and ash were considered by Wilding and Drees (1971) to be more susceptible because they have a large surface area compared to those from American beech. Grasses produce phytoliths in much greater quantities than dicotyledonous plants and conifers, which produce relatively small amounts. The size ranges of the individual phytoliths vary considerably; Wilding and Drees (1971) estimated that 75 per cent or more of the phytoliths from forest species were in sizes less than two microns (the upper limit for clay). Verma and Rust (1969) found that in a typic hapludoll many phytoliths were in the 5 to 20 micron fraction and the largest amounts were derived from bur oak and ironwood. They noted that, where increases in concentration of the phytoliths were found at a buried surface, it could reflect either a grass vegetation and landscape or windblown deposits of loess containing phytoliths. The determination of amounts of phytoliths in a soil can be used to obtain age estimates for the period that vegetation, particularly grasses, may have existed on a soil. Witty and Knox (1964), for example, studied a transition zone between grass and ponderosa pine forest in north central Oregon and, using an estimate of grass opal accumulation of 22 to 34 kg/ha/yr, concluded that the vegetation boundary between these two formations had been relatively stable for 4,000 to 7,000 years. Soil drainage may affect the amount of phytoliths since it also will alter plant growth (Jones and Beavers, 1964).

Figure 16.3 Root development of young eucalyptus (*E. saligna*) growing in soil replaced after strip mining. Note restriction of roots to replaced soil. Scale divisions in decimetres. Photo courtesy of S.R. Shea

CULTURAL

A general review of man's influence on soil development has been given by Bidwell and Hole (1965). They considered actions primarily in terms of additions or losses either in absolute terms or in relation to intensity of processes associated with the principal soil-forming processes of parent material, relief, climate, organisms, and time.

Large-scale strip mining operations provide ample evidence of the effect of man in exposing new surfaces for soil development. Attempts at amelioration of the rooting conditions by placement of some form of top soil on the surface may enhance growth temporarily, but in the long run often creates further difficulties for tree growth, since root development may be restricted once the top soil has been exploited (Figure 16.3). In some areas, large-scale additions of fertilizers, especially of elements not subject to loss from the system, increase the fertility. In local situations, fall-out of particulate matter as, for example, heavy metals in the vicinity of smelting operations can result in large concentrations of copper, nickel, and zinc in some soils. The effect of these metals is often masked because of the presence of other materials such as sulphur dioxide, which inhibit plant growth. The regular and complete removal of vegetation results in loss of many elements from the soil and certain aspects of this have already been discussed (Chapter 14). What is generally less appreciated are the effects that early man may have had in many areas, even in North America. Although it is commonly believed that the forests were relatively undisturbed until the coming of the white man, there would appear to be scant evidence for this and considerable counter evidence such as that presented by Day (1953). Abler (1970) summarizes further historical records for Iroquoian villages of the seventeenth century and a recurring theme is the extent to which agricultural crop production and firewood use changed the surrounding forest landscape. The soils used were primarily the coarser-textured, well-drained, post-glacial deposits of outwash sands common in much of northeastern North America. Undoubtedly, some lower, moister soils may have been cultivated, particularly if they were subject to regular alluvial deposition, but it is evident that primitive 'hoe culture' on a shifting basis is only effective in a forest-soil system where fertility maintenance is based on a forest crop rotation and short-term soil fertility enhancement is dependent on the ashes from the burned forest. The fire-disturbed forest soil is also much more readily subject to hoe culture than grassland soils of finer texture. These latter, which are the best suited for modern arable use in temperate regions, had to wait for the technology of the plow before they could be cultivated. The enhancement of soil fertility by wood ash has as its corollary the lowering of fertility if the ash is removed. While no complete proof is possible, the occurrence of potassium deficient soils in the Warrensburg area of New York already described (Chapter 14) may well have been intensified by the removal of the wood ash when the forests were cleared and burned by white settlers. At that time there was a strong market for potash, a source of cash income to the farmers and Warrensburg was a centre for this trade.

Locations of settlement by early man have been detected or confirmed by increases in soil phosphorus, particularly in and about old habitation sites both in northern Europe and North America (Lutz, 1951; Deitz, 1957). The use of burial mounds in conjunction with pollen analysis by Dimbleby (1962) to provide an explanation of heathlands has already been mentioned. In North America, soil profiles at archaeological sites have been studied to determine relative time for profile development. Parsons et al. (1962) concluded that in a series of mounds in Iowa the A1 horizon formed under deciduous forest reached its maximum development at 1,000 years or less and clay film development at 1,000 to 2,500 years.

The type of vegetation that a soil can support may be greatly changed by drainage. Drainage is costly and usually undertaken for agricultural or urban purposes; however, soil and vegetation in the same watershed will be affected. The most obvious examples are to be found along highways where the natural drainage sequences have been entirely disrupted both during construction and after road completion. Less obvious are the effects on farm woodlots, often located at a distance from a road. When arable land is drained, water movement from the soils of woodlands which lie in the drainage basin is increased. The changes in species composition and soil properties, although less perceptible, are still inexorable and, because they are gradual, may go unrecorded. When organic soils are drained, the net result over a period of years is the lowering of the surface as increased decomposition occurs. Conversely, impedance of drainage can result in the development of organic soils in areas that were previously well-drained, and as a result the surface level of the soil is elevated. Although man is often a major factor in creating changes in drainage patterns, there are other vectors. Destruction of vegetation by fire in an area where the water-table is within the rooting zone can result in a rise in the level of the water so that subsequent vegetation is better adapted to poorer drainage than the previous one. Natural erosion may result in damming of drainage channels. In many North American forests beaver are responsible for dramatic and long-lasting changes affecting soil and vegetation. An example of such a change is shown in Figure 16.4, where a spodosol (podzol) had developed in a deep well-drained sand but an impedance in drainage has now resulted in the development of a histosol approximately 2 m thick. Much of the iron in the B horizon has been removed in the reducing conditions following saturation. The lack of change or evidence of disturbance in the upper portion of the A2 horizon indicates that the raising of the water-table was a sudden one such as would occur by the building of a beaver dam. Change in drainage may occur gradually and results from changes in the soil profile which may in turn have been caused by other factors. An example of this is given by Proudfoot (1958) for a former forest area in Ireland. Following glaciation, a solum developed under forest in a till deposit over chalk. During Neolithic time (3000 to 2500 BC) the forest was cleared and there was evidence that it became a ritual site which was abandoned and then subjected to some form of tillage. During this period, more intensive weathering and leaching occurred and a spodosol profile is evident. By 2200 BC an impermeable iron pan had developed in the B horizon and as a consequence waterlogging occurred and peat formation commenced. The pattern of this profile development varied with differences in use. Cultivation and periodic burning would have occurred in the area and resulted in a range of intensity of the iron pan, being least where the

Figure 16.4 Peat over a buried spodosol (podzol) profile which had developed in a well-drained sand prior to peat formation. Note Ae (A2) immediately beneath lower boundary of peat. Scale: 1 division = 10 cm. Photo by K.A. Armson

soil was cultivated. Eventually there was a general abandonment of fields in the area. The particular site studied was occupied by huts in 1500 AD and during this period there was evidence of further development of the B horizon. From 1700 AD increasing peat accumulation buried previous evidence of man's occupation. While in retrospect there has been a most dramatic change in the landscape from forest on well-drained soil to a treeless peat bog over a time period of three or four millennia, the landscape appeared stable during the lifespan of any one person. Yet the major landscape change was essentially brought about by the cumulation of man's 'short time' cultural actions.

The New Forest – a case history

It is inevitable that most of the evidence of past actions by man, particularly as they affect the soil, are either fragmentary or must be inferred from the greater and more obvious evidence related to his activities of war, trade, and general living. The lack of evidence is particularly great for forested areas, due primarily to

Figure 16.5 Location of the New Forest in England.

the transient and low intensity of cultural activities in such an environment. What makes the New Forest in southern England of especial interest is the fact that from the Domesday Book (1086 AD) to the present there has been a set of written records for the area and if to this are added the archaeological records, a picture emerges of changing landscape and processes which provide a valuable microcosm. In addition, the various pieces of evidence have been brought together and made available by Tubbs (1968).

Figure 16.6 Diagram of cross-section through main geological deposits of the New Forest area. The surficial gravels do not exist as a continuous covering.

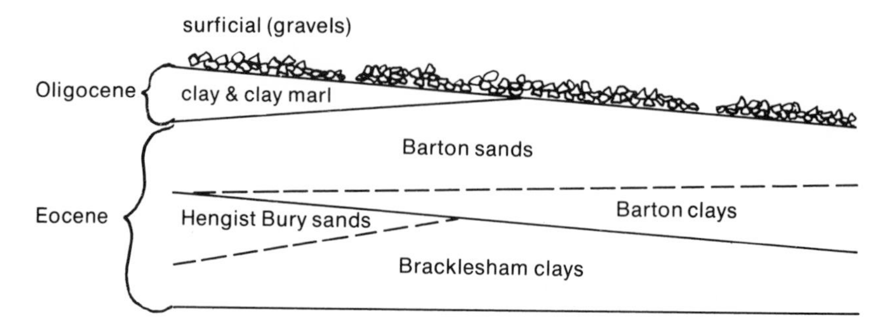

LOCATION, AREA, AND GEOLOGY

The New Forest is located in the south of England (Figure 16.5) and comprises an area of approximately 38,000 hectares, of which 27,000 or 71 per cent are Crown Land (government owned) and the remainder is privately owned but subject to specific legislation because it lies within the New Forest. Although it was outside of the area covered by ice during the most recent glaciation, there is evidence that the surficial geological material was subject to periglacial weathering and some portion consists of waterlaid deposits of glacial origin. Gravel deposits, either the more recent ones in valleys or the older Plateau gravels, cap a sequence of fine and coarse-textured beds of Oligocene and Eocene origin. The upper (Oligocene) finer-textured clays and clay marls are exposed mainly in the southern portion of the forest. Beneath are the thicker beds of Eocene age, an upper zone of Barton sands, and a lower one of Barton clay; however, Eocene deposits, primarily of clays (Bracklesham beds) lie below the Barton clay. The general sequence is illustrated in Figure 16.6.

SETTLEMENT, CULTURE, VEGETATION, AND SOILS

The evidence, prior to written records, is archaeological. Although Neolithic man may have been present, no evidence for settlement appears to exist and the most obvious and general remains are burial mounds left by Bronze Age culture. From studies, particularly by Dimbleby and associated workers, it seems clear that what was an area of extensive deciduous woodland was largely removed either prior to or during this period and that the landscape as a result probably consisted of open grass land, heath, and some cultivated areas, together with open woodland. Tubbs (1968) suggests that the burial mounds, which are on the well-drained coarser-textured soils, were perhaps in fact located on abandoned agri-

Figure 16.7 Heathland spodosol (podzol) developed in gravel in the New Forest. Photo by K.A. Armson

cultural land. The cause of abandonment on these soils could well be a result of lowered fertility. Certainly the dominant feature by the close of the Bronze Age was the existence of relatively open areas of heath and grassland. Pollen analyses and examination of buried soils indicate that the deciduous forest grew on acid brown forest soils but that clearing and a combination of agriculture, fire, and grazing resulted in reduced fertility and a heathland vegetation which initiated the spodosol (podzol) profile development (Figure 16.7). Where pollen analysis by Dimbleby and Gill (1955) discloses no evidence of previous heath occupation on the Plateau gravels and beech has not been introduced, the oak forest retains a rich flora over a brown forest soil.

The Iron Age following the Bronze Age is not well represented archaeologically in the New Forest except for several forts, of which only one in the south just inland from the coast was of any size. It can be inferred that this represented a consolidation of the population into larger centres on the areas of finer-textured soils. During this period the open grass-heathland areas within the forest were undoubtedly maintained by a combination of fire and grazing. With the arrival of the Romans little development took place except for a pottery industry located on clays mainly in the northwest portion of the forest. There was perhaps some pastoral use, but it would seem reasonable to assume that the importation of technology by the Romans would make possible large-scale cultivation of clay soils and development of road systems elsewhere than in the New Forest.

The record from the time of the departure of the Romans to the arrival of the Normans is not complete and provides an opportunity for further study. The area was thinly populated with small villages, largely on the finer-textured soils most suited for cultivation. From the twelfth century on, the pattern of land use within the area became regulated because of its status as a royal forest. The small-scale rural economy combining limited cultivation with pastoral rights in the forest was in fact sustained indirectly by legislation.

It is of interest that where former heathland has been set aside for afforestation, under certain tree species, particularly oak, there is a rehabilitation of the infertile heathland spodosol and the development of a mull humus which paradoxically may be found over a strongly leached A2. Pollen analysis (Dimbleby and Gill, 1955) clearly confirms that the heath vegetation was related to the spodosol occurrence.

The change in the forest composition, its exploitation during the seventeenth century followed by the establishment of new forests into the present century and the related social and economic conditions provide a background for alteration in the faunal and floral populations, but these have not changed the essential soil-vegetation relationships described. In this example of the New Forest it is the associations of established features that have then been placed in a historical context. The nature of the processes involved can only be inferred from other studies of soil-forming processes.

L'Envoi

There seems little doubt that soil and the landscape of which it is a part occupy a special place in the perceptions of mankind. The studies of soil scientists have focused on the task of exposing and understanding soil properties and processes. As might be expected, those soil properties which affect crop growth and are capable of modification by agricultural practices have generally received the greatest attention. Increasingly, there are opportunities for the application of soils' knowledge and expertise, not only in the management of forest lands but also within a much broader range of studies such as history and archaeology, in which soil science provides technical knowledge and methodology and an ecological awareness of the interactions between man and soil.

The scars, degradation, and general destruction that the land and soil bear, often but not always as a result of man's activities, frequently receive emotional attention if not much else. Soils and the land will often show an unexpected resilience to destructive forces. Stone (1971) referred to the 'conservative nature of soils,' particularly the ability of developed soils to respond to the impact of treatments and stabilize around a new 'steady state.' This new condition may be quite different from the prior one and does not always result in a return to the soil's previous state. It is when the soil cannot rebound from the impact of treatments that our knowledge of soils can be especially useful in preventing or modifying both the treatment and its action on the soil. The conditions, under which seemingly small soil changes can set off a change in direction and intensity of soil processes with far-reaching effects, provide a major challenge to those who study soil.

APPENDIX 1

Procedures for soil profile description and sampling

In order to examine a soil profile it must be exposed, usually by digging a pit. Often it seems easier to use road cuts or excavations made for other purposes, but while these may be useful, it should always be kept in mind that, because they were made for other purposes, they often ignore, modify, or destroy a portion of the soil. Thus in road cuts the upper layers are much modified and in excavations for buildings the surface layers are usually destroyed.

The digging of a soil pit is frequently regarded as an irksome, tedious, and tiring piece of labour. Yet it is in the physical process of digging that many interesting attributes are seen or felt. Changes in compactness are sensed immediately at the handle of a shovel or spade. The presence of animals and their holes and the forms of roots are best seen when first exposed. Often these are destroyed as the pit is enlarged. Mechanical excavators such as backhoes and bulldozers are sometimes very useful in preparing excavations so that hand removal of soil on a large profile can be done more easily. Soil augers and probes are of use mainly in exploratory work to determine the general nature of soil materials in an area. Soil profile descriptions should never be based on the results of examining a soil with an auger only.

There are a number of features which should be kept in mind when a soil pit is located:

1 / In an exploratory study, the pit should be located so that it is representative of the general situation. It is good practice to place it so that not only the general root distribution is exposed but also the roots of a dominant tree.

2 / Position the pit so that the primary face, the one to be described, will receive the maximum natural light. In the northern hemisphere this side should face in a southerly direction.

3 / Commence the digging of a soil pit at least 50 cm away from the position of the prime face and make sure that the soil excavated is thrown well back from the pit location.

4 / Take great care to ensure that no soil which is dug out is inadvertently dropped on the surface near the prime face of the pit. Do not walk on the soil surface in the vicinity of the prime face or lay down shovels or other equipment on top of it.

5 / If the soil pit is to remain open, it should be properly fenced or otherwise protected so that humans and animals will not injure themselves. The pit should have a sloping ramp access so that small animals are not trapped in it.

6 / Normally soil pits are filled in after they have been used for sampling and description. To facilitate this, and also to leave the location with minimal disturbance, it is wise to lay down one or two tarpaulins or large plastic sheets on the ground back from the pit location and to place the excavated soil on these sheets. During the excavation of the pit it is preferable to separate the upper surface horizons from the lower soil materials. On filling the pit these surface materials may then be replaced last.

7 / Never underestimate the final size of your soil pit. The surface area should be a minimum of one square metre and, if the pit is to be deeper than one metre, the surface area should be increased. Large soil pits are necessary if photographic records are to be made, particularly of the lower soil.

A list of equipment and material usually required in excavating and describing a soil is as follows: 1 spade (short handle), 1 shovel (round mouth with either a long or short handle), 1 pickaxe, 1 swede saw, 1 trowel, 1 knife (stiff blade and blunt, an old table knife is suitable), 1 knife (large pocket knife or hunting knife - sharp), 1 bottle of dilute acid (HCl), 1 pail with 2m length of rope, 1 pair secateurs, 1 handlens (X8 or X10), 1 metric tape, 1 field book (preferably with waterproof paper - surveyors' type books are usually suitable), 1 soil description manual (if available) for the region, 1 Munsell Soil Color book,* plastic sample bags and labels, 1 permanent ink marker pen, 1 roll plastic, coloured seismic tape ('flagging').

Soil profile description

When a soil pit has been located and excavated, time and care should be taken in making a complete examination of both the soil and the surrounding site so that there is a clear understanding of the type of information to be recorded.

It is useful to include with the written record a drawing (to scale) illustrating the gross morphology of the profile. Obviously the date, location, topographic position, nature of the vegetation, together with the name of the investigator and other necessary information should be recorded first.

* Munsell Soil Color Charts, Munsell Color Company Inc., Baltimore, Maryland, USA

For the description of the profile itself, it is convenient to use a fixed sequence in noting the properties of each horizon, beginning with the uppermost, as follows:

1 / Horizon thickness - usually measured in centimetres; the upper boundary of mineral soil is taken as datum zero. Organic horizons are measured down to this datum.

2 / Colour: A Munsell colour notation based on the colour of moist soil is always given. Although it is preferable to note the colour of the soil both moist and dry, the latter is not always possible to determine in the field. If a horizon has several colours and is mottled, the colours of the mottles should be described separately.

3 / Texture: This property refers to relative amounts of particles less than 2 mm in diameter. It is described by terms such as sandy loam, clay, silt loam, etc.

4 / Structure: Often the primary particles are held together in aggregates which have a distinct shape and arrangement. They may be described as medium granular, coarse angular blocky, etc.

5 / Consistence: Refers to the attributes of soil materials that are related to the degree and kind of adhesion and cohesion of the particles or aggregates or by their resistance to deformation and rupture. This property varies greatly with the moisture condition of the soil and is usually expressed at three moisture conditions - dry, moist, and wet. Terms such as hard, firm, sticky, and plastic are used to describe consistence. Cementation and its degree are also noted under this heading.

6 / Mottles: This term refers to patches of soil where one colour appears against a background of another, often as streaks or blotches. The most common reason is that of variable oxidation - reduction states of iron and manganese compounds giving rise to blue gray-yellow-red mottles. They are described in terms of abundance, size, and colour.

7 / Roots and organisms: A description of the abundance, size, orientation, and distribution of roots in each horizon focuses attention on the specific location of these organs and on the interrelations between the soil and the vegetation it supports. Descriptions of organisms or the results of their activities (burrows and mounds) are often important in evaluating soil processes, particularly those which relate to the transformation and movement of organic material.

8 / Clay films and concretions: The occurrence of clay films is described in terms of their frequency of occurrence, thickness, and location. For concretions, the colour, hardness, and relative abundance are recorded. The nature of the cementing agent or dominant material in the concretion should be noted if possible.

9 / Horizon boundaries: The lower boundary of each horizon is described in terms of its distinctness and form.

10 / Stoniness: A description of stones and pebbles, their abundance, size, shape, and orientation. If possible the geological nature of the coarse fragments should be recorded.

11 / Soil reaction: Frequently a portable pH meter is used to determine soil reaction in the field. If a meter is not used, then dilute acid should be taken into the field so that the presence of free carbonates can be detected.

Soil sampling procedures

In many exploratory soil studies, slavish attention to a rigid sampling system that has the virtue of statistical correctness should not be an excuse for precluding observation and measurement of the unusual and aberrant. Accuracy of description and reliability of quantitative data are desired, but these depend as much upon a high level of professional knowledge, experience, and judgment as upon a statistical sampling program.

METHODS OF SAMPLING

There are three basic sampling procedures. For surface soil layers (the forest floor) a sample of known area may be cut, usually with a steel sampling tube or frame. Sample size varies, but frequently areas of 50 to 1000 cm^2 are used; larger sizes are most common.

Mineral horizons may be sampled either by compositing samples taken from the face of a soil profile exposed by digging or by taking samples with a sampling tube or auger by insertion vertically into a soil, without digging a pit. Although their study was on agricultural soils, Welch and Fitts (1956) could find no significant differences between certain soil properties for samples collected with tube, spade, or trowel but did find lower values for soil pH and organic matter when samples were taken with augers. The writer's experience is that augers are unreliable and of minimal use for sampling forest soils.

It is most important when samples are being taken that samples for bulk density determinations are obtained. This will ensure that other properties may be expressed on a soil volume as well as a weight basis.

NUMBERS OF SAMPLES

The number of samples which should be taken depends on the variation of the specific property in the soil population. This variation depends not only on the property itself - more samples are usually required to determine exchangeable calcium to a desired level than are required for soil pH determination - but also on the horizon sampled. In general, for a given level of accuracy, a larger number of samples are needed with increasing depth. It may also be found that the younger the soil, the greater the variation. Harridine (1950) found this to be true for bulk density, colloidal clay, soil pH, and total nitrogen in certain California soils.

Forest floor

Stutzbach et al. (1972) found coefficients of variation for weight measurements of forest floors under red pine plantations to be 20 to 30 per cent, using sample areas of 930 cm^2. In natural pine stands Van Wagner (1963), using the same sample area, determined that (for $p = 0.05$) the numbers of samples for ± 10 per cent accuracy were 48 for the L layer and 81 for the F-H layer. Mader and Lull (1968) determined that 20 to 25 forest floor samples were necessary to give standard errors of ± 10 to 15 per cent of the mean ($p = 0.05$) for estimates of forest floor weights; a similar number of samples gave moisture content values that were 20 to 25 per cent of the mean. In an old undisturbed forest of yellow birch and spruce, McFee and Stone (1965), using a small sampling area of 59 cm^2, found that coefficients of variation for horizon thickness averaged 53 per cent and concluded that approximately 50 observations of a forest floor were required to give standard errors of the mean of ±10 per cent. In the Pacific northwest, Gessel and Balci (1965) found coefficients of variation for forest floor weights to average 41 per cent with a range of 30 to 79 per cent.

It is always preferable to have data on variability for the particular soils that are studied but failing this it can be concluded that a minimum of 20 and preferably 50 observations of forest floors are to be made if standard errors approaching ±10 per cent of the mean are desired.

Mineral soil horizons

Only general guidelines will be suggested because of the wide variation in soil properties that may be encountered. Measurements from a single soil pit should be considered unreliable. Ike and Clutter (1968) in the southeastern United States estimated that at least two to four soil pits were necessary and McFee and Stone (1965) concluded that between 12 and 14 separate cores to determine horizon thickness were necessary to give reliable estimates. In a study of soils developed on clay loam calcareous tills, Wilding et al. (1964) found that the number of profile samples needed to provide values ±10 per cent of the mean ($p = 0.05$) varied considerably with soil property, being fewest (3) for pH values and greatest for depth to mottling (64 to 74) and organic matter (44 to 49). They concluded that, while only 15 or fewer profiles need be sampled in order to estimate values for the less variable properties, at least 20 to 50 profiles are necessary for the determination of the more variable properties to the same accuracy.

TABLE 1
Tentative guidelines for sampling intensity. Where ranges are indicated they are for ± 10 per cent accuracy (standard error of mean) at 95% level of probability (p = 0.05)

No. profiles to be sampled - 2 to 50 Soil horizon	Soil property	Approximate number of samples required
Forest floor:		
1,000 cm^2	Weight - kg/ha	20 - 50
sample area	Thickness - cm	
	Nitrogen - %	15
Mineral horizon: A		
50 - 100 cm^3	Organic matter %	20 - 50
sample volume	Bulk density g/cm^3	2 - 5
	Texture %	5 - 50
	pH	1 - 5
	Total nitrogen %	15 - 40
	Phosphorus mg/100g	> 20
	Exchangeable potassium meq/100g	> 20
	Exchangeable calcium meq/100g	>> 50
	Exchangeable magnesium meq/100g	> 50
Mineral horizon: B		
50 - 100 cm^3	Organic matter %	20 - 50
sample volume	Bulk density g/cm^3	2 - 5
	Texture %	5 - 50
	pH	1 - 5
	Total nitrogen %	> 20
	Phosphorus mg/100g	> 30
	Exchangeable potassium meq/100g	> 30
	Exchangeable calcium meq/100g	>> 50
	Exchangeable magnesium meq/100g	> 30

The range in number of samples required to give the same level of accuracy for chemical properties can be extremely wide. Ike and Clutter (1968) determined that to give ±10 per cent standard error of the mean in the A2 horizon using two soil pits per plot, 20 samples should be taken to determine available phosphorus but that 438 were needed for exchangeable calcium determinations and only 19 and 85 were needed for exchangeable potassium and magnesium, respectively. Metz et al. (1966) made a comprehensive study of both soil and plant tissue values in a loblolly pine plantation and found that in the upper 15 cm of soil, to obtain ±10 per cent accuracy ($p = 0.05$) between 90 and 158 determinations were required for soil phosphorus and 67 to 135 for exchangeable calcium In the B horizon the sample requirements for phosphorus ranged from 27 to 150 and for calcium from 29 to 46. In a detailed study of two forest soils in Sweden, Troedsson and Tamm (1969) concluded that, if 15 to 20 samples were taken, each equidistant in a rectangular lattice arrangement at 1 m spacing, then mixed to provide a composite sample, the coefficient of variation would be less than 10 per cent for nitrogen and loss-on-ignition values in the upper organic horizons. Ten samples collected in this way were considered satisfactory for phosphorus and potassium determinations throughout the soil profile.

In Table 1 an attempt has been made to set out tentative guidelines for sampling intensity. In addition to the literature already mentioned, other data on which the table is based are taken from studies by Ball and Williams (1971), Hammond et al. (1958), and Mader (1963).

APPENDIX 2

Common and scientific names of trees and shrubs

American beech *Fagus grandifolia* Ehrn.
American elm *Ulmus americana* L.

balsam fir *Abies balsamea* (L.) Mill.
bamboo *Arundinaria alpina*
basswood *Tilia americana* L.
beaked hazel *Corylus cornuta* Marsh.
bell-heather (ling) *Calluna vulgaris* L.
bigleaf maple *Acer macrophyllum* Pursh.
bitterbrush *Purdia tridentata* (Pursh.) DC.
black alder *Alnus glutinosa* (L.) Gaertn.
black ash *Fraxinus nigra* Marsh.
black locust *Robinia pseudoacacia* L.
black spruce *Picea mariana* (Mill.) B.S.P.
black walnut *Juglans nigra* L.
bur oak *Quercus macrocarpa* Michx.

chestnut oak *Quercus prinus* L.
Corsican pine *Pinus nigra* var. *calabrica*
cottonwood *Populus deltoides* Bartr.
cypress *Cupressus macrocarpa* Hartw.

Douglas-fir *Pseudotsuga menziesii* (Mirb.) Franco.

eastern hemlock *Tsuga canadensis* (L.) Carr.
eastern white cedar *Thuja occidentalis* L.
eastern white pine *Pinus strobus* L.

elderberry *Sambucus* spp.
English holly *Ilex aquifolium* L.
English oak *Quercus robur* L.
European beech *Fagus sylvatica* L.
European birch *Betula pendula* Roth.
European buckthorn *Rhamnus cathartica* L.
European larch *Larix decidua* Mill.
European silver fir *Abies alba* Mill.

grand fir *Abies grandis* (Dougl.) Lindl.
grey birch *Betula populifolia* Marsh.

hoop-pine *Araucaria cunninghamii* (Sweet.)
hop-hornbeam (ironwood) *Ostrya virginiana* (Mill.) K.Koch

jack pine *Pinus banksiana* Lamb.
Japanese larch *Larix leptolepis* (Sieb. and Zucc.)
jarrah *Eucalyptus marginata* Sm.

kauri *Agathis australis* Salisb.

largetooth aspen *Populus grandidentata* Michx.
loblolly pine *Pinus taeda* L.
lodgepole pine *Pinus contorta* Dougl.
longleaf pine *Pinus palustris* Mill.

Monterey pine *Pinus radiata* D.Don.
mountain alder *Alnus tenuifolia* Nutt.
mountain laurel *Kalmia latifolia* L.
mesquite *Prosopsis juliflora* (Swartz) DC.

Norfolk Island pine *Araucaria heterophylla* (Salisbury) Franco.
Norway spruce *Picea abies* (L.) Karst.
northern pin oak *Quercus ellipsoidalis* E.J. Hill

paper (white) birch *Betula papyrifera* Marsh.
pitch pine *Pinus rigida* Mill.
ponderosa pine *Pinus ponderosa* Laws.

red alder *Alnus rubra* Bong.
red maple *Acer rubrum* L.

red mulberry *Morus rubra* L.
red oak *Quercus rubra* L.
red pine *Pinus resinosa* Ait.
red spruce *Picea rubens* Sarg.
rhododendron *Rhododendron maximum* L.

scarlet oak *Quercus coccinea* Muench
Scots pine *Pinus Sylvestris* L.
shortleaf pine *Pinus echinata* Mill.
shumard oak *Quercus shumardii* Bodd.
silver maple *Acer saccharinum* L.
Sitka spruce *Picea sitchensis* (Bong.) Carr.
slash pine *Pinus elliottii* Engelm.
snowbrush *Ceanothus velutinus* Dougl.
speckled alder *Alnus rugosa* (Du Roi.) Spreng.
sugar maple *Acer saccharum* Marsh.
sugi *Cryptomeria japonica* (L.F.) Don.
sweet fern *Myrica asplenifolia* L.

tamarack *Larix laricina* (Du Roi) K.Koch
trembling aspen *Populus tremuloides* Michx.

Virginia pine *Pinus virginiana* Mill.

western hemlock *Tsuga heterophylla* (Raf.) Sarg.
western larch *Larix occidentalis* Nutt.
western red cedar *Thuja plicata* Donn.
western white pine *Pinus monticola* Dougl.
white ash *Fraxinus americana* L.
white fir *Abies concolor* (Gord. and Glend.) Lindl.
white oak *Quercus alba* L.
white spruce *Picea glauca* (Moench) Voss.

yellow birch *Betula alleghaniensis* Britton

References

Introduction

Aaltonen, V.T. 1948. Boden und Wald. Paul Parey, Berlin

Bornebusch, C.H. 1930. The Fauna of Forest Soil. Nielson and Lydriche, Copenhagen

Duchaufour, P. 1960. Précis de pédologie. Masson, Paris

Ebermayer, E. 1876. Die gesammte Lehre der Waldstreu, mit Rücksicht auf die chemische Statik des Waldbaues, unter Zugrundlegung der in den Königlichen Staatsforsten Bayerns angestellten Untersuchungen. Julius Springer, Berlin

Henry, G. 1908. Les Sols forestiers. Berger-Leviault, Paris

Heyer, G. 1856. Lehrbuch der forstlichen Bodenkunde und Klimatologie. Erlangen

Lutz, H.J., and Chandler, R.F. Jr. 1946. Forest Soils. John Wiley and Sons, New York

Müller, P.E. 1887. Studien über die natürlichen Humusformen und deren Einwirkung auf Vegetation und Boden. Julius Springer, Berlin

Ramann, E. 1893. Forstliche Bodenkunde und Standortslehre. Julius Springer, Berlin

Tamm, O. 1950. Northern Coniferous Forest Soils. Transl. M.L. Anderson. Scrivener Press, Oxford

Wilde, S.A. 1946. Forest Soils and Forest Growth. Chronica Botanica, Waltham, Mass.

\- 1958. Forest Soils - Their Properties and Relation to Silviculture. Ronald Press, New York

\- 1962. Forstliche Bodenkunde. Transl. Th. Keller, F. Richard, and H.H. Krause. Paul Parey, Berlin

1 Forest soil: what it is and how to describe it

Canada Department of Agriculture. 1974. The System for Soil Classification for Canada. Publ. No. 1455, Ottawa
Soil Science Society of America. 1973. Glossary of Soil Science Terms. Madison
Soil Survey Staff. 1960. Soil Classification: A Comprehensive System, Seventh Approximation. USDA Soil Conservation Service, Washington
- 1975. Soil Taxonomy. Handbook No. 436, USDA Soil Conservation Service, Washington

2 The architecture of soil: texture, structure, and porosity

Arnold, R.W., and Cline, M.G. 1961. Origin of a surficial deposit in soils of East Fulton County, New York. Soil Sci. Soc. Amer. Proc. 25:240-3
Boelter, D.H. 1964. Water storage characteristics of several peats in situ. Soil Sci. Soc. Amer. Proc. 28:433-5
Boelter, D.H., and Blake, G.R. 1964. Importance of volumetric expression of water contents of organic soils. Soil Sci. Soc. Amer. Proc. 28:176-8
Canada Dept. of Agriculture. 1974. The System of Soil Classification for Canada. Publ. No. 1455 rev. Queen's Printer, Ottawa
Curtis, R.O., and Post, B.W. 1964. Estimating bulk density from organic matter content in some Vermont forest soils. Soil Sci. Soc. Amer. Proc. 28:285-6
Gurr, C.G. 1962. Use of gamma rays in measuring water content and permeability in unsaturated columns of soil. Soil Sci. 94:224-9
Hoyle, M.C. 1973. Nature and properties of some forest soils in the White Mountains of New Hampshire. USDA Forest Service Res. Paper NE-260, Northeastern Forest Expt. Station, Upper Darby
Lyford, W.H. 1964. Coarse fragments in the Gloucester soils of the Harvard Forest. Harvard Univ., Harvard Forest Paper No. 9
Reigner, I.C., and Phillips, J.J. 1964. Variations in bulk density and moisture content within two New Jersey coastal plain soils, Lakeland and Lakehurst sands. Soil Sci. Soc. Amer. Proc. 28:287-9
Rennie, P.J. 1957. Routine determinations of the solids, water and air volumes within soil clods of natural structure. Soil Sci. 84:351-65
Wooldridge, D.D. 1968. An air pycnometer for forest and range soils. USDA Forest Service Pacific Northwest Forest and Range Expt. Station, Portland, Ore.

3 Colour, temperature, and aeration

Armson, K.A., and Struik, H. 1969. The effects of oxygen levels on the growth of black spruce and red pine seedlings. Univ. Toronto Faculty of Forestry Glendon Hall Ann. Rep. 1967-8

Birkle, D.E., Letey, J., Stolzy, L.H., and Szuszkiewicz, T.E. 1964. Measurement of oxygen diffusion rates with the platinum microelectrode. II. Factors influencing measurement. Hilgardia 35 (20):555-66

Bouyoucos, G.J. 1916. Soil Temperature. Mich. Agric. College Expt. Station Tech. Bull. No. 26, East Lansing

Cochran, P.H. 1969. Thermal properties and surface temperatures of seedbeds - a guide for foresters. USDA Forest Service Pacific Northwest Forest and Range Expt. Station, Portland, Ore.

Federer, C.A. 1973. Annual cycles of soil and water temperatures at Hubbard Brook. USDA Forest Service Res. Note NE-167, Northeastern Forest Expt. Station, Upper Darby

Hart, G., and Lull, H.W. 1963. Some relationships among air, snow and soil temperatures and soil frost. USDA Forest Service Res. Note NE-3, Northeastern Forest Expt. Station, Upper Darby

Lees, J. 1972. Soil aeration and sitka spruce seedling growth in peat. J. Ecol. 60:343-9

Lemon, E.R., and Erickson, A.E. 1952. The measurement of oxygen diffusion in the soil with a platinum microelectrode. Soil Sci. Soc. Amer. Proc. 16:160-3

- 1955. Principle of the platinum microelectrode as a method of characterizing soil aeration. Soil Sci. 79:383-92

Letey, J., and Stolzy, L.H. 1964. Measurement of oxygen diffusion rates with the platinum microelectrode. I. Theory and equipment. Hilgardia 35(20): 545-54

Leyton, L., and Rousseau, L.Z. 1958. Root growth of seedlings in relation to aeration. *In* The Physiology of Forest Trees, ed. K.V. Thimann, pp. 467-75. Ronald Press, New York

Li, T.T. 1926. Soil Temperature as Influenced by Forest Cover. Yale Univ. School of Forestry Bull. No. 18

Pierce, R.S., Lull, H.W., and Storey, H.C. 1958. Influence of land use and forest condition on soil freezing and snow depth. For. Sci. 4:246-63

Schelling, J. 1955. Stuifzandgronden: inland-dune sand soils. Forest Research Station T.N.O. Wageningen Rep. 2:1:1-58

Striffler, W.D. 1959. Effects of forest cover on soil freezing in northern lower Michigan. USDA Forest Service Station Paper No. 76, Lake States Forest Expt. Station, St Paul, Minn.

Troedsson, T. 1956. Marktemperaturen i ytsteniga jordarter. Kgl. Skoghögskolans Skr. No. 25

Will, G.M. 1962. Soil moisture and temperature studies under radiata pine, Kaingaroa State Forest 1956-8. N.Z. J. Agric. Res. 5:111-20

4 Soil water: the lifeblood of soil

Bouyoucos, G.J., and Mick, A.H. 1940. An electrical resistance method for the continuous measurement of soil moisture under field conditions. Mich. Agric. Expt. Station Tech. Bull. 172, East Lansing

Colman, E.A., and Hendrix, T.M. 1949. Fiberglass electrical soil moisture instrument. Soil Sci. 67:425-38

Eagleman, J.R., and Jamison, V.C. 1962. Soil layering and compaction effects on unsaturated moisture movement. Soil Sci. Soc. Amer. Proc. 26:519-25

Grover, B.L., Cahoon, G.A., and Hotchkiss, C.W. 1964. Moisture relations of soil inclusions of a texture different from a surrounding soil. Soil Sci. Soc. Amer. Proc. 28:692-5

Gurr, C.G. 1962. Use of gamma rays in measuring water content and permeability in unsaturated columns of soil. Soil Sci. 94:224-9

Hendrickson, A.H., and Veihmeyer, F.J. 1945. Permanent wilting percentage of soils obtained from field and laboratory trials. Plant Physiol. 20:517-39

Hillel, D. 1971. Soil and Water. Academic Press, New York

Mack, A.R., and Brach, E.J. 1966. Soil moisture measurement with ultrasonic energy. Soil Sci. Soc. Amer. Proc. 30:544-8

Metz, L.J., and Douglass, J.E. 1959. Soil moisture depletion under several Piedmont cover types. USDA Forest Service Tech. Bull. No. 1207

Miller, D.E. 1969. Flow and retention of water in layered soils. USDA Agric. Res. Service Conservation Rep. No. 13, Washington

Miller, D.E., and Gardner, W.R. 1962. Water infiltration into stratified soil. Soil Sci. Soc. Amer. Proc. 26:115-19

McClain, K.M. 1973. Growth responses of some conifer seedlings to regimes of soil moisture and fertility under greenhouse and field nursery conditions. Univ. Toronto M Sc F thesis, Toronto

Nielsen, D.R., Kirkham, D., and Perrier, E.R. 1960. Soil capillary conductivity: comparison of measured and calculated values. Soil Sci. Soc. Amer. Proc. 24:157-60

Nielsen, D.R., Davidson, J.M., Biggar, J.W., and Miller, R.J. 1964. Water movement through Panoche clay loam soil. Hilgardia 35(17):491-506

Pierpoint, G. 1966. Measuring surface soil moisture with a neutron depth probe and a surface shield. Soil Sci. 101:189-92

Reginato, R.J., and van Bavel, C.H.M. 1964. Soil water measurement with gamma attenuation. Soil Sci. Soc. Amer. Proc. 28:721-4

Richards, L.A., and Richards, S.J. 1957. Soil moisture. *In* Soil - The Yearbook of Agriculture, pp. 49-60. USDA, Washington

Sartz, R.S., and Curtis, W.R. 1961. Field calibration of a neutron-scattering soil moisture meter. USDA Forest Service Station Paper No. 91, Lake States Expt. Station, St Paul, Minn.
Visvalingam, M., and Tandy, J.D. 1972. The neutron method for measuring soil moisture content - a review. J. Soil Sci. 23:499-511
Walkotten, W.J. 1972. A recording soil moisture tensiometer. USDA Forest Service Research Note PNW-180, Pacific Northwest Forest and Range Expt. Station, Portland, Ore.
Wilson, R.G. 1970. Methods of measuring soil moisture. National Research Council International Field Year for the Great Lakes Tech. Manual No. 1

5 **Soil Organic matter**

Alexander, M. 1961. An Introduction to Soil Microbiology. John Wiley and Sons, New York
Alway, F.J., and Zon, R. 1930. Quantity and nutrient contents of pine leaf litter. J. For. 28:715-27
Alway, F.J., Methley, W.J., and Younge, O.R. 1933. Distribution of volatile matter, lime, and nitrogen among litter, duff and leaf mould under various forest types. Soil Sci. 36:399-407
Bocock, K.L. 1963. Changes in the amount of nitrogen in decomposing leaf litter of sessile oak (*Quercus petraea*). J. Ecol. 51:555-66
- 1964. Changes in the amounts of dry matter, nitrogen carbon and energy in decomposing woodland leaf litter in relation to the activities of the soil fauna. J. Ecol. 52:273-84
Bocock, K.L., and Gilbert, O.J.W. 1957. The disappearance of leaf litter under different woodland conditions. Plant and Soil 9:179-85
Bornebusch, C.H., and Heiberg, S.O. 1936. Proposal to the Third International Congress of Soil Science, Oxford, England, 1935, for the nomenclature of forest humus layers. Trans. Third Internat. Congr. Soil Sci. 3:250-61
Bornebusch, C.H., and Holstener-Jørgensen, H. 1953. Laboratorie forsøg til belysning af regnormenes biologi. Dansk Skovforeningens Tids. 9:557-9
Bray, J.R., and Gorham, E. 1964. Litter production in forests of the world. *In* Advances in Ecological Research, 2:101-57. Academic Press, New York
Coulson, C.B., Davies, R.I., and Lewis, D.A. 1960a. Polyphenols in plant, humus and soil. I. Polyphenols of leaves, litter and superficial humus from mull and mor sites. J. Soil Sci. 11:20-9
- 1960b. Polyphenols in plant, humus and soil. II. Reduction and transport by polyphenols of iron in model soil columns. J. Soil Sci. 11:30-44

Daubenmire, R., and Prusso, D.C. 1963. Studies of the decomposition rates of tree litter. Ecology 44:589-92

Ebermayer, E. 1876. Die gesammte Lehre der Waldstreu mit Rücksicht auf die chemische Statik des Waldbaues, unter Zugrundlegung der in den Königlichen Staatsforsten Bayerns angestellten Untersuchungen. Julius Springer, Berlin

Gessel, S.P., and Balci, A.N. 1965. Amount and composition of forest floors under Washington coniferous forests. *In* Forest-Soil Relationships in North America, ed. C.T. Youngberg, pp. 11-23. Oregon State Univ. Press, Corvallis

Handley, W.R.C. 1954. Mull and mor formation in relation to forest soils. UK For. Comm. Bull. No. 23, London

Hart, G., Leonard, R.F., and Pierce, R.S. 1962. Leaf fall, humus depth and soil frost in a northern hardwood forest. USDA Forest Service Res. Note No. 131, Northeastern Forest Expt. Station, Upper Darby

Hatch, A.B. 1955. The influence of plant litter on the jarrah forest soils of the Dwellingup region, Western Australia. Austral. For. Bur. Leaflet 70

Hesselman, H. 1926. Studier över barrskogens humustäcke dess egenskaper och bervende av skogsvården. Statens. Skogsforsoks Medd. 22:169-552

Heyward, F., and Barnette, R.M. 1936. Field characteristics and partial chemical analysis of the humus layer of longleaf pine forest soils. Florida Agric. Expt. Station Bull. No. 302

Hoover, M.D., and Lunt, H.A. 1952. A key for the classification of forest humus types. Soil Sci. Soc. Amer. Proc. 16:368-70

Hurd, R.M. 1971. Annual tree litter production by successional forest stands, Juneau, Alaska. Ecology 52:881-4

Kawada, H. 1961. A study on approximate organic matter composition of leaf litters and their transformation during decomposing process. Japan For. Expt. Station, Tokyo Bull. 128:115-44

Kubiena, W.L. 1953. The Soils of Europe. Thomas Murby and Co., London

Lutz, H.J., and Chandler, R.F. Jr. 1946. Forest Soils. John Wiley and Sons, New York

Mader, D.L., and Lull, H.W. 1968. Depth, weight and water storage in white pine stands in Massachusetts. USDA Forest Service Res. Paper No. 109, Northeastern Forest Expt. Station, Upper Darby

Madge, D.S. 1965. Leaf fall and litter disappearance in tropical forest. Pedobiologia 5:273-88

McFee, W.W., and Stone, E.L. 1965. Quantity, distribution and variability of organic matter and nutrients in a forest podozol in New York. Soil Sci. Soc. Amer. Proc. 29:432-6

Metz, L.J., Wells, C.G., and Kormanik, P.P. 1970. Comparing the forest floor and surface soil beneath four pine species in the Virginia Piedmont. USDA

Forest Service Res. Paper No. SE-55, Southeastern Forest Expt. Station, Asheville
Mikola, P. 1960. Comparative experiment on decomposition rates of forest litter in southern and northern Finland. Oikos 11:161-6
Moir, W.H., and Grier, H. 1969. Weight and nitrogen, phosphorus, potassium and calcium content of forest floor humus of lodgepole pine stands in Colorado. Soil Sci. Soc. Amer. Proc. 33:137-40
Mork, E. 1942. Om strofallet i vare skoger. Medd. Norske Skorsforsoksv. 29:297-365
Nykvist, N. 1959a. Leaching and decomposition of litter. 1. Experiments on leaf litter of *Fraxinus excelsior.* Oikos 10:190-201
- 1959b. Leaching and decomposition of litter. 2. Experiments on needle litter of *Pinus sylvestris.* Oikos 10:212-24
- 1960a. Leaching and decomposition of litter. 3. Experiments on leaf litter of *Betula verrucosa.* Oikos 11: 249-63
- 1960b. Leaching and decomposition of litter. 4. Experiments on needle litter of *Picea abies.* Oikos 11:264-79
- 1962. Leaching and decomposition of litter. 5. Experiments on *Alnus glutinosa, Fagus silvatica* and *Quercus robur.* Oikos 12:232-48
Odum, E.P. 1971. Fundamentals of Ecology 3rd ed. W.B. Saunders Company, Philadelphia
Olson, J.S. 1963. Energy storage and the balance of producers and decomposers in ecological systems. Ecology 44:322-31
Ovington, J.D. 1962. Quantitative ecology and woodland ecosystem. *In* Advances in Ecological Research, vol. 1, ed. J.B. Cragg, pp. 103-92. Academic Press, London and New York
Reiners, W.A. 1972. Structure and energetics of three Minnesota forests. Ecology 42:71-94
Reiners, W.A., and Reiners, N.M. 1970. Energy and nutrient dynamics of forest floors in three Minnesota forests. J. Ecol. 58:497-519
Reukema, D.L. 1964. Litterfall in a young Douglas-fir stand as influenced by thinning. USDA Forest Service Res. Note PNW-14, Pacific Northwest Forest and Range Expt. Station, Portland, Ore.
Rodin, L.E., and Brazilevich, N.J. 1967. Production and Mineral Cycling in Terrestrial Vegetation. Oliver and Boyd, Edinburgh
Romell, L.G. 1935. Ecological problems of the humus layer in the forest. Cornell Univ. Agric. Expt. Station Mem. 170
Romell, L.G., and Heiberg, S.O. 1931. Types of humus layer in the forests of northeastern United States. Ecology 12:567-608
Scott, D.R.M. 1955. Amount and chemical composition of the organic matter

contributed by overstory and understory vegetation to forest soil. Yale Univ. School of Forestry Bull. No. 62

Tamm, C.O. 1953. Growth, yield and nutrition in carpets of a forest moss (*Hylocomium splendens*). Medd. Statens skogsforskning. Inst. 43(1):1–140

Tarrant, R.F., and Miller, R.E. 1963. Accumulation of organic matter and soil nitrogen beneath a plantation of red alder and Douglas-fir. Soil Sci. Soc. Amer. Proc. 27:231–4

Tarrant, R.F., Isaac, L.A., and Chandler, R.F. Jr. 1951. Observations on litter fall and foliage nutrient content of some Pacific northwest tree species. J. For. 49:914–15

Van Cleve, K., and Noonan, L.L. 1971. Physical and chemical properties of the forest floor in birch and aspen stands in interior Alaska. Soil Sci. Soc. Amer. Proc. 35:356–60

Wallwork, J.A. 1970. Ecology of Soil Animals. McGraw-Hill, London

Wells, C.G., and Davey, C.B. 1966. Cation-exchange characteristics of forest floor materials. Soil Sci. Soc. Amer. Proc. 30:399–402

Wilde, S.A. 1946. Forest Soils and Forest Growth. Chronica Botanica, Waltham, Mass.

Will, G.M. 1967. Decomposition of *Pinus radiata* litter on the forest floor. Pt. I. Changes in dry matter and nutrient content. N.Z. Jr. Sci. 10:1030–44

Williams, C.B. Jr., and Dyrness, C.T. 1967. Some characteristics of forest floors and soils under true-fir-hemlock stands in the Cascade Range. USDA Forest Service Res. Paper PNW-37, Pacific Northwest Forest and Range Expt. Station, Portland, Ore.

Witkamp, M. 1963. Microbial populations of leaf litter in relation to environmental conditions and decomposition. Ecology 44:370–7

Witkamp, M., and Olson, J.S. 1963. Breakdown of confined and non-confined litter. Oikos 14:138–47

Wollum, A.G. II. 1973. Characterization of the forest floor in stands along a moisture gradient in southern New Mexico. Soil Sci. Soc. Amer. Proc. 37:637–44

Wollum, A.G. II., and Davey, C.B. 1975. Nitrogen accumulation, transformation and transport in forest soils. *In* Forest Soils and Forest Land Management, ed. B. Bernier and C.H. Winget, pp. 67–106. Les Presses de l'université Laval, Québec

Wooldridge, D.D. 1970. Chemical and physical properties of forest litter layers in central Washington. *In* Tree Growth and Forest Soils, ed. C.T. Youngberg and C.B. Davey, pp. 327–37. Oregon Sate Univ. Press, Corvallis

Youngberg, C.T. 1966. Forest floors in Douglas-fir forests: I. Dry weight and chemical properties. Soil Sci. Soc. Amer. Proc. 30:406–9

Zavitkovski, J., and Newton, M. 1971. Litterfall and litter accumulation in red alder stands in western Oregon. Plant and Soil 35:257-68

6 **Soil biology: organisms and processes**

Alexander, M. 1961. An Introduction to Soil Microbiology. John Wiley and Sons, New York

Bernier, B., and Roberge, M.R. 1962. Etude in vitro sur la minéralisation de l'azote organique dans les humus forestiers. 1. Influence de litières forestières. Université Laval, Fonds de Recherches forestières Contribution No. 9, Québec

Bjorkman, E. 1942. Über die Bedingungen der Mykorrhizabildung bei Kiefer und Fichte. Symb. Bot. Upsalienses 6:2-191

Bond, G. 1970. Fixation of nitrogen in non-legumes with Alnus-type root nodules. *In* Nitrogen Nutrition of the Plant, ed. E.A. Kirkby, pp. 1-8. Univ. Leeds, Leeds

Bornebusch, C.H. 1930. The Fauna of Forest Soil. Nielsen and Lydriche, Copenhagen

Bornebusch, C.H., and Holstener-Jørgensen, H. 1953. Laboratorie forsøg til belysning af regnormenes biologi. Dansk. Skovforeningens Tids. 9:557-79

Burges, A. 1963. The microbiology of a podzol profile. *In* Soil Organisms, ed. J. Doeksen and J. van der Drift, pp. 151-7. North Holland, Amsterdam

Burges, A., and Raw, F., eds. 1967. Soil Biology. Academic Press, London and New York

Chase, F.E., and Baker, G. 1954. A comparison of microbial activity in an Ontario forest soil under pine, hemlock and maple cover. Can. J. Microbiol. 1:45-54

Clark, F.E. 1967. Bacteria in soil. *In* Soil Biology, ed. A. Burges and F. Raw, pp. 15-49. Academic Press, London and New York

Corke, C.T. 1958. Nitrogen transformations in Ontario forest podsols. *In* Proc. First North American Soils Conference, pp. 116-21. Michigan State Univ. Agric. Expt. Station, East Lansing

Crocker, R.L., and Major, J. 1955. Soil development in relation to vegetation and surface age at Glacier Bay, Alaska. J. Ecol. 43:422-48

Davidson, W.H. 1970. Deer prefer pine seedlings growing near black locust. USDA Forest Res. Note NE-111, Northeastern Forest Expt. Station, Upper Darby

Doeksen, J., and van der Drift, J., eds. 1963. Soil Organisms. North Holland, Amsterdam

Eaton, T.H. Jr., and Chandler, R.F. Jr. 1942. The fauna of the forest humus layers in New York. Cornell Agric. Expt. Station, Mem. 247

Garrett, S.D. 1963. Soil Fungi and Soil Fertility. Pergamon Press, London

Handley, W.R.C. 1954. Mull and mor formation in relation to forest soils. UK For. Comm. Bull. No. 23, London

Harley, J.L. 1969. The Biology of Mycorrhiza. 2nd ed. Leonard Hill, London

Hatch, A.B. 1937. The Physical Basis of Mycotrophy in *Pinus*. Black Rock For. Bull. No. 6

Hesse, P.R. 1957. Sulphur and nitrogen changes in forest soils of East Africa. Plant and Soil 9:86–96

Ike, A.F., and Stone, E.L. 1958. Soil nitrogen accumulation under black locust. Soil Sci. Soc. Amer. Proc. 22:346–9

Ivarson, K.C., and Katznelson, H. 1960. Studies on the rhizosphere microflora of yellow birch seedlings. Plant and Soil 12:30–40

Jacot, A.P. 1939. Reduction of spruce and fir litter by minute animals. J. For. 37:858–60

Jurgensen, M.F., and Davey, C.B. 1971. Nonsymbiotic nitrogen-fixing micro-organisms in forest and tundra soils. Plant and Soil 34:341–56

Katznelson, H. 1965. Nature and importance of the rhizosphere. *In* Ecology of Soil-Borne Plant Pathogens, ed. K.F. Baker and W.C. Snyder, pp. 187–209. Univ. California Press, Berkeley

Kevan, D.K.McE., ed. 1955. Soil Zoology. Butterworths, London

– 1962. Soil Animals. H.F. and G. Witherby, London

Knowles, R. 1965. The significance of non-symbiotic nitrogen fixation. Soil Sci. Soc. Amer. Proc. 29:223

Kubiena, W.L. 1938. Micropediology. Collegiate Press, Ames, Iowa

Kühnelt, W. 1961. Soil Biology – With Special Reference to the Animal Kingdom. Transl. N. Walker. Faber and Faber, London

Langmaid, K.K. 1964. Some effects of earthworm invasion in virgin podzols. Can. J. Soil Sci. 44:34–7

Lunt, H.A., and Jacobson, H.G.M. 1944. The chemical composition of earth-worm casts. Soil Sci. 58:367–75

Lyford, W.H. 1963. Importance of ants to brown podzolic soil genesis in New England. Harvard Univ. For. Paper No. 7

Marx, D.H., and Bryan, W.C. 1975. The significance of mycorrhizae to forest trees. *In* Forest Soils and Forest Land Management, ed. B. Bernier and C.H. Winget, pp. 107–17. Les Presses de l'université Laval, Québec

Marx, D.H., and Davey, C.B. 1969a. The influence of ectotrophic mycorrhizal fungi on the resistance of pine roots to pathogenic infections. III. Resistance of aseptically formed mycorrhizae to infection by *Phytophthora cinnamomi*. Phytopathology 59:549

- 1969b. The influence of ectotrophic mycorrhizal fungi on the resistance of pine roots to pathogenic infections. IV. Resistance of naturally occurring mycorrhizae to infections by *Phytophthora cinnamomi.* Phytopathology 59:559

Murphy, P.W., ed. 1962. Progress in Soil Zoology. Butterworths, London

Neal, J.L. Jr., Bollen, W.B., and Zak, B. 1964. Rhizosphere microflora associated with mycorrhizae of Douglas fir. Can. J. Microbiol. 10:259-65

Newell, P.F. 1967. Mollusca. *In* Soil Biology, ed. A. Burges and F. Raw, pp. 413-33. Academic Press, London and New York

Nielson, G.A., and Hole, F.D. 1964. Earthworms and the development of coprogenous A1 horizons in forest soils of Wisconsin. Soil Sci. Amer. Proc. 28:426-30

Nye, P.H. 1954. Some soil-forming processes in the humid tropics. 1. A field study of a catena in a West African forest. J. Soil Sci. 5:7-21

- 1955. Some soil-forming processes in the humid tropics. IV. The action of the soil fauna J. Soil Sci. 6:73-83

O'Connor, F.B. 1957. An ecological study of the enchytraeid worm population of a coniferous forest soil. Oikos 8:161-99

Poole, T.B. 1961. An ecological study of the Collembola in a coniferous forest soil. Pedobiol. 1:113-37

Powers, W.L., and Bollen, W.B. 1935. The chemical and biological nature of certain forest soils. Soil Sci. 40:321-31

Rayner, M.C. 1927. Mycorrhiza. Wheldon and Wesley Ltd., London

Reid, C.P.P., and Woods, F.W. 1969. Translocation of C14-labeled compounds in mycorrhizae and its implications in interplant nutrient cycling. Ecology 50:179-87

Salem, M.Z., and Hole, F.D. 1968. Ant (*Formica exsectoides*) pedoturbation in a forest soil. Soil Sci. Amer. Proc. 32:563-7

Satchell, J.E. 1967. Lumbricidae. *In* Soil Biology, ed. A. Burges and F. Raw, pp. 259-322. Academic Press, London and New York

Slankis, V. 1958. The role of auxin and other exudates in mycorrhizal symbiosis of forest trees. *In* Physiology of Forest Trees, ed. K.V. Thimann, pp. 427-43. Ronald Press, New York

Stegman, L.C. 1960. A preliminary survey of earthworms of the Tully Forest in central New York. Ecology 41:779-82

Stone, E.L. 1950. Some effects of mycorrhizae on the phosphorus nutrition of Monterey pine seedlings. Soil Sci. Soc. Amer. Proc. 14:340-5

Stout, J.D., and Heal, O.W. 1967. Protozoa. *In* Soil Biology, ed. A. Burges and F. Raw, pp. 149-95. Academic Press, London and New York

Tarrant, R.F., and Trappe, J.M. 1971. The role of *Alnus* in improving the forest environment. Plant and Soil Spec. Vol. :335-48
Wallwork, J.A. 1970. Ecology of Soil Animals. McGraw-Hill, London
Went, J.C. 1963. Influence of earthworms on the number of bacteria in the soil. *In* Soil Organisms, ed. J. Doeksen and J. van der Drift, pp. 260-5. North Holland, Amsterdam
Wiegert, R.G. 1970. Energetics of the nest-building termite, *Nasutitermes costalis* (Holmgren) in a Puerto Rican forest. *In* A Tropical Rain Forest, a Study of Irradiation and Ecology at El Verde, Puerto Rico, ed. H.T. Odum and R.F. Pigeon. Office of Inform. Services, U.S. Atomic Energy Comm., Washington
Youngberg, C.T., and Wollum, A.G. II. 1970. Non leguminous symbiotic nitrogen fixation. *In* Tree Growth and Forest Soils, ed. C.T. Youngberg and C.B. Davey, pp. 383-95. Oregon State Univ. Press, Corvallis
Zavitkovski, J., and Newton, M. 1968. Ecological importance of snowbrush, *Ceonothus velutinus* in the Oregon Cascades. Ecology 49:1134-45
Zentmyer, G.A. 1961. Chemotaxis of zoospores for root exudates. Science 133:1595-6

7 Soil chemistry

Association of Official Agricultural Chemists. 1955. Official Methods of Analysis. 8th ed. Washington
Ball, D.F., and Beaumont, P. 1972. Vertical distribution of extractable iron and aluminium in soil profiles from a brown earth - peaty podzol association. J. Soil Sci. 23:298-308
Black, C.A., ed. 1965. Methods of Soil Analysis. Agronomy Series No. 9, Pt. 1 and Pt. 2. Amer. Soc. Agronomy, Madison, Wis.
Black, C.A. 1968. Soil-Plant Relationships. 2nd ed. John Wiley and Sons, New York
Bloomfield, C. 1957. The possible significance of polyphenols in soil formation. J. Sci. Fd. Agric. 8:389-92
Burger, D. 1969. Relative weatherability of calcium-containing minerals. Can. J. Soil Sci. 49:21-8
Chapman, H.D., and Pratt, P.F. 1961. Methods of Analysis for Soils, Plants and Waters. Univ. California Div. Agr. Sci.
Hesse, P.R. 1971. A Textbook of Soil Chemical Analysis. Chemical Publishing Co., New York
Jackson, M.L. 1958. Soil Chemical Analysis. Prentice-Hall, Englewood Cliffs

- 1964. Chemical composition of soils. *In* Chemistry of the Soil, ed. F.E. Bear, 2nd ed., pp. 71-141. Reinhold, New York

Kilian, W.H.E., and Lumbe, C. 1972. Different P-uptake of seedlings (*Picea excelsa, Pinus silvestris, Alnus glutinosa* and *Secale cereale*). *In* Isotopes and Radiation in Soil-Plant Relationships Including Forestry, pp. 301-12. Internat. Atomic Energy Agency, Vienna

Krause, H.H. 1965. Effect of pH on leaching losses of potassium applied to forest nursery soils. Soil Sci. Soc. Amer. Proc. 29:613-15

Lafond, A. 1950. Oxidation-reduction potential as a characteristic of forest humus types. Soil Sci. Soc. Amer. Proc. (1949) 14:337-40

Larsen, S. 1967. Soil phosphorus. *In* Advances in Agronomy, vol. 19, ed. A.G. Norman, pp. 151-210. Academic Press, New York and London

Lees, H., and Quastel, J.H. 1946. Biochemistry of nitrification in soil. I. Kinetics of, and effects of poisons on, soil nitrification as studied by a soil perfusion technique. Biochem. J. 40:803-15

Leyton, L. 1952. The effect of pH and form of nitrogen on the growth of Sitka spruce seedlings. Forestry 25:32-40

McFee, W.W., and Stone, E.L. 1968. Ammonium and nitrate as nitrogen sources for *Pinus radiata* and *Picea glauca*. Soil Sci. Soc. Amer. Proc. 32:879-84

Metson, A.J. 1961. Methods of chemical analysis for soil survey samples. New Zealand Dept. Sci. Ind. Res. Bull. No. 12

Rennie, P.J. 1966. The use of micropedology in the study of some Ontario podzolic profiles. J. Soil Sci. 17:99-106

Richards, L.A., ed. 1954. The Diagnosis and Improvement of Saline and Alkali Soils. USDA Handbook No. 60, Washington

Sticker, H., and Bach R. 1966. Fundamentals in the chemical weathering of silicates. Soils and Fertilizers 29:321-5

Swan, H.S.D. 1960. The mineral nutrition of Canadian pulpwood species. I. The influence of nitrogen, phosphorus, potassium and magnesium deficiencies on the growth and development of white spruce, black spruce, jack pine and western hemlock seedlings grown in a controlled environment. Pulp Paper Inst. Can. Tech. Rep. No. 168

Tamm, C.O., and Pettersson, A. 1969. Studies on nitrogen mobilisation in forest soils. Stud. Forestal. Suec. Skoghögskolan, Stockholm

Troedsson, T. 1952. Den geologiska miljons inverkan på grundvattnets halt av lösta växtnäringsämmen. Kgl. Skogshögskolans Skr. No. 10

Van Groenewoud, H. 1961. Variation in pH and buffering capacity of the organic layer of grey wooded soils. Soil Sci. 92:100-5

Vézina, P.E. 1965. Methods of determination and seasonal pH fluctuations in Quebec forest humus. Ecology 46:752-5

Waito, R., and Gregory, R.C. 1973. The effect of copper on the germination and growth of jack pine seedlings. Univ. Toronto Faculty of Forestry, mimeo.

Wilde, S.A., Voigt, G.K., and Iyer, J.G. 1972. Soil and Plant Analysis for Tree Culture. 4th ed. (ed. G. Chesters). Oxford and IBH Publ. Co., New Delhi

8 **Soil fertility**

Armson, K.A. 1959. An example of the effect of past use of land on fertility levels and growth of Norway spruce (*Picea abies* L. Karst.). Univ. Toronto Faculty of Forestry. Tech. Rep. No. 1

- 1966. The growth and absorption of nutrients by fertilized and unfertilized white spruce seedlings. For. Chron. 42:127-36

- 1968. The effects of fertilization and seedbed density on the growth and nutrient content of white spruce and red pine seedlings. Univ. Toronto Faculty of Forestry. Tech. Rep. No. 10

- 1972. Fertilizer distribution and sampling techniques in the aerial fertilization of forests. Univ. Toronto Faculty of Forestry. Tech. Rep. No. 11

Armson, K.A., and Sadreika, V. 1974. Forest Tree Nursery Soil Management and Related Practices. Ontario Ministry of Natural Resources, Toronto

Baule, H., and Fricker, C. 1970. The Fertilizer Treatment of Forest Trees. BLV, Munich

Bengtson, G.W., and Voigt, G.K. 1962. A greenhouse study of relations between nutrient movement and conversion in a sandy soil and the nutrition of slash pine seedlings. Soil Sci. Soc. Amer. Proc. 26:609-12

Bevege, D.I., and Richards, B.N. 1972. Principles and practice of foliar analysis as a basis for crop-logging in pine plantations. II. Determination of critical phosphorus levels. Plant and Soil 37:159-69

Carrow, J.R., and Betts, R.E. 1973. Effects of different foliar applied nitrogen fertilizers on balsam woolly aphid. Can. J. For. Res. 3:122-39

Chapman, H.D., ed. 1966. Diagnostic Criteria for Plants and Soils. Univ. California Div. Agr. Sci.

Gladstone, W.T., and Gray, R.L. 1973. Effects of forest fertilization on wood quality. *In* Forest Fertilization, pp. 167-73. USDA Forest Service Gen. Tech. Rep. NE-3, Northeastern Forest Expt. Station, Upper Darby

Goodall, D.W., and Gregory, F.C. 1947. Chemical composition of plants as an index of their nutritional status. Imp. Bur. Hort. Plant Crops Tech. Comm. No. 17

Goyer, R.A., and Benjamin, D.M. 1972. Influence of soil fertility on infestation of jack pine plantations with the pine root weevil. For. Sci. 18:139-47

Heiberg, S.O., and Loewenstein, H. 1958. Depletion and rehabilitation of a sandy outwash plain in northern New York. *In* Proc. First North American Forest Soils Conf., pp. 172-80. Mich. State Univ. Agric. Expt. Station, East Lansing

Heiberg, S.O., and White, D.P. 1951. Potassium deficiency of reforested pine and spruce stands in northern New York. Soil Sci. Soc. Amer. Proc. 15:369-76

Hesterberg, G.A., and Jurgensen, M.F. 1972. The relation of forest fertilization to disease incidence. For. Chron. 48:92-6

Hughes, R.H., Bengtson, G.W., and Harrington, T.A. 1971. Forage response to nitrogen and phosphorus fertilization in a 25-year-old plantation of slash pine. USDA Forest Service Res. Paper SE-82, Southeastern Forest Expt. Station, Asheville

Jacks, G.V. 1956. The influence of man on soil fertility. Address to the British Association for the Advancement of Science. Sheffield

Klem, G.S. 1972. Virkningen av gjødsling av gran (*Picea abies* L. Karst.) og furu (*Pinus sylvestris* L.) på virkets sommerved-prosent, volumvekt og ekstraktinnhold. Medd. Norske Skogsforsøksv 30:4:286-305

Klemmedson, J.O., and Ferguson, R.B. 1969. Response of bitterbrush seedlings to nitrogen and moisture in a granite soil. Soil Sci. Amer. Proc. 33:962-6

Lavallée, A. 1972. Etat pathologique d'une plantation d'épinette blanche après fertilisation à Grand' Mère. Can. For. Service Centre de Recherch. Forêt des Laurentides Rapp. d'Info Q-F-X3, Québec

Leaf, A.L. 1968. K, Mg and S deficiencies in forest trees. *In* Forest Fertilization Theory and Practice, pp. 88-122. Tennessee Valley Authority, Muscle Shoals

- 1973. Plant analysis as an aid in fertilizing forests. *In* Soil Testing and Plant Analysis, ed. L.M. Walsh and J.D. Beaton, pp. 427-54. Soil Sci. Soc. Amer., Madison

Leaf, A.L., and Madgwick, H.A.I. 1960. Evaluation of chemical analysis of soils and plants as aids in intensive soil management. Proc. 5th World For. Congr. 1:554-6

Leyton, L. 1954. The growth and mineral nutrition of spruce and pine in heathland plantations. Oxford Univ. Imp. For. Inst. Paper No. 31

Leyton, L., and Armson, K.A. 1955. Mineral composition of the foliage in relation to the growth of Scots pine. For. Sci. 1:210-18

Li, C.Y., Lu, K.C., Trappe, J.M., and Bollen, W.B. 1967. Selective nitrogen assimilation by *Poria weirii.* Nature (London) 213(5078):814

Madgwick, H.A.I. 1964. Variations in the chemical composition of red pine (*Pinus resinosa* Ait.) leaves: a comparison of well-grown and poorly-grown trees. Forestry 37:87-94

Mitchell, H.L. 1939. The growth and nutrition of white pine (*Pinus strobus* L.) seedlings in cultures with varying nitrogen, phosphorus, potassium and calcium (with observations on the relation of seed weight to seedling yield). Black Rock For. Bull. 9

Mitchell, H.L., and Chandler, R.F. Jr. 1939. The nitrogen nutrition and growth of certain deciduous trees of northeastern United States. Black Rock For. Bull. 11

Morrison, I.K. 1974. Mineral nutrition of conifers with special reference to nutrient status interpretation: a review of literature. Canadian Forestry Service, Ottawa, Publ. No. 1343

Nelson, E. 1970. Effects of nitrogen fertilizer on survival of *Poria weirii* and populations of soil fungi and aerobic actinomycetes. Northwest Sci. 44:102-6

Oh, J.H., Jones, M.B., Longhurst, W.M., and Connolly, G.E. 1970. Deer browsing and rumen microbial fermentation of Douglas-fir as affected by fertilization and growth stage. For. Sci. 16:21-7

Pritchett, W.L. 1968. Progress in the development of techniques and standards for soil and foliar diagnosis of phosphorus deficiency in slash pine. *In* Forest Fertilization Theory and Practice, pp. 81-7. Tennessee Valley Authority, Muscle Shoals

Rennie, P.J. 1955. The uptake of nutrients by mature forest growth. Plant and Soil 7:49-95

Richards, B.N., and Bevege, D.I. 1971. Principles and practice of foliar analysis for crop logging in pine plantations. I. Basic considerations. Plant and Soil 36:109-19

Saucier, J.R., and Ike, A.F. 1969. Effects of fertilization on selected wood properties of sycamore. For. Prod. J. 19:93-6

Smirnoff, W.A., and Bernier, B. 1973. Increased mortality of the Swaine jack-pine sawfly and foliar nitrogen concentrations after urea fertilization. Can. J. For. Res. 3:112-21

Soil Science Society of America. 1973. Glossary of Soil Science Terms. Soil Sci. Soc. Amer., Madison

Sprague, H.B., ed. 1964. Hunger Signs in Crops. David McKay Co., New York

Steenbjerg, F. 1954. Manuring plant production and the chemical composition of the plant. Plant and Soil 5:226-42

Stone, E.L. 1968. Micro element nutrition of forest trees: A review. *In* Forest Fertilization Theory and Practice, pp. 132-75. Tennessee Valley Authority, Muscle Shoals

Tisdale, S.L., and Nelson, W.L. 1966. Soil Fertility and Fertilizers. Macmillan, New York

Voigt, G.K. 1966. Phosphorus uptake in young pitch pine (*Pinus rigida* Mill.). Soil Sci. Soc. Amer. Proc. 30:403-6

Walker, L.C. 1955. Foliar analysis as a method of indicating potassium-deficient soils for reforestation. Soil Sci. Soc. Amer. Proc. 19:233-6

Wright, T.W. 1957. Abnormalities in nutrient uptake by Corsican pine growing on sand dunes. J. Soil Sci. 8:150-7

Xydias, G.K., and Leaf, A.L. 1964. Weevil infestation in relation to fertilization of white pine. For. Sci. 10:428-31

Young, H.E., and Guinn, V.P. 1966. Chemical elements in complete mature trees of seven species in Maine. Tappi 49:190-7

9 **Soil classification**

Basinski, J.J. 1959. The Russian approach to soil classification and its recent development. J. Soil Sci. 10:14-26

Bidwell, O.W., and Hole, F.D. 1964. Numerical taxonomy and soil classification. Soil Sci. 97:58-62

Bunting, B.T. 1965. The Geography of Soil. Hutchinson Univ. Library, London

Buol, S.W., Hole, F.D., and McCracken, R.J. 1973. Soil Genesis and Classification. Iowa State Univ. Press, Ames

Canada Department of Agriculture. 1974. The System of Soil Classification for Canada. Publ. 1455 rev. Queen's Printer, Ottawa

Grigal, D.F., and Arnemann, H.F. 1969. Numerical classification of some forested Minnesota soils. Soil Sci. Soc. Amer. Proc. 33:433-8

Hole, F.D., and Hironaka, M. 1960. An experiment in ordination of some soil profiles. Soil Sci. Soc. Amer. Proc. 24:309-12

Jenny, H. 1941. Factors of Soil Formation. McGraw-Hill, New York

Kubiena, W.L. 1953. The Soils of Europe. Thomas Murby and Co., London

Manil, G. 1959. General considerations on the problem of soil classification. J. Soil Sci. 10:5-13

Northcote, K.H. 1965. A factual key for the recognition of Australian soils. CSIRO Australia, Div. Soils, Rep. 2/65

Rayner, J.H. 1966. Classification of soils by numerical methods. J. Soil Sci. 17:79-92

Robinson, G.W. 1936. Soils, Their Origin, Constitution and Classification. Thomas Murby and Co., London

Simonson, R.W. 1968. Concept of soil. *In* Advances in Agronomy, ed. A.G. Norman, 20:1-47. Academic Press, New York and London

Sneath, P.H.A., and Sokal, R.R. 1962. Numerical taxonomy. Nature 193:855-60

Soil Survey Staff. 1960. Soil Classification: A Comprehensive System, Seventh Approximation. 1964 and 1967 Supplements. USDA Soil Conservation Service, Washington
- 1975. Soil Taxonomy. Handbook No. 436, USDA Soil Conservation Service, Washington

10 **Soil surveys**

Clarke, G.R. 1957. The Study of the Soil in the Field. Clarendon Press, Oxford

Coile, T.S. 1948. Relation of soil characteristics to site index of loblolly and short leaf pines in the lower Piedmont region of North Carolina. Duke Univ. School of Forestry Bull. No. 13

Day, J.H., Farstad, L., and Laird, D.G. 1959. Soil survey of southeast Vancouver Island and Gulf Islands, British Columbia. Canada Dept. Agric. Res. Br. Rep. No. 6, B.C. Soil Survey

DeMent, J.A., and Stone, E.L. 1968. Influence of soil and site on red pine plantations in New York. II. Soil type and physical properties. Cornell Univ. Agric. Expt. Station Bull. 1020

Gibbons, F.R., and Downes, R.G. 1964. A Study of the Land in Southwestern Victoria. Soil Conservation Authority, Victoria

Hayashi, S., and Shimoide, A. 1954. Soils of the Yogawa National Forest. *In* Forest Soils of Japan, Rep. 4, pp. 87-116. Government For. Expt. Station, Tokyo

Jurdant, M., Beaubien, J., Belair, J.L., Dionne, J.C., and Gerardin, V. 1972. Carte écologique de la région du Saguenay – Lac St Jean. Can. For. Service Centre de Recherch. Forêt des Laurentides Rapp. d'Info. QF-X-31, Québec

Kizaki, T., and Watanabe, T. 1954. Soils of the Tokyo National Forest. *In* Forest Soils of Japan, Rep. 4, pp. 61–86. Gov. For. Expt. Station, Tokyo

Lacate, D.S., ed. 1967. Guidelines for bio-physical land classification as applied to forest and associated wildlands. Canada Dep. For. and Rur. Develop., Victoria, B.C

Montgomery, P.H., and Edminster, F.C. 1966. Use of soil surveys in planning for recreation. *In* Soil Surveys and Land Use Planning, eds. L.J. Bartelli, A.A. Klingebiel, J.V. Baird, and M.R. Heddleson, pp. 104–12. Soil Sci. Soc. Amer. and Amer. Soc. Agron., Madison

Richards, N.A., Morrow, R.R., and Stone, E.L. 1962. Influence of soil and site on red pine plantations in New York. I. Stand development and site index curves. Cornell Univ. Agric. Expt. Station Bull. No. 977

Schelling, J. 1955. Stuifzandgronden: inland-dune sand soils. Forest Research Station T.N.O. Wageningen, Rep. 2:1:1–58

Soil Science Society of America. 1973. Glossary of Soil Science Terms. Soil Sci. Soc. Amer., Madison

Soil Survey Staff. 1951. Soil Survey Manual. USDA Handbook 18, Washington

- 1967. Soil Survey of Scotland County, North Carolina. USDA Soil Conservation Service, Washington

Steinbrenner, E.C. 1965. The influence of individual soil and physiographic factors on the site index of Douglas-fir in western Washington. *In* Forest-Soil Relationships in North America, ed. C.T. Youngberg, pp. 261–77. Oregon State Univ. Press, Corvallis

Valentine, K.W.G., Lord, T.M., Watt, W., and Bedwany, A.L. 1971. Soil mapping accuracy from black and white, colour and infrared aerial photography. Can. J. Soil Sci. 51:461-9

Zinke, P.J., and Colwell, W.L. Jr. 1965. Some general relationships among California forest soils. *In* Forest-Soil Relationships in North America, ed. C.T. Youngberg, pp. 353–65. Oregon State Univ. Press, Corvallis

11 **Roots and soil**

Armson, K.A., and van den Driessche, R. 1959. Natural root grafts in red pine (*Pinus resinosa,* Ait.). For. Chron. 35:232–41

Armson, K.A., and Williams, J.R.M. 1960. The root development of red pine (*Pinus resinosa,* Ait.) seedlings in relation to various soil conditions. For. Chron. 36:14–17

Biswell, H.H. 1935. Effects of environment upon the root habits of certain deciduous forest trees. Bot. Gaz. 96:676-708

Blevins, R.L., Holowaychuk, N., and Wilding, L.P. 1970. Micromorphology of soil fabric at tree root-soil interface. Soil Sci. Soc. Amer. Proc. 34:460–5

Büsgen, M., and Münch, E. 1929. The Structure and Life of Forest Trees. Transl. T. Thomson. Chapman and Hall, London

Coile, T.S. 1937. Distribution of forest tree roots in North Carolina Piedmont soils. J. For. 35:247–57

Cook, D.B., and Welch, D.S. 1957. Backflash damage to residual stands incident to chemi-peeling. J. For. 55:265–7

Cornforth, I.S. 1968. Relationship between soil volume used by roots and nutrient accessibility. J. Soil Sci. 19:291-301

Day, W.R. 1955. Forest Hygiene in Great Britain. Univ. Toronto, Faculty of Forestry Bull. No. 4

- 1959. Observations on eucalypts in Cypress. Pt. II. Emp. For. Rev. 38:186–97

- 1962. Notes on the development of the root system with Sitka spruce. Scott. Forestry 16:72–83

DeByle, N.V., and Place, I.C.M. 1959. Rooting habits of oak and pine on Plainfield sands in central Wisconsin. Univ. Wisconsin For. Res. Note 44, Madison

Eis, S. 1974. Root system morphology of western hemlock, western red cedar and Douglas-fir. Can. J. For. Res. 4:28-38

Fayle, D.C.F. 1965. Rooting habit of sugar maple and yellow birch. Canada Dept. Forestry Publ. No. 1120, Ottawa

Fisher, R.F., and Stone, E.L. 1969. Increased availability of nitrogen and phosphorus in the root zone of conifers. Soil Sci. Soc. Amer. Proc. 33:956-61

Gersper, P.L., and Holowaychuk, N. 1970a. Effects of stemflow water on a Miami soil under a beech tree. I. Morphological and physical properties. Soil Sci. Soc. Amer. Proc. 34:779-86

- 1970b. Effects of stemflow water on a Miami soil under a beech tree. II. Chemical properties. Soil Sci. Soc. Amer. Proc. 34:786-94

Graham, B.F., and Bormann, F.H. 1966. Natural root grafts. Bot. Rev. 32:255-92

Heyward, F. 1933. The root system of longleaf pine on the deep sands of western Florida. Ecology 14:136-48

Horton, K.W. 1958. Rooting habits of lodgepole pine. Canada Dept. Northern Affairs Nat. Res. For. Branch Tech. Note 67, Ottawa

Hoyle, M.C. 1971. Effects of the chemical environment on yellow birch, root development and top growth. Plant and Soil 35:623-33

Jeffrey, W.W. 1959. White spruce rooting modifications on the fluvial deposits of the lower Peace River. For. Chron. 35:304-11

Kalela, E.K. 1950. [The horizontal roots of pine and spruce stands] Acta For. Fenn. 57:1-79

Kohmann, K. 1972a. Rotøkologiske undersøkelser på furu. I. Metodiske problemer og generelle rotforhold. Det Norske Skogsforsøksv, Ås 30:325-57

- 1972b. Rotøkologiske undersøkelse på furu. II. Rotsystemets reaksjon på gjødsling. Det Norske Skogsforsøksv. Ås 30:359-66

Ladefoged, K. 1939. Untersuchungen über die Periodizität im Ausbruch und Längenwachstum der Wurzeln bei einigen unserer gewöhnlichsten Waldbaume. Det forstl. Forsøgsv. Danmark 16

Leaphart, C.D., and Wicker, E.F. 1966. Explanation of pole blight from responses of seedlings grown in modified environment. Can. J. Bot. 44:121-37

Little, S., and Somes, H.A. 1964. Root systems of direct-seeded and variously planted loblolly, shortleaf and pitch pines. USDA Forest Service Res. Paper NE-26, Northeastern Forest Expt. Station, Upper Darby

Lorio, P.L., Jr., and Hodges, J.D. 1971. Microrelief, soil water regime and loblolly pine growth on a wet, mounded site. Soil Sci. Soc. Amer. Proc. 35:795-800

Lorio, P.L., Jr., Howe, V.K., and Martin, C.N. 1972. Loblolly pine rooting varies with microrelief on wet sites. Ecology 53:1134-40

Lutz, H.J., Ely, J.B., and Little, S. 1937. The influence of soil profile horizons on root distribution of white pine (*Pinus strobus* L.). Yale Univ. School of Forestry Bull. 44, New Haven

Lyford, W.H., and Wilson, B.F. 1964. Development of the root system of *Acer rubrum* L. Harvard Univ. Forest Paper No. 10

- 1966. Controlled growth of forest tree roots: technique and application. Harvard Univ. Forest Paper No. 16

McMinn, R.G. 1963. Characteristics of Douglas-fir root systems. Can. J. Bot. 41:105-22

McQuilken, W.E. 1935. Root development of pitch pine with some comparative observations on shortleaf pine. USDA J. Agr. Res. 51:983-1016

Morrison, I.K. 1974. Dry-matter and element content of roots of several natural stands of *Pinus banksiana* Lamb. in Northern Ontario. Can. J. For. Res. 4:61-4

Nye, P.H. 1961. Organic matter and nutrient cycles under moist tropical forest. Plant and Soil 13:333-46

- 1968. Processes in the root environment. J. Soil Sci. 19:205-15

O'Loughlin, C.L. 1974. A study of tree root strength deterioration following clearfelling. Can. J. For. Res. 4:107-13

Paavilainen, E. 1961. [The effect of fertilization on the root systems of swamp pine stands] Folia for. Inst. For. Fenn No. 31

Phillips, W.S. 1963 Depth of roots in soil. Ecology 44:424

Reynolds, E.R.C. 1970. Root distribution and the cause of its spatial variability in *Pseudotsuga taxifolia* (Poir.) Britt. Plant and Soil 32:501-17

Richards, B.N. 1962. Increased supply of soil nitrogen brought about by *Pinus*. Ecology 62:538-41

- 1973. Nitrogen fixation in the rhizosphere of conifers. Soil Biol. Biochem. 5:149-52

Richardson, S.D. 1956. Studies of root growth in *Acer saccharinum* L. IV. The effect of differential shoot and root temperatures on root growth. Koninkl. Nederl. Akad. v Wet. 59c:428-38

Safford, L O., and Bell, S. 1972. Biomass of fine roots in a white spruce plantation. Can. J. For. Res. 2:169-72

Schultz, J.D. 1969. The vertical rooting habit of black spruce, white spruce and balsam fir. PhD dissertation, Univ. Michigan. University Microfilms Inc., Ann Arbor

Schultz, R.P. 1972. Root development of intensively cultivated slash pine. Soil Sci. Soc. Amer. Proc. 36:158-62

- 1973. Site treatment and planting method alter root development of slash pine. USDA Forest Service Res. Paper SE-109, Southeastern Forest Expt. Station, Asheville

Schuurman, J.J., and Goedewaagen, M.A.J. 1965. Methods for the Examination of Root Systems and Roots. Centre Agric. Publ. and Documents, Wageningen

Shea, S.R. 1973. Growth and development of jack pine (*Pinus banksiana*, Lamb.) in relation to edaphic factors in northeastern Ontario. Univ. Toronto, PhD dissertation

Stanek, W. 1961. The properties of certain peats in northern Ontario. Univ. Toronto M.Sc.F thesis

Stout, B.B. 1956. Studies of the root systems of deciduous trees. Black Rock For. Bull. 15

Sutton, R.F. 1969. Form and Development of Conifer Root Systems. Comm. Agric. Bur. Tech. Comm. No. 7, Farnham Royal

Tucker, R.E., Jarvis, J.M., and Waldron, R.M. 1968. Early survival and growth of white spruce plantations. Riding Mountain National Park. Canada Dept. For. and Rural Develop. For. Branch Publ. 1239, Ottawa

Voigt, G.K. 1960. Distribution of rainfall under forest stands. For. Sci. 6:2-10

Voigt, G.K., Richards, B.N., and Mannion, E.C. 1964. Nitrogen utilization by young pitch pine. Soil Sci. Soc. Amer. Proc. 28:707-9

Wagg, J.W.B. 1967. Origin and development of white spruce root forms. Canada Dept. For. and Rural Develop. For. Branch Publ. 1192, Ottawa

White, E.H., Pritchett, W.L., and Robertson, W.K. 1971. Slash pine root biomass and nutrient concentrations. *In* Forest Biomass Studies, pp. 165-76. Univ. Maine, Life Sciences and Agric., Orono

Will, G.M. 1966. Root growth and dry-matter production in a high-producing stand of *Pinus radiata*. N.Z. For. Service Res. Note 44

Yeatman, C.W. 1955. Tree root development on upland heaths. UK For. Comm. Bull. 21

Yorke, J.S. 1968. A review of techniques for studying root systems. Canada Dept. For. and Rural Develop. Info. Rep. N-X-20, St John's

Youngberg, C.T. 1959. The influence of soil conditions following tractor logging on the growth of planted Douglas-fir seedlings. Soil Sci. Soc. Amer. Proc. 23:76-8

Zak, B. 1961. Aeration and other soil factors affecting southern pines as related to littleleaf disease. USDA Forest Service Tech. Bull. No. 1248, Washington

12 **Fire and soil**

Ahlgren, I.F., and Ahlgren, C.E. 1960. Ecological effects of forest fires. Bot. Rev. 26:483–533

– 1965. Ecology of prescribed burning on soil microorganisms in a Minnesota jack pine forest. Ecology 46:304–10

Armson, K.A., Taylor, J.McG., and Astley, E. 1973. The effect of fire on organic layers of spodosols in the boreal forests of Ontario. Agron. Abst. 1973:137

Batchelder, R.B. 1967. Spatial and temporal patterns of fire in the tropical world. *In* Proc. Tall Timbers Fire Ecology Conf. No. 6, pp. 171–92. Tall Timbers Research Station, Tallahasse

Beaton, J.D. 1959. The influence of burning on the soil in the timber range of Lac Le Jeune, British Columbia. I. Physical properties. II. Chemical properties. Can. J. Soil Sci. 39:1–11

Berndt. H.W. 1971. Early effects of forest fire on streamflow characteristics. USDA Forest Service Res. Note PNW-148, Pacific Northwest Forest and Range Expt. Station, Portland, Ore.

Burns, P.Y. 1952. Effects of fire on forest soils in the pine barren regions of New Jersey. Yale Univ. School of Forestry Bull. No. 57

De Bano, L.F., Mann, L D., and Hamilton, D.A. 1970. Translocation of hydrophobic substances into soil by burning organic litter. Soil Sci. Soc. Amer. Proc. 34:130–3

De Byle, N.V., and Packer, P.E. 1972. Plant nutrient and soil losses in overland flow from burned forest clearcuts. *In* Watersheds in Transition, pp. 296–307. Proc. Amer. Water Resources Assoc.

Diebold, C.H. 1942. Effect of fire and logging upon the depth of the forest floor in the Adirondack region. Soil Sci. Soc. Amer. Proc. (1941) 6:409–13

Dyrness, C.T , and Youngberg, C.T. 1957. The effect of logging and slash-burning on soil structure. Soil Sci. Soc. Amer. Proc. 21:444–7

Fellin, D.G., and Kennedy, P.C. 1972. Abundance of arthropods inhabiting duff and soil after prescribed burning in forest clear cuts in northern Idaho. USDA Forest Service Res. Note INT-162, Intermountain Forest and Range Expt. Station, Ogden

Fuller, W.H., Shannon, S., and Burgess, P.S. 1955. Effect of burning on certain forest soils of northern Arizona. For. Sci. 1:44–50

Gagnon, J.D. 1963. Effect of fire on the availability of nitrogen and growth of black spruce (*Picea mariana* (Mill.) B.S.P.) in Gaspé, Québec. Canada Dept. Forestry For. Res. Br. Que. Dist. Rep. 63-Q-1

Helvey, J.D. 1972. First-year effect of wildfire on water yield and stream temperature in north-central Washington. *In* Watersheds in Transition, pp. 308–12. Proc. Amer. Water Resources Assoc.

Heyward, F., and Barnette, R.M. 1934. Effect of frequent fires on chemical composition of forest soils in the longleaf pine region. Univ. Florida Agric. Expt. Station Bull. 265

Heyward, F., and Tissot, A.N. 1936. Some changes in the soil fauna associated with fires in the longleaf pine region. Ecology 17:569–666

Humphreys, F.R., and Lambert, M. 1965. Soil temperature profiles under slash and log fires of various intensities. Austral. For. Res. 1:4:23–9

Jorgensen, J.R., and Hodges, C.S., Jr. 1971. Effects of prescribed burning on the microbial characteristics of soil. *In* Prescribed Burning Symposium Proc., pp. 107–11. USDA Forest Service Southeastern Forest Expt. Station, Asheville

Kennedy, P.C., and Fellin, D.G. 1969. Insects affecting western white pine following direct seeding in northern Idaho. USDA Forest Service Res. Note INT-106, Intermountain Forest and Range Expt. Station, Ogden

Knight, H. 1966. Loss of nitrogen from the forest floor by burning. For. Chron. 42:149–52

Komarek, E.V., Sr. 1967. Fire and the ecology of man. *In* Proc. Tall Timbers Fire Ecology Conference No. 6, pp. 143–70. Tall Timbers Research Station, Tallahasse

Krammes, J.S., and De Bano, L.F. 1965. Soil wettability: a neglected factor in watershed management. Water Resources Res. 1:283–6

Lunt, H.A. 1951. Liming and twenty years of litterraking and burning under red (and white) pine. Soil Sci. Soc. Amer. Proc. 15:381–90

Lutz, H.J., and Griswold, F.S. 1939. The influence of tree roots on soil morphology. Amer. J. Sci. 237:389–400

McNamara, P.J. 1955. A preliminary investigation of the fauna of humus layers in the jarrah forest of Western Australia. Austral. For. Bur. Leaflet No. 71, Canberra

Metz, L.J., and Farrier, M.H. 1971. Prescribed burning and soil mesofauna on the Santee Experimental Forest. *In* Prescribed Burning Symposium Proc., pp. 100–6. USDA Forest Service Southeastern Forest Expt. Station, Asheville

Moehring, D.M., Grano, C.X., and Bassett, J.R. 1966. Properties of forested loess soils after repeated prescribed burns. USDA Forest Service Res. Note SO-40, Southern Forest Expt. Station, New Orleans

Neal, J.L., Wright, E., and Bollen, W.B. 1965. Burning Douglas-fir slash; physical, chemical and microbial effects in the soil. Oregon State Univ. For. Res. Lab. Res. Paper

Pase, C.P., and Lindenmuth, A.W., Jr. 1971. Effects of prescribed fire on vegetation and sediment in oak-mountain mahogany chaparral. J. For. 69:800-5
Severson, R.C., and Arnemann, H.F. 1973. Soil characteristics of the forest-prairie ecotone in northern Minnesota. Soil Sci. Soc. Amer. Proc. 37:593-9
Stephens, E.P. 1956. The uprooting of trees: a forest process. Soil Sci. Soc. Amer. Proc. 20:113-16
Sweeney, J.R., and Biswell, H.H. 1961. Quantitative studies of the removal of litter and duff by fire under controlled conditions. Ecology 42:572-5
Tarrant, R.F. 1956. Effects of slash burning on some soils of the Douglas-fir region. Soil Sci. Soc. Amer. Proc. 20:408-11
Van Wagner, C.E. 1963. Prescribed burning experiments: red and white pine. Canada Dept. Forestry Publ. No. 1020, Ottawa
Wells, C.G. 1971. Effects of prescribed burning on soil chemical properties and nutrient availability. *In* Prescribed Burning Symposium Proc., pp. 86-97. USDA Forest Service Southeastern Forest Expt. Station, Asheville
Wright, E., and Bollen, W.B. 1961. Microflora of a Douglas-fir forest soil. Ecology 42:825-8
Wright, E., and Tarrant, R.F. 1957. Microbiological soil properties after logging and slash burning. USDA Forest Service Res. Note 157, Pacific Northwest Forest and Range Expt. Station, Portland, Ore.
Zoltai, S.C., and Pettapiece, W.W. 1973. Studies of vegetation, landform and permafrost in the Mackenzie Valley: terrain, vegetation and permafrost relationships in the northern part of the Mackenzie Valley and northern Yukon. Information Canada, Ottawa, Cat. No. R72-7973

13 **The hydrologic cycle**

Barrett, J.W., and Youngberg, C.T. 1965. Effect of tree spacing and understory vegetation on water use in a pumic soil. Soil Sci. Soc. Amer. Proc. 29:472-5
Bassett, J.R. 1964. Tree growth as affected by soil moisture availability. Soil Sci. Soc. Amer. Proc. 28:436-7
Bernard, J.M. 1963. Forest floor moisture capacity of the New Jersey pine barrens. Ecology 44:574-6
Clements, J.R. 1971. Evaluating summer rainfall through a multilayered large-tooth aspen community. Can. J. For. Res. 1:20-31
Cole, D.W. 1958. Alundum tension lysimeter. Soil Sci. 85:293-6
– 1966. The forest soil-retention and flow of water. Proc. Soc. Amer. Foresters, Washington, pp. 150-4

Cole, D.W., Gessel, S.P., and Held, E.E. 1961. Tension lysimeter studies of ion and moisture movement in glacial till and coral atoll soils. Soil Sci. Soc. Amer. Proc. 25:321–5

Cowan, I.R. 1965. Transport of water in the soil-plant-atmosphere system. J. Appl. Ecol. 2:221–39

Croft, A.R., and Monninger, L.V. 1953. Evapotranspiration and other losses on some aspen forest types in relation to water available for streamflow. Trans. Amer. Geophys. Union 34:563–74

Dahms, W.G. 1973. Tree growth and water use response to thinning in a 47-year old lodgepole pine stand. USDA Forest Service Res. Note, PNW-194, Pacific Northwest Forest and Range Expt. Station, Portland, Ore.

Douglass, J.E. 1967. Effects of species and arrangements of forests on evapotranspiration. *In* Forest Hydrology, ed. W.E. Sopper and H.W. Lull, pp. 451–61. Pergamon Press, Oxford

Douglass, J.E., and Swank, W.T. 1972. Streamflow modification through management of eastern forests. USDA Forest Service Res. Paper SE-94, Southeastern Forest Expt. Station, Asheville

Dyrness, C.T. 1969. Hydrologic properties of soils on three small watersheds in the western Cascades of Oregon. USDA Forest Service Res. Note PNW-111, Pacific Northwest Forest and Range Expt. Station, Portland, Ore.

Ekern, P.C. 1964. Direct interception of cloud water on Lanaihale, Hawaii. Soil Sci. Soc. Amer. Proc. 28:419–21

Federer, C.A. 1970. Measuring forest evapotranspiration – theory and problems. USDA Forest Service Res. Paper NE-165, Northeastern Forest Expt. Station, Upper Darby

– 1973. Forest transpiration greatly speeds streamflow recession. Water Resources Res. 9:1599–1604

Fletcher, P.W., and McDermott, R.E. 1957. Moisture depletion by forest cover in a seasonally saturated Ozark ridge soil. Soil Sci. Soc. Amer. Proc. 21:547–50

Golding, D.L., and Stanton, C.R. 1972. Water storage in the forest floor of subalpine forests of Alberta. Can. J. For. Res. 2:1–6

Goodell, B.C. 1959. Management of forest stands in western United States to influence the flow of snow-fed streams. Internat. Assoc. Sci. Hydrology Publ. 48, pp. 49–58

Hanks, R.J., and Bowers, S.A. 1963. Influence of variations in the diffusivity – water content relation in infiltration. Soil Sci. Soc. Amer. Proc. 27:263–5

Heiberg, S.O. 1942. Silvicultural significance of mull and mor. Soil Sci. Soc. Amer. Proc. (1941) 6:405–8

Helvey, J.D., and Patric, J.H. 1965. Canopy and litter interception of rainfall by hardwoods of eastern United States. Water Resources Res. 1:193–206

Hewlett, J.D., and Hibbert, A.R. 1967. Factors affecting the response of small watersheds to precipitation in humid areas. *In* Forest Hydrology, ed. W.E. Sopper and H.W. Lull, pp. 275–90. Pergamon Press, Oxford

Hibbert, A.R. 1967. Forest treatment effects on water yield. *In* Forest Hydrology, ed. W.E. Sopper and H.W. Lull, pp. 527–43. Pergamon Press, New York

Holstener-Jørgensen, H. 1961. Undersøgelse af troearts-og aldersindflydelsen på grundvandstanden i skovtraebevoksninger på bregentved. Det. forstlige Forsøgsv. Danmark 27:237–480

Holstener-Jørgensen, H., Eiselstein, L.M., and Johansen, M.B. 1968. En undersølgelse af variationerne i grundvandstanden i maj maned i en 90-årig bøg-bevoksning på leret moraene med højstaende grundvand. Det. forstlige Forsøgsv. Danmark 31:15–30

Hoover, M.D. 1950. Hydrologic characteristics of South Carolina piedmont forest soils. Soil Sci. Soc. Amer. Proc. (1949) 14:353–8

Johnson, E.A., and Kovner, J.L. 1956. Effect on streamflow of cutting a forest understory. For. Sci. 2:82–91

Kittredge, J. 1948. Forest Influences. McGraw-Hill, New York

Kline, J R., Stewart, M.L., Jordan, C.F., and Kovac, P. 1972. Use of tritiated water for determination of plant transpiration and biomass under field conditions. *In* Isotopes and Radiation in Soil-Plant Relationships Including Forestry, pp. 419–37. Internat. Atomic Energy Agency, Vienna

Klock, G.O. 1972. Snowmelt temperature influence on infiltration and soil water retention. J. Soil Water Conser. 27:12–14

Kramer, P.J. 1969. Plant and Soil Water Relationships: A Modern Synthesis. McGraw-Hill, New York

Kramer, P.J., and Coile, T.S. 1940. An estimation of the volume of water made available by root extension. Plant Physiol. 15:743–7

Lambert, J.L., Gardner, W.R., and Boyle, J.R. 1971. Hydrologic response of a young pine plantation to weed removal. Water Resources Res. 7:1013–19

Lull, H.W., and Reinhart, K.G. 1967. Increasing water yield in the northeast by management of forested watersheds. USDA Forest Service Res. Paper NE-66, Northeastern Forest Expt. Station, Upper Darby

– 1972. Forests and floods in the eastern United States. USDA Forest Service Res. Paper NE-226, Northeastern Forest Expt. Station, Upper Darby

Lyford, W.H. 1964. Watertable fluctuations in periodically wet soils of central New England. Harvard Univ. Forest Paper No. 8

Mader, D.L., and Lull, H.W. 1968. Depth, weight and water storage in white pine stands in Massachusetts. USDA Forest Service Res. Paper NE-109, Northeastern Forest Expt. Station, Upper Darby

Mahendrappa, M.K. 1974. Chemical composition of stemflow of some eastern Canadian tree species. Can. J. For. Res. 4:1-7

Meeuwig, R.O. 1971. Infiltration and water repellency in granitic soils. USDA Forest Service Res. Paper INT-111, Intermountain Forest and Range Expt. Station, Ogden

Megahan, W.F. 1972. Logging, erosion, sedimentation - are they dirty words? J. For. 70:403-7

Meginnis, H.G. 1959. Increasing water yields by cutting forest vegetation. *In* Woodlands and Water - Lysimeters Symp. Proc. No. 48, L'Association International d'Hydrologie scientifique, pp. 59-68

Moehring, D.M., and Ralston, C.W. 1967. Diameter growth of loblolly pine related to available soil moisture and rate of soil moisture loss. Soil Sci. Soc. Amer. Proc. 31:560-2

Moldenhauer, W.C., and Long, D.C 1964. Influence of rainfall energy on soil loss and infiltration rates: I. Effect over a range of texture. Soil Sci. Soc. Amer. Proc. 28:813-17

Nicolson, J.A., Thorud, D.B., and Sucoff, E.I. 1968. The interception-transpiration relationship of white spruce and white pine. J. Soil Water Conserv. 23:181-4

Ovington, J.D. 1962. Quantitative ecology and woodland ecosystem. *In* Advances in Ecological Research, vol. 1, ed. J.B. Cragg, pp. 103-92. Academic Press, London and New York

Patric, J.H. 1961. A forester looks at lysimeters. J. For. 59:889-93

- 1966. Rainfall interception by mature coniferous forests of southeast Alaska. J. Soil Water Conserv. 21:229-31

Patric, J.H., Douglas, J.E., and Hewlett, J.D. 1965. Soil water absorption by mountain and piedmont forests. Soil Sci. Soc. Amer. Proc. 29:303-8

Penman, H.L. 1948. Natural evaporation from open water, bare soil and grass. Proc. Roy. Soc. London (A) 193:120-45

- 1956. Estimating evaporation. Trans. Amer. Geophys. Union 37:43-50

- 1963. Vegetation and hydrology. UK Comm. Agric. Bur. Soils Tech. Comm. No. 53

Pereira, H.C., and Hosegood, P.H. 1962. Comparative water-use of solftwood plantations and bamboo forest. J. Soil Sci. 13:301-13

Pierpoint, G., and Farrar, J.L. 1962. The water balance of the University of Toronto forest. Univ. Toronto Faculty of Forestry Tech. Rep. 3

Reynolds, E.R.C , and Henderson, C.S. 1967. Rainfall interception by beech, larch and Norway spruce. Forestry 40:165-84

Rothacher, J. 1963. Net precipitation under a Douglas-fir forest. For. Sci. 9:423-9

- 1970. Increases in water yield following clear-cut logging in the Pacific Northwest. Water Resources Res. 6:653-8

Rowe, P.B. 1955. Effects of the forest floor on disposition of rainfall in pine stands. J. For. 53:342-8

Rutter, A.J. 1967. An analysis of evaporation from a stand of Scots pine. *In* Forest Hydrology, ed. W.E. Sopper and H.W. Lull, pp. 403-17. Pergamon Press, New York

Sartz, R.S. 1967. Winter thaws can raise ground water levels in driftless area. USDA Forest Service Res. Note NC-35, North Central Forest Expt. Station, St Paul

Slatyer, R.O. 1967. Plant-Water Relationships. Academic Press, New York

Smith, R.M., and Stamey, W.L. 1965. Determining the range of tolerable erosion. Soil Sci. 100:414-24

Soil Science Society of America. 1973. Glossary of Soil Science Terms. Soil Sci. Soc. Amer., Madison

Stoeckeler, J.H., and Weitzman, S. 1960. Infiltration rates in frozen soils in northern Minnesota. Soil Sci. Soc. Amer. Proc. 24:137-9

Stone, E.C., and Fowells, H.A. 1955. Survival value of dew under laboratory conditions with *Pinus ponderosa.* For. Sci. 1:183-8

Stransky, J.J., and Wilson, D.H. 1964. Terminal elongation of loblolly and short leaf pine seedlings under soil moisture stress. Soil Sci. Soc. Amer. Proc. 28:439-40

Swanston, D.N. 1971. Judging impact and damage of timber harvesting to forest soils in mountainous regions of western North America. Proc. Western Reforest. Comm., Western Forestry and Conserv. Assoc., Portland, Ore.

Tackle, D. 1962. Infiltration in a western larch-Douglas-fir stand following cutting and slash treatment. USDA Forest Service Res. Note 89, Intermountain Forest and Range Expt. Station, Ogden

Taylor, S.A. 1972. Physical Edaphology. Rev. ed., ed. G.L. Ashcroft. W.H. Freeman and Co., San Francisco

Thornthwaite, C.W. 1948. An approach toward a rational classification of climate. Geog. Rev. 38:85-94

Thornthwaite, C.W., and Mather, J.R. 1957. Instructions and tables for computing potential evaportranspiration and the water balance. Drexel Inst. Tech. Publ. in Climatology 10:184-311

Toumey, J.W., and Kienholz, R. 1931. Trenched plots under forest canopies. Yale Univ. School of Forestry Bull. No. 30

Troendle, C.A. 1970. A comparison of soil moisture loss from forested and clear cut areas in West Virginia. USDA Forest Service Res. Note NE-120, Northeastern Forest Expt. Station, Upper Darby

Trousdell, K.B., and Hoover, M.D. 1955. A change in ground-water level after clear cutting of loblolly pine in the Coastal Plain. J. For. 53:493–8

Voigt, G.K. 1960. Distribution of rainfall under forest stands. For. Sci. 6:2–10

White, D.P., and Wood, R.S. 1958. Growth variations in a red pine plantation influenced by a deep-lying fine soil layer. Soil Sci. Soc. Amer. Proc. 22:174–7

Wright, T.W. 1955. Profile development in the sand dunes of Culbin Forest, Morayshire. I. Physical properties. J. Soil Sci. 6:270–83

Zahner, R. 1955. Soil water depletion by pine and hardwood stands during a dry season. For. Sci. 1:258–64

14 **Nutrient cycling**

Abee, A., and Lavender, D. 1972. Nutrient cycling in throughfall and litterfall in 450-year-old Douglas-fir stands. *In* Proc. Research on Coniferous Forest Ecosystems, ed. J.F. Franklin, L.J. Dempster, and R.H. Waring, pp. 133–43. USDA Forest Service Pacific Northwest Forest and Range Expt. Station, Portland, Ore.

Armson, K.A. 1965. Seasonal patterns of nutrient absorption by forest trees. *In* Forest-Soil Relationships in North America, ed. C.T. Youngberg, pp. 65–75. Oregon State Univ. Press, Corvallis

Attiwell, P.M. 1966. The chemical composition of rain water in relation to cycling of nutrients in mature Eucalyptus forest. Plant and Soil 24:390–406

Bernier, B., Brazeau, M., and Winget, C.H. 1972. Gaseous losses of ammonia following urea application in a balsam fir forest. Can. J. For. Res. 2:59–62

Bormann, F.H., Likens, G.E., Fisher, D.W., and Pierce, R.S. 1968. Nutrient loss accelerated by clear-cutting of a forest ecosystem. *In* Primary Productivity and Mineral Cycling in Natural Ecosystems, pp. 187–96. Univ. Maine, Orono

Boyle, J.R., and Ek, A.R. 1972. An evaluation of some effects of bole and branch pulpwood harvesting on site macronutrients. Can. J. For. Res. 2:407–12

Carlisle, A., Brown, A.H.F., and White, E.J. 1967. The nutrient content of rainfall and its role in the forest nutrient cycle. Proc. Internat. Union For. Res. Org. 14th Congr., Sect. 21:145–58

Cole, D.W., and Ballard, T.M. 1970. Mineral and gas transfer in a forest floor - a phase model approach. *In* Tree Growth and Forest Soils, ed. C.T. Youngberg and C.B. Davey, pp. 347-58. Oregon State Univ. Press, Corvallis

Cole, D.W., and Gessel, S.P. 1965. Movement of elements through a forest soil as influenced by tree removal and fertilizer additions. *In* Forest-Soil Relationships in North America, ed. C.T. Youngberg, pp. 95-104. Oregon State Univ. Press, Corvallis

Cole, D.W., Gessel, S.P., and Dice, S.F. 1968. Distribution and cycling of nitrogen, phosphorus, potassium and calcium in a second-growth Douglas-fir ecosystem. *In* Primary Productivity and Mineral Cycling in Natural Ecosystems, pp. 197-232. Univ. Maine, Orono

Cole, D.W., Gessel, S.P., and Held, E.E. 1961. Tension lysimeter studies of ion and moisture movement in glacial till and coral atoll soils. Soil Sci. Soc. Amer. Proc. 25:321-5

Curlin, J.W. 1967. Clonal differences in yield response of *Populus deltoides* to nitrogen fertilization. Soil Sci. Soc. Amer. Proc. 31:276-80

Duvigneaud, P., and Denaeyer-De Smet, S. 1970. Biological cycling of minerals in temperate deciduous forests. *In* Analysis of Temperate Forest Ecosystem, ed. D.E. Reichie, pp. 199-225. Springer-Verlag, New York and Berlin

Edmisten, J. 1970. Preliminary studies of the nitrogen budget of a tropical rain forest. *In* A Tropical Rain Forest - A Study of Irradiation and Ecology at El Verde, Puerto Rico, ed. H.T. Odum and R.F. Pigeon, Book 3, pp. 211-15. U.S. Atomic Energy Comm., Office of Information Services, Washington

Emanuelsson, A., Ericksson, E., and Egner, H. 1954. Composition of atmospheric precipitation in Sweden. Tellus 6:261-7

Fober, H., and Giertych, M. 1971. Variation among Norway spruce of Polish provenances in seedling growth and mineral requirements. Arbor. Körnichie 16:107-20

Foster, N.W. 1974. Annual macroelement transfer from *Pinus banksiana* Lamb. forest to soil. Can. J. For. Res. 4:470-6

Fredriksen, R.L. 1972. Nutrient budget of a Douglas-fir forest on an experimental watershed in western Oregon. *In* Proc. Research on Coniferous Forest Ecosystems, ed. J.F. Franklin, L.J. Dempster, and R.H. Waring, pp. 115-31. USDA Forest Service Pacific Northwest Forest and Range Expt. Station, Port-Portland, Ore.

Gadgil, R.L. 1971. The nutritional role of *Lupinus arboreus* in coastal sand dune forestry. Plant and Soil 35:113-26

Greenland, D.J., and Kowal, J.M.L. 1960. Nutrient content of the moist tropical forest of Ghana. Plant and Soil 12:154-74

Grier, C.G., and Cole, D.W. 1972. Elemental transport changes occurring during development of a second-growth Douglas-fir system. *In* Proc. Research on Coniferous Forest Ecosystems, ed. J.F. Franklin, L.J. Dempster, and R.H. Waring, pp. 103-13. USDA Forest Service Pacific Northwest Forest and Range Expt. Station, Portland, Ore.

Heiberg, S.O., Madgwick, H.A.I., and Leaf, A.L. 1964. Some long-time effects of fertilization on red pine plantations. For. Sci. 10:17-23

Jordan, C.F. 1970. A progress report on studies of mineral cycles at El Verde. *In* A Tropical Rain Forest - A Study of Irradiation and Ecology at El Verde, Puerto Rico, ed. H.T. Odum and R.F. Pigeon, 3:217-19. U.S. Atomic Energy Comm., Office of Information Services, Washington

Kenworthy, J.B. 1971. Water and nutrient cycling in a tropical rain forest. *In* Trans. 1st Aberdeen-Hull Symp. on Malesian Ecology, ed. J.R. Flenley, pp. 49-59. Univ. Hull, Dept. Geog. Misc. Ser. No. 11

Likens, G.E., Bormann, F.H., Johnson, N.M., Fisher, D.W., and Pierce, R.S. 1970. Effects of forest cutting and herbicide treatment on nutrient budgets in the Hubbard Brook Watershed ecosystem. Ecol. Mon. 40:23-47

Mahendrappa, M.S. 1974. Chemical composition of stemflow of some eastern Canadian tree species. Can. J. For. Res. 4:1-7

McColl, J.G. 1972. Dynamics of ion transport during moisture flow from a Douglas-fir forest floor. Soil Sci. Soc. Amer. Proc. 36:668-74

Morrison, I.K. 1973. Distribution of elements in aerial components of several natural jack pine stands in northern Ontario. Can. J. For. Res. 3:170-9

Nye, P.H. 1961. Organic matter and nutrient cycles under moist tropical forest. Plant and Soil 13:333-46

Overrein, L.N. 1968. Lysimeter studies on tracer nitrogen in forest soil: 1. Nitrogen losses by leaching and volatilization after addition of urea-N15. Soil Sci. 106:280-90

\- 1969. Lysimeter studies on tracer nitrogen in forest soil: 2. Comparative losses of nitrogen through leaching and volatilization after the addition of urea - ammonium - and nitrate N15. Soil Sci. 107:149-59

Ovington, J.D. 1962. Quantitative ecology and woodland ecosystem. *In* Advances in Ecological Research, vol. 1, ed. J.B. Cragg, pp. 103-92. Academic Press, London and New York

\- 1965. Woodlands. English Universities Press, London

Ovington, J.D., Forrest, W.G., and Armstrong, J.E. 1968. Tree biomass estimation. *In* Primary Productivity and Mineral Cycling in Natural Ecosystems, pp. 4-31. Univ. Maine, Orono

Pritchett, W.L., and Goddard, R.E. 1967. Differential responses of slash pine progeny lines to some cultural practices. Soil Sci. Soc. Amer. Proc. 31:280-4

Richards, B.N., and Voigt, G.K. 1965. Nitrogen accretion in coniferous forest ecosystems. *In* Forest-Soil Relationships, ed. C.T. Youngberg, pp. 105-16. Oregon State Univ. Press, Corvallis

Rodin, L.E., and Brazilevich, N.J. 1967. Production and Mineral Cycling in Terrestrial Vegetation. Oliver and Boyd, Edinburgh

Smith, W.H., and Goddard, R.E. 1973. Effects of genotype on the response to fertilizer. Doc. F/73/28, FAO/IUFRO Internat. Symp. on Forest Fertilization, Paris, mimeo.

Smith, W.H., Nelson, L.E., and Switzer, G.L. 1971. Development of the shoot system of young loblolly pine. II. Dry matter and nitrogen accumulation. For. Sci. 17:55-62

Stone, E.L., and Leaf, A.L. 1967. Potassium deficiency and response in young conifer forests in eastern North America. *In* Proc. Coll. on Forest Fertilization, Finland 1967, pp. 217-19. Internat. Potash Inst., Berne

Switzer, G.L., and Nelson, L.E. 1972. Nutrient accumulation and cycling in loblolly pine (*Pinus taeda* L.) plantation ecosystems: the first twenty years. Soil Sci. Soc. Amer. Proc. 36:143-7

\- 1973. Maintenance of productivity under short rotations. Doc. F/73/15, FAO/IUFRO Internat. Symp. on Forest Fertilization, Paris, mimeo.

Switzer, G.L., Nelson, L.E., and Smith, W.H. 1968. The mineral cycle in forest stands. *In* Forest Fertilization - Theory and Practice, pp. 1-9. Tennessee Valley Authority, Muscle Shoals

Tamm, C.O., and Troedsson, T. 1955. An example of the amounts of plant nutrients supplied to the ground in road dust. Oikos 6:61-70

Ulrich, B., and Mayer, R. 1972. Systems analysis of mineral cycling in forest ecosystems. *In* Isotopes and Radiation in Soil-Plant Relationships Including Forestry, pp. 329-39. Internat. Atomic Energy Agency, Vienna

Van Cleve, K., Viereck, L.A., and Schlentner, R.L. 1971. Accumulation of nitrogen in alder (*Alnus*) ecosystems near Fairbanks, Alaska. Arctic and Alpine Res. 3:101-14

Van den Driessche, R. 1973. Foliar nutrient concentration differences between provenances of Douglas-fir and their significance to foliar analysis interpretation. Can. J. For. Res. 3:323-8

Walker, L.C., and Hatcher, R.D. 1965. Variation in the ability of slash pine progeny groups to absorb nutrients. Soil Sci. Soc. Amer. Proc. 29:616-21

Weetman, G.F., and Webber, B. 1972. The influence of wood harvesting on the nutrient status of two spruce stands. Can. J. For. Res. 2:351-69

Will, G.M. 1959. Nutrient return in litter and rainfall under some exotic-conifer stands in New Zealand. N.Z. J. Agric. Res. 2:719-34

- 1968. The uptake, cycling and removal of mineral nutrients by crops of *Pinus radiata.* Proc. N.Z. Ecol. Soc. 15:20–24

Woodwell, G.M., and Whittaker, R.H. 1968. Primary production and the cation budget of the Brookhaven Forest. *In* Primary Productivity and Mineral Cycling in Natural Ecosystems, pp. 151–66. Univ. Maine, Orono

Wright, T.W. 1956. Profile development in the sand dunes of Culbin Forest, Morayshire. II. Chemical properties. J. Soil Sci. 7:33–42

- 1957. Abnormalities in nutrient uptake by Corsican pine growing on sand dunes. J. Soil Sci. 8:150–7

15 **Forest soil development**

Acquaye, D.E., and Tinsley, J. 1965. Soluble silica in soils. *In* Experimental Pedology, ed. E.G. Hallsworth and D.V. Crawford, pp. 126–48. Butterworths, London

Armson, K.A., and Fessenden, R.J. 1973. Forest windthrows and their influence on soil morphology. Soil Sci. Soc. Amer. Proc. 37:781–3

Arnold, R.W. 1965. Multiple working hypothesis in soil genesis. Soil Sci. Soc. Amer. Proc. 29:717–24

Ball, D.F., and Beaumont, P. 1972. Vertical distribution of extractable iron and aluminium in soil profiles from a brown earth-peaty podzol association. J. Soil Sci. 23:298–308

Bates, J.A.R. 1960. Studies on a Nigerian forest soil. I. The distribution of organic matter in the profile and in various soil fractions. J. Soil Sci. 11:246–56

Bloomfield, C. 1953a. A study of podzolization. I. The mobilization of iron and aluminium by Scots pine needles. J. Soil Sci. 4:3–16

- 1953b. A study of podzolization. II. The mobilization of iron and aluminium by the leaves and bark of *Agathis australis* (Kauri). J. Soil Sci. 4:17–23

- 1957. The possible significance of polyphenols in soil formation. J. Sci. Fd. Agric. 8:389–92

Blume, H.P., and Schwertmann, U. 1969. Genetic evaluation of profile distribution of aluminium, iron and manganese oxides. Soil Sci. Soc. Amer. Proc. 33:438–44

Buol, S.W., and Hole, F.D. 1961. Clay skin genesis in Wisconsin soils. Soil Sci. Soc. Amer. Proc. 25:377–9

Buol, S.W., Hole, F.D., and McCracken, R.J. 1973. Soil Genesis and Classification. Iowa State Univ. Press, Ames

Coulson, C.B., Davies, R.I., and Lewis, D.A. 1960a. Polyphenols in plant, humus and soil. I. Polyphenols of leaves, litter and superficial humus from mull and mor sites. J. Soil Sci. 11:20–9

- 1960b. Polyphenols in plant, humus and soil. II. Reduction and transport by polyphenols of iron in model soil columns. J. Soil Sci. 11:30–44

Davies, R.I., Coulson, C.B., and Lewis, D.A. 1964a. Polyphenols in plant, humus and soil. III. Stabilization of gelatin by polyphenol tanning. J. Soil Sci. 15:299–309

- 1964b. Polyphenols in plant, humus and soil. IV. Factors leading to increase in biosynthesis of polyphenol in leaves and their relationship to mull and mor formation. J. Soil Sci. 15:310–18

Delong, W.A., and Schnitzer, M. 1955. Investigations on the mobilization and transport of iron in forested soils: I. The capacities of leaf extracts and leachates to react with iron. Soil Sci. Soc. Amer. Proc. 19:360–3

Dickson, B.A., and Crocker, R.L. 1953. A chronosequence of soils and vegetation near Mt. Shasta, California. II. The development of the forest floors and the carbon and nitrogen profiles of the soils. J. Soil Sci. 4:142–54

Douglas, L.A., and Tedrow, J.C.F. 1959. Organic matter decomposition rates in Arctic soils. Soil Sci. 88:305–12

Duchaufour, Ph. 1973. Processus de formation des sols biochimie et géochimie. Univ. de Nancy Centre de Pédologie CNRS, Nancy

Gersper, P.L., and Holowaychuk, N. 1970a. Effects of stemflow water on a Miami soil under a beech tree. I. Morphological and physical properties. Soil Sci. Soc. Amer. Proc. 34:779–86

- 1970b. Effects of stemflow water on a Miami soil under a beech tree. II. Chemical properties. Soil Sci. Soc. Amer. Proc. 34:786–94

- 1971. Some effects of stemflow from forest canopy trees on chemical properties of soils. Ecology 52:691–702

Handley, W.R.C. 1954. Mull and mor formation in relation to forest soils. UK For. Comm. Bull. No. 23, London

Holstener-Jørgensen, H. 1960. Indfygning af jord i en plantages Vestrand. Forstlige Forsøgsv. Danmark 26:391–7

Jenny, H. 1941. Factors of Soil Formation. McGraw-Hill, New York

Kyuma, K., and Kawaguchi, K. 1964. Oxidative changes of polyphenols as influenced by allophane. Soil Sci. Soc. Amer. Proc. 28:371–4

Langmaid, K.K. 1964. Some effects of earthworm invasion in virgin podzols. Can. J. Soil Sci. 44:34–7

Lutz, H.J. 1940. Disturbance of forest soil resulting from the uprooting of trees. Yale Univ. School of Forestry Bull. 45

Lutz, H.J., and Griswold, F.S. 1939. The influence of tree roots on soil morphology. Amer. J. Sci. 237:389–400

Malcolm, R.L., and McCracken, R.J. 1968. Canopy drip: a source of mobile soil organic matter for mobilization of iron and aluminium. Soil Sci. Soc. Amer. Proc. 32:834–8

McCaleb, S.B. 1959. The genesis of the red-yellow podzolic soils. Soil Sci. Soc. Amer. Proc. 23:164–8

McFee, W.W., and Stone, E.L. 1965. Quantity, distribution and variability of organic matter and nutrients in a forest podzol in New York. Soil Sci. Soc. Amer. Proc. 29:432–6

Milfred, C.J., Olsen, G.W., and Hole, F.D. 1967. Soil resources and forest ecology of Menominee County, Wisconsin. Univ. Wisconsin, Geol. and Nat. Hist. Survey, Soil Survey Div. Bull. 85, Soil Ser. No. 60, Madison

Mueller, O.P., and Cline, M.G. 1959. Effects of mechanical soil barriers and soil wetness on rooting of trees and soil-mixing by blow-down in central New York. Soil Sci. 88:107–11

Nye, P.H. 1954. Some soil-forming processes in the humid tropics. 1. A field study of a catena in a west African forest. J. Soil Sci. 5:7–21

– 1955. Some soil-forming processes in the humid tropics. IV. The action of the soil fauna. J. Soil Sci. 6:73–83

Nykvist, N. 1959a. Leaching and decomposition of litter. 1. Experiments on leaf litter of *Fraxinus excelsior.* Oikos 10:190–201

– 1959b. Leaching and decomposition of litter. 2. Experiments on needle litter of *Pinus sylvestris.* Oikos 10:212–24

– 1960a. Leaching and decomposition of litter. 3. Experiments on leaf litter of *Betula verrucosa.* Oikos 11:249–63

– 1960b. Leaching and decomposition of litter. 4. Experiments on needle litter of *Picea abies.* Oikos 11:264–79

– 1962. Leaching and decomposition of litter. 5. Experiments on *Alnus glutinosa, Fagus silvatica* and *Quercus robur.* Oikos 12:232–48

Schnitzer, M. 1959. Interaction of iron with rainfall leachates. J. Soil Sci. 10:300–8

Schramm, J.R. 1958. The mechanism of frost heaving of tree seedlings. Proc. Amer. Phil. Soc. 102:4:333–50

Simonson, R.W. 1959. Outline of a generalized theory of soil genesis. Soil Sci. Soc. Amer. Proc. 23:152–6

Stephens, E.P. 1956. The uprooting of trees: a forest process. Soil Sci. Soc. Amer. Proc. 20:113–16

Strang, R.M. 1973. The rate of silt accumulation in the Lower Peel River, Northwest Territories. Can. J. For. Res. 3:457–8

Swindale, L.D., and Jackson, M.L. 1956. Genetic processes in some residual podzolized soils of New Zealand. Trans. 6th Internat. Congr. Soil Sci. 233–9

Syers, J.K., Adams, J.A., and Walker, T.W. 1970. Accumulation of organic matter in a chronosequence of soils developed in windblown sand in New Zealand. J. Soil Sci. 21:146–53

Tamm, C.O., and Holmen, H. 1967. Some remarks on soil organic matter turnover in Swedish podzol profiles. Det. Norske Skogsforsøksv. 23:69-88

van Schuylenborgh, J. 1971. Weathering and soil-forming processes in the tropics. *In* Soils and Tropical Weathering, pp. 39-50. Unesco, Paris

de Villiers, J.M. 1965. Present soil-forming factors and processes in tropical and subtropical regions. Soil Sci. 99:50-7

16 **Soils and changing landscapes and use**

Abler, T.S. 1970. Longhouse and palisade: northeastern Iroquoian villages of the 17th century. Ont. Hist. 62:17-40

Bailey, L.W., Odell, R.T., and Boggess, W.R. 1964. Properties of selected soils developed near the forest-prairie border in east-central Illinois. Soil Sci. Soc. Amer. Proc. 28:257-63

Bidwell, O.W., and Hole, F.D. 1965. Man as a factor of soil formation. Soil Sci. 99:65-72

Bushue, L.J., Fehrenbacher, J.B., and Ray, B.W. 1970. Exhumed paleosols and associated modern till soils in western Illinois. Soil Sci. Soc. Amer. Proc. 34:665-9

Daniels, R.B., and Gamble, E.E. 1967. The edge effect in some ultisols in the North Carolina coastal plain. Geoderma 1:117-24

Daniels, R.B., Gamble, E.E., and Cady, J.G. 1970. Some relations among coastal plain soils and geomorphic surfaces in North Carolina. Soil Sci. Soc. Amer. Proc. 34:648-51

- 1971. The relations between geomorphology and soil morphology and genesis. *In* Advances in Agronomy, ed. N.C. Brady, vol. 23, pp. 51-88. Academic Press, New York

Day, G.M. 1953. The Indian as an ecological factor in the northeastern forest. Ecology 34:329-46

Deitz, R. 1957. Phosphorus accumulation in soil on an Indian habitation site. Amer. Antiquity 22:405-9

Dimbleby, G.W. 1961. Soil pollen analysis. J. Soil Sci. 12:1-11

- 1962. The development of British heathlands and their soils. Oxford Univ. For. Mem. No. 23

Dimbleby, G.W., and Gill, J.W. 1955. The occurrence of podzols under deciduous woodland in the New Forest. Forestry 28:95-106

Finney, H.R., Holowaychuk, N., and Heddleson, M.R. 1962. The influence of microclimate on the morphology of certain soils of the Alleghany plateau of Ohio. Soil Sci. Soc. Amer. Proc. 26:287-92

Flint, R.F. 1957. Glacial and Pleistocene Geology. John Wiley and Sons Inc., New York

Gimingham, C.H. 1972. Ecology of Heathlands. Chapman and Hall, London
Jessup, R.W. 1960a. The lateritic soils of the southeastern portion of the Australian arid zone. J. Soil Sci. 11:106-13
- 1960b. The stony tableland soils of the southeastern portion of the Australian arid zone and their evolutionary history. J. Soil Sci. 11:188-96
Jones, R.L., and Beavers, A.H. 1964. Aspects of catenary and depth distribution of opal phytoliths in Illinois soils. Soil Sci. Soc. Amer. Proc. 28:413-16
Lutz, H.J. 1951. The concentration of certain chemical elements in the soils of Alaskan archaeological sites. Amer. J. Sci. 246:925-8
Mulcahy, M.J. 1960. Laterites and lateritic soils in southwestern Australia. J. Soil Sci. 11:206-25
Parsons, R.B., Scholtes, W.H., and Reicken, F.F. 1962. Soils of Indian mounds in northeastern Iowa as benchmarks for studies of soil genesis. Soil Sci. Soc. Amer. Proc. 26:491-6
Proudfoot, V.B. 1958. Problems of soil history. Podzol development at Goodland and Torr. Townlands, Co. Antrim, Northern Ireland. J. Soil Sci. 9:186-98
Ruhe, R.V. 1956. Geomorphic surfaces and the nature of soils. Soil Sci. 82:441-55
Severson, R.C., and Arnemann, H.F. 1973. Soil characteristics of the forest-prairie ecotone in northwestern Minnesota. Soil Sci. Soc. Amer. Proc. 37:593-9
Stone, E.L. 1971. Effects of prescribed burning on long-term productivity of Coastal Plain soils. *In* Proc. Prescribed Burning Symposium, pp. 115-27. USDA Forest Service Southeastern Forest Expt. Station, Asheville
Thorp, J., Johnson, W.M., and Reed, E.C. 1951. Some post-pliocene buried soils of central United States. J. Soil Sci. 2:1-19
Tubbs, C.R. 1968. The New Forest: An Ecological History. David and Charles, Newton Abbot
Verma, S.D., and Rust, R.H. 1969. Observations on opal phytoliths in a soil biosequence in southeastern Minnesota. Soil Sci. Soc. Amer. Proc. 33:749-51
Wilding, L.P., and Drees, L.R. 1971. Biogenic opal in Ohio soils. Soil Sci. Amer. Proc. 35:1004-10
Witty, J.E., and Knox, E.G. 1964. Grass opal in some chestnut and forested soils in north central Oregon. Soil Sci. Soc. Amer. Proc. 28:685-8

Appendix 1 **Procedures for soil profile description**

Ball, D.F., and Williams, W.M. 1971. Further studies on the variability of soil chemical properties: efficiency of sampling programmes on an uncultivated brown earth. J. Soil Sci. 22:60-8

Gessel, S.P., and Balci, A.N. 1965. Amount and composition of forest floors under Washington coniferous forests. *In* Forest-Soil Relationships in North America, ed. C.T. Youngberg, pp. 11-23. Oregon State Univ. Press, Corvallis

Hammond, L.C., Pritchett, W.L., and Chew, V. 1958. Soil sampling in relation to soil heterogeneity. Soil Sci. Soc. Amer. Proc. 22:548-52

Harridine, F.F. 1950. The variability of soil properties in relation to stage of profile development. Soil Sci. Soc. Amer. Proc. (1949) 14:302-11

Ike, A.F., and Clutter, J.L. 1968. The variability of forest soils in the Georgia Blue Ridge mountains. Soil Sci. Soc. Amer. Proc. 32:284-8

Mader, D.L. 1963. Soil variability - a serious problem in soil-site studies in the Northeast. Soil Sci. Soc. Amer. Proc. 27:707-9

Mader, D.L., and Lull, H.W. 1968. Depth, weight and water storage in white pine stands in Massachusetts. USDA Forest Service Res. Paper NE-109, Northeastern Forest Expt. Station, Upper Darby

McFee, W.W., and Stone, E.L. 1965. Quantity, distribution and variability of organic matter and nutrients in a forest podzol in New York. Soil Sci. Soc. Amer. Proc. 29:432-6

Metz, L.J., Wells, C.G., and Swindel, B.F. 1966. Sampling soil and foliage in a pine plantation. Soil Sci. Soc. Amer. Proc. 30:397-9

Soil Survey Staff. 1960. Soil Classification: A Comprehensive System, Seventh Approximation. 1964, 1967 Supplement. USDA Soil Conservation Service, Washington

Stutzbach, J.J., Leaf, A.L., and Leonard, R.E. 1972. Variation in forest floor under a red pine plantation. Soil Sci. 114:24-8

Troedsson, T., and Tamm, C.O. 1969. Small-scale spatial variation in forest soil properties and its implications for sampling procedures. Stud. Forestal. Suec. No. 74, Skogshögskolan, Stockholm

Van Wagner, C.E. 1963. Prescribed burning experiments: red and white pine. Canada Dept. Forestry Publ. No. 1020, Ottawa

Welch, C.D., and Fitts, J.W. 1956. Some factors affecting soil sampling. Soil Sci. Soc. Amer. Proc. 20:54-6

Wilding, L.P., Schafer, G.M., and Jones, R.B. 1964. Morley and Blount soils: a statistical summary of certain physical and chemical properties of some selected profiles from Ohio. Soil Sci. Soc. Amer. Proc. 28:674-9

Index